Methods in Enzymology

Volume 417
FUNCTIONAL GLYCOMICS

METHODS IN ENZYMOLOGY

EDITORS-IN-CHIEF

John N. Abelson   Melvin I. Simon

DIVISION OF BIOLOGY
CALIFORNIA INSTITUTE OF TECHNOLOGY
PASADENA, CALIFORNIA

FOUNDING EDITORS

Sidney P. Colowick and Nathan O. Kaplan

Methods in Enzymology

Volume 417

Functional Glycomics

EDITED BY

Minoru Fukuda

GLYCOBIOLOGY PROGRAM
CANCER RESEARCH CENTER
THE BURNHAM INSTITUTE FOR MEDICAL RESEARCH
LA JOLLA, CALIFORNIA

AMSTERDAM • BOSTON • HEIDELBERG • LONDON
NEW YORK • OXFORD • PARIS • SAN DIEGO
SAN FRANCISCO • SINGAPORE • SYDNEY • TOKYO
Academic Press is an imprint of Elsevier

ELSEVIER

Academic Press is an imprint of Elsevier
525 B Street, Suite 1900, San Diego, California 92101-4495, USA
84 Theobald's Road, London WC1X 8RR, UK

This book is printed on acid-free paper.

For information on all Elsevier Academic Press publications
visit our Web site at www.books.elsevier.com

ISBN-13: 978-0-12-182822-6
ISBN-10: 0-12-182822-0

PRINTED IN THE UNITED STATES OF AMERICA
06 07 08 09 9 8 7 6 5 4 3 2 1

Working together to grow
libraries in developing countries

www.elsevier.com | www.bookaid.org | www.sabre.org

ELSEVIER   BOOK AID
             International   Sabre Foundation

Table of Contents

Section I. N-Glycan Function Revealed by Gene Inactivation

Section II. Neural Cell Function

Section VII. Newly Developed Field

Contributors to Volume 417

Article numbers are in parentheses following the names of contributors.
Affiliations listed are current.

KIYOHIKO ANGATA (3), *Glycobiology Program, Cancer Center, Burnham Institute for Medical Research, La Jolla, California*

ANNA ARNQVIST (20), *Department of Medical Biochemistry and Biophysics, Umeå University, Umeå, Sweden*

RAFFI V. AROIAN (21), *Section of Cell and Developmental Biology, La Jolla, California*

MARINA ASPHOLM (20), *Department of Molecular Biosciences, University of Oslo, Oslo, Norway*

BRAD D. BARROWS (21), *Section of Cell and Developmental Biology, La Jolla, California*

LINDA G. BAUM (17), *Department of Pathology and Laboratory Medicine, UCLA School of Medicine, Los Angeles, California*

DOUGLAS E. BERG (20), *Department of Molecular Microbiology, Washington University Medical School, St. Louis, Missouri*

IRA J. BLADER (23), *Department of Microbiology and Immunology, University of Oklahoma Health Sciences Center, Oklahoma City, Oklahoma*

THOMAS BORÉN (20), *Departments of Medical Biochemistry and Biophysics, Umeå University, Umeå, Sweden*

OLENA BUKALO (5), *Zentrum fuer Molekulare Neurobiologie, Universitaetsklinikum Hamburg-Eppendorf, Hamburg, Germany*

ADRIANA CARVALHO DE SOUZA (16), *Department of Bio-Organic Chemistry, Bijvoet Center, Padualaan 8, CH Utrecth, The Netherlands*

PAM CHEUNG (1), *Samuel Lunenfeld Research Institute, Mount Sinai Hospital, Toronto, Ontario*

DAN DANIELSSON (20), *Department of Clinical Microbiology, Örebro University Hospital, Örebro, Sweden*

JAMES W. DENNIS (1), *Samuel Lunenfeld Research Institute, Mount Sinai Hospital, Toronto, Ontario, Canada*

ALEXANDER DITYATEV (5), *Neurophsiologie and Pathophysiologie, Universitaetsklinikum Hamburg-Eppendorf, Hamburg, Germany*

ANDRE DUBOIS (20), *Laboratory of Gastrointestinal and Liver Studies, University of the Health Sciences, Bethesda, Maryland*

TAMAO ENDO (11), *Glycobiology Research Group, Tokyo Metropolitan Institute of Gerontology Foundation for Research on Aging and Promotion of Human Welfare, Tokyo, Japan*

MINORU FUKUDA (3), *Glycobiology Program, Cancer Research Center, The Burnham Institute for Medical Research, La Jolla, California*

KEIKO FURUKAWA (4), *Department of Biochemistry II, Nagoya University School of Medicine, Nagoya, Aichi, Japan*

KOICHI FURUKAWA (4), *Department of Biochemistry II, Nagoya University School of Medicine, Nagoya, Aichi, Japan*

MARKUS GERHARD (20), *Uppsalalaan 8, Hubrecht Lab/NIOB, Utrecht, The Netherlands*

JOEL S. GRIFFITTS (21), *Department of Biological Sciences, Stanford University, Stanford, California*

JIANGUO GU (2), *Department of Biochemistry, Osaka University Graduate School of Medicine, Osaka, Japan*

ROBERT S. HALTIWANGER (8), *Department of Biochemistry and Cell Biology, SUNY at Stony Brook, Stony Brook, New York*

JIALE HE (17), *Department of Pathology and Laboratory Medicine, UCLA School of Medicine, Los Angeles, California*

YOSHIO HIRABAYASHI (12), *Hirabayashi Laboratory Unit and CREST, Neuronal Cicruit Mechanisms Research Group, Brain Science Institute RIKEN, Saitama, Japan*

KOICHI HONKE (2), *Department of Molecular Genetics, Kochi Medical School, Kochi, Japan*

DANIEL K. HSU (18), *Department of Dermatology, University of California, Davis School of Medicine, Sacramento, California*

AWDHESH KALIA (20), *Departments of Biology and Microbiology and Immunology, University of Louisville, Louisville, Kentucky*

JOHANNIS P. KAMERLING (16), *Department of Bio-Organic Chemistry, Bijvoet Center, Padualaan 8, CH Utrecth, The Netherlands*

KAREN KAUCIC (13), *Children's National Medical Center, Center for Cancer and Immunology Research, Washington, District of Columbia*

MITCHELL KRONENBERG (14), *10355 Science Center Dr, La Jolla Institute for Allergy and Immunology, San Diego, California*

STEPHAN LADISCH (13), *Children's National Medical Center, Center for Cancer and Immunology Research, Washington, District of Columbia*

WENJAU LEE (3), *Glycobiology Program, Cancer Research Center, The Burnham Institute for Medical Research, La Jolla, California*

WOO-KON LEE (20), *Department of Molecular Microbiology, Washington University Medical School, St. Louis, Missouri*

SARA LINDÉN (20), *Aubigny place, level 3, Mater Medical Research Institute, South Brisbane, Australia*

FU-TONG LIU (18), *Department of Dermatology, University of California, Davis School of Medicine, Sacramento, California*

YIHUI LIU (13), *Children's National Medical Center, Center for Cancer and Immunology Research, Washington, District of Columbia*

JEFFREY M. LONG (6), *Department of Biochemistry, University of Texas Southwestern Medical Center, Dallas, Texas*

PABLO H. H. LOPEZ (15), *Department of Neurology, Johns Hopkins University School of Medicine, Baltimore, Maryland*

LINCHAO LU (10), *Department of Cell Biology, Albert Einstein College of Medicine, Bronx, New York*

HIROSHI MANYA (11), *Glycobiology Research Group, Tokyo Metropolitan Institute of Gerontology Foundation for Research on Aging and Promotion of Human Welfare, Tokyo, Japan*

JAMEY D. MARTH (3), *Department, of Cellular and Molecular Medicine, Howard Hughes Medical Institute, La Jolla, California*

TSUKASA MATSUDA (9), *Applied Molecular Biosciences, Nagoya University Graduate School of Bioagricultural Sciences, Nagoya, Japan*

JUNYA MITOMA (3), *Glycobiology Program, Cancer Research Center, The Burnham Institute for Medical Research, La Jolla, California*

EIJI MIYOSHI (2), *Department of Biochemistry, Osaka University Graduate School of Medicine, Osaka, Japan*

YASUKO NAGATSUKA (12), *Hirabayashi Laboratory Unit and CREST, Neuronal Circuit Mechanisms Research Group, Brain Science Institute RIKEN, Saitama, Japan*

SUSUMU NAKAHARA (19), *Karmanos Cancer Institute, Wayne State University, Detroit, Michigan*

ALEKSANDRA NITA-LAZAR (8), *Department of Biochemistry and Cell Biology, SUNY at Stony Brook, Stony Brook, New York*

TETSUYA OKUDA (4), *Nagoya University School of Medicine, Nagoya, Aichi, Japan*

TETSUYA OKAJIMA (9), *Department of Applied Molecular Biosciences, Nagoya University Graduate School of Bioagricultural Sciences, Nagoya, Aichi, Japan*

KODEESWARAN PARAMESHWARAN (7), *Pharmacal Sciences, Auburn University, Auburn, Alabama*

EMILY A. PARTRIDGE (1), *Samuel Lunenfeld Research Institute, Mount Sinai Hospital, Toronto, Ontario*

AVRAHAM RAZ (19), *Karmanos Cancer Institute, Wayne State University, Detroit, Michigan*

STEFAN RUHL (20), *Department of Operative Dentistry and Periodontology, University of Regensburg, Regensburg, Germany*

HARRY SCHACHTER (22), *Program in Structural Biology and Biochemistry, Hospital for Sick Children, Toronto, Ontario, Canada*

STAFFAN SCHEDIN (20), *Department of Applied Physics and Electronics, Umeå University, Umeå, Sweden*

RONALD L. SCHNAAR (15), *Departments of Pharmacology and Neuroscience, Johns Hopkins University School of Medicine, Baltimore, Maryland*

CRISTINA SEMINO-MORA (20), *Laboratory of Gastrointestinal and Liver Studies, University of the Health Sciences, Bethesda, Maryland*

HUI SHI (22), *Program in Structural Biology and Biochemistry, Hospital for Sick Children, Toronto, Ontario, Canada*

ROLF SJÖSTRÖM (20), *Departments of Medical Biochemistry and Biophysics, Umeå University, Umeå, Sweden*

PAMELA STANLEY (10), *Department of Cell Biology, Albert Einstein College of Medicine, Bronx, New York*

VISHNU SUPPIRAMANIAM (7), *Pharmacal Sciences, Auburn University, Auburn, Alabama*

JENNY TAN (22), *Program in Structural Biology and Biochemistry, Hospital for Sick Children, Toronto, Ontario, Canada*

NAOYUKI TANIGUCHI (2), *Department of Biochemistry, Osaka University Graduate School of Medicine, Osaka, Japan*

SUSANN TENEBERG (20), *Institute of Medical Biochemistry, Göteborg University, Göteborg, Sweden*

HIROMASA TOJO (12), *Department of Biochemistry and Molecular Biology, Osaka University Graduate School of Medicine, Suita, Osaka, Japan*

EMMANUEL TUPIN (14), *La Jolla Institute for Allergy and Immunology, San Diego, California*

MAGNUS UNEMO (20), *Department of Clinical Microbiology, Örebro University Hospital, Örebro, Sweden*

THIRUMALINI VAITHIANATHAN (7), *Pharmacal Sciences, Auburn University, Auburn, Alabama*

HANKE VAN DER WEL (23), *Department of Biochemistry and Molecular Biology, Oklahoma Center for Medical Glycobiology, Oklahoma City, Oklahoma*

XIANGCHUN WANG (2), *Department of Biochemistry, Osaka University Graduate School of Medicine, Osaka, Japan*

CHRISTOPHER M. WEST (23), *Department of Biochemistry and Molecular Biology, Oklahoma Center for Medical Glycobiology, Oklahoma City, Oklahoma*

RI-YAO YANG (18), *Department of Dermatology, University of California, Davis School of Medicine, Sacramento, California*

Preface

In the past decade, we have seen an explosion of progress in understanding the roles of carbohydrates in biological systems. This explosive progress was made with the efforts in determining the roles of carbohydrates in immunology, neurobiology and many other disciplines, examining each unique system and employing new technology. Thanks to Academic Press Editorial Management, particularly to Ms. Cindy Minor, three books, namely Glycobiology (vol. 415), Glycomics (vol. 416), and Functional Glycomics (vol. 417) in the series of Methods in Enzymology have been dedicated to disseminate information on methods in determining the biological roles of carbohydrates. These books are designed to provide an introduction of new methods to a large variety of readers who would like to participate in and contribute to the advancement of glycobiology. The methods covered include structural analysis of carbohydrates, biological and chemical synthesis of carbohydrates, expression and determination of ligands for carbohydrate-binding proteins, gene expression profiling including microarray, and generation of gene knockout mice and their phenotype analyses. The book also covers recent advances in special topics such as chaperones for glycosyltransferase, the roles of glycosylation in signal transduction, chemokine and cytokine binding, muscle development, glycolipids in development and cell-cell interaction. I believe that we have a collection of outstanding contributors, who represent their respective expertise and field.

The current Functional Glycomics vol. 417 covers methods on N-glycan functions revealed by gene knockout, analysis of neural cell function, Notch signaling and muscular dystrophy, glycolipid function and interaction, galectin function and emerging fields such as Helicobacter adhesion to carbohydrates. I believe that this book together with two previous volumes 415, Glycobiology and 416, Glycomics will be useful to a wide variety of readers from graduate students, researchers in academics and industry, to those who would like to teach glycobiology at various levels.

The volume 417 also concludes the current series of Methods in Enzymology on Glycobiology and Glycomics. We hope that these three books will contribute to explosive progress in Glycobiology and Glycomics, necessitating a new series on these fields in a near future. Finally I would like to acknowledge our mentors, our colleagues, our parents, our families and the staff at the Academic Press for their generous support and encouragement.

MINORU FUKUDA
June 21, 2006

METHODS IN ENZYMOLOGY

VOLUME 72. Lipids (Part D)
Edited by JOHN M. LOWENSTEIN

VOLUME 73. Immunochemical Techniques (Part B)
Edited by JOHN J. LANGONE AND HELEN VAN VUNAKIS

VOLUME 74. Immunochemical Techniques (Part C)
Edited by JOHN J. LANGONE AND HELEN VAN VUNAKIS

VOLUME 75. Cumulative Subject Index Volumes XXXI, XXXII, XXXIV–LX
Edited by EDWARD A. DENNIS AND MARTHA G. DENNIS

VOLUME 76. Hemoglobins
Edited by ERALDO ANTONINI, LUIGI ROSSI-BERNARDI, AND EMILIA CHIANCONE

VOLUME 77. Detoxication and Drug Metabolism
Edited by WILLIAM B. JAKOBY

VOLUME 78. Interferons (Part A)
Edited by SIDNEY PESTKA

VOLUME 79. Interferons (Part B)
Edited by SIDNEY PESTKA

VOLUME 80. Proteolytic Enzymes (Part C)
Edited by LASZLO LORAND

VOLUME 81. Biomembranes (Part H: Visual Pigments and Purple Membranes, I)
Edited by LESTER PACKER

VOLUME 82. Structural and Contractile Proteins (Part A: Extracellular Matrix)
Edited by LEON W. CUNNINGHAM AND DIXIE W. FREDERIKSEN

VOLUME 83. Complex Carbohydrates (Part D)
Edited by VICTOR GINSBURG

VOLUME 84. Immunochemical Techniques (Part D: Selected Immunoassays)
Edited by JOHN J. LANGONE AND HELEN VAN VUNAKIS

VOLUME 85. Structural and Contractile Proteins (Part B: The Contractile Apparatus and the Cytoskeleton)
Edited by DIXIE W. FREDERIKSEN AND LEON W. CUNNINGHAM

VOLUME 86. Prostaglandins and Arachidonate Metabolites
Edited by WILLIAM E. M. LANDS AND WILLIAM L. SMITH

VOLUME 87. Enzyme Kinetics and Mechanism (Part C: Intermediates, Stereo-chemistry, and Rate Studies)
Edited by DANIEL L. PURICH

VOLUME 88. Biomembranes (Part I: Visual Pigments and Purple Membranes, II)
Edited by LESTER PACKER

VOLUME 89. Carbohydrate Metabolism (Part D)
Edited by WILLIS A. WOOD

Section I

N-Glycan Function Revealed by Gene Inactivation

[1]   Cytokine Sensitivity and N-Glycan Processing Mutations

By EMILY A. PARTRIDGE, PAM CHEUNG, and JAMES W. DENNIS

Abstract

The EGF and TGF-β families of cytokines are critical regulators of cell proliferation, morphogenesis, and tissue repair. The signaling pathways downstream of EGF and TGF-β receptors also contribute to cancer growth and metastasis. Cytokine receptors are glycoproteins, and we have recently shown that GlcNAc-branching of N-glycans enhances their cell surface residency and contributes to the growth autonomy of cancer cells. Ligand-induced dimerization of EGF receptors leads to phosphorylation of Erk1/2, whereas TGF-β binding to its receptors stimulates phosphorylation of Smad2/3. Activated Erk1/2 and Smad2/3 translocate independently into the nucleus and regulate gene expression. Here we describe a sensitive and robust method to quantify TGF-β and EGF signaling in cancer cells and primary cells from mice by quantitative fluorescence imaging.

Introduction

Many glycosyltransferases are developmentally regulated and generate tissue-specific patterns of protein glycosylation. Activation of the Ras-Erk-Ets oncogene pathway stimulates Mgat5 transcription (Chen *et al.*, 1998; Kang *et al.*, 1996), a gene encoding Golgi β1, 6 N-acetylglucosaminyltransferase V in the N-glycan processing pathway. Although Mgat5$^{-/-}$ mice develop normally, they display a number of adult phenotypes associated with altered sensitivity to trophic factors in T lymphocytes, macrophages, and fibroblasts (Demetriou *et al.*, 2001; Partridge *et al.*, 2004). The Mgat5-deficient background suppresses mammary tumor growth and metastasis in polyoma middle T oncogene (PyMT) transgenic mice (Granovsky *et al.*, 2000). The PyMT oncoprotein is a cytosolic scaffold that promotes Src, PI3 kinase, and Shc/Ras/Erk activation (Webster *et al.*, 1998). However, our results indicate that full activation of PyMT signaling depends on glyco-protein glycosylation. Indeed, Mgat5$^{-/-}$PyMT tumor cells display a loss of sensitivity to multiple cytokines including EGF, IGF-1, PDGF, FGF, and TGF-β, a defect that can be rescued by infection of the cells with a retro-viral vector for Mgat5 expression (Partridge *et al.*, 2004). At the molecular level, Mgat5-modified N-glycans on EGF and TGF-β receptors bind to

METHODS IN ENZYMOLOGY, VOL. 417
0076-6879/06 $35.00
DOI: 10.1016/S0076-6879(06)17001-9

galectins that cross-link and retain the receptors at the cell surface in a lattice that delays receptor removal by constitutive endocytosis. Activation of receptor tyrosine kinases in the EGFR family induces phosphorylation and translocation of Erk1/2 into the nucleus, where Erk1/2 substrates are activated to stimulate gene expression (Chen et al., 1992). Stimulation of the Erk1/2 pathway increases autocrine TGF-β signaling (Oft et al., 2002) with nuclear localization of Smad2/3-Smad4 complexes (Macias-Silva et al., 1996), and both Erk and Smad pathways are required for epithelial-mesenchymal transition (EMT), a characteristic of invasive carcinomas. These requirements for Erk1/2 and Smad2/3 activation in carcinoma cells are also dependent on Mgat5 gene expression (Partridge et al., 2004). Genetic deficiencies in N-glycan processing cause a variety of phenotypes observed in genetically engineered mice (Haltiwanger and Lowe, 2004; Lowe and Marth, 2003) and in human patients with congenital deficiencies of glycosylation (Freeze, 2002). Quantitative fluorescence imaging provides a sensitive method well suited to comparative analysis of signal transduction in various cell types from mutant and wild-type mice.

Materials

1. Dulbecco's Modified Eagle's Medium (DMEM), high glucose (33 mM), supplemented with 10% fetal bovine serum (FBS) (Gibco/BRL, Bethesda, MD).
2. EGF cytokine (Sigma, St. Louis, MO. Cat# E9644) is dissolved at 100 μg/ml in phosphate-buffered saline (PBS) (0.137 M NaCl, 2.7 mM KCl, 10 mM Na$_2$PO$_4$, 1.8 mM KH$_2$PO$_4$, pH 7.4), stored in single use aliquots at $-80°$, and diluted as required.
3. TGF-β1 cytokine (R&D Systems, Minneapolis, MN) is dissolved at 1 mg/ml in PBS, stored in single-use aliquots at $-80°$, and diluted as required.
4. Mouse anti-Smad2/3 antibodies (Transduction Laboratories, Cat# S66220).
5. Mouse monoclonal antibodies against dual-phosphorylated Erk1/2 at Thr183/Tyr185 (abbreviated Erk-P) (Sigma, Cat# M-8159).
6. Secondary antibodies are mouse monoclonal anti-mouse Ig labeled with AlexaFlour 488 (Molecular Probes, Eugene, OR. Cat# A-11029; and Hoechst 33342; Molecular Probes, Cat# 33342).

Methods

A number of commercial systems for automated and quantitative fluorescence imaging of cells are available. We have used the Scan Array by

Cellomics (Pittsburgh, PA), a fixed end-point reader with software supplied by the company. The cytoplasmic-nuclear translocation algorithm identifies objects as cells by nuclear staining with Hoechst 33342. Erk-P or Smad2/3 are quantified by use of primary mouse monoclonal antibodies specific for these proteins, and a secondary florescence AlexaFluor 488-tagged anti-mouse antibody. By demarcating a mask around the Hoechst 33342–stained nucleus and a ring in the cytoplasm at a set distance from the nuclear mask, the fluorescence intensity in the nucleus minus that in the cytoplasmic ring is calculated for each cell. Generally, 200 individual cells per well are imaged, and each well represents an experimental condition. Our data indicate that tumor cells respond to increasing doses of either EGF or TGF-β cytokines as a single population. However, if required, cell populations can be identified and represented as the fraction of responding and nonresponding cells. Nuclear-cytoplasmic differences measured for Erk-P and Smad2/3 in 200 cells routinely have standard errors $<5\%$ of the mean.

Cells from Mutant Mice

Mammary Tumor Cells from Polyomavirus Middle T Transgenic Mice

1. Polyomavirus middle T (PyMT) transgenic mice on a 129sv $\times$ FVB background with either Mgat5$^{+/+}$ or Mgat5$^{-/-}$ genotypes display mammary tumors at 8–16 wk of age (Granovsky *et al.*, 2000). PyMT gene expression is under the control of the mouse mammary tumor virus long terminal repeat (MMTV) promoter. Mice with tumors from 0.5–2 cm in diameter are sacrificed with CO_2, the mice are dipped in 75% ethanol, and placed in a tissue culture hood. The fur and skin are removed with surgical scissors, exposing the underlying tissue and mammary tumors.

2. Each tumor is excised, dipped rapidly in 70% ethanol, and placed in a 10-cm tissue culture dish, keeping the tumors from each mammary fat fad separate.

3. The tumors are "diced" into small fragments with a scalpel directly in the tissue culture dish on which they will grow. Fragment size varies from 2–7 mm in diameter.

4. 20 ml of DMEM (high glucose, 33 mM) plus 20%FBS is added gently, such that tissue fragments remained where they were cut. The tissue culture dish will show scratch marks from dicing the tumors, and this roughening of the surface seems to provide points of adherence for the tumor cells, because they tend to grow out from the cut abrasions.

5. Plates are left undisturbed for 1–2 days in a standard tissue culture incubator with 5% CO_2 at 37°. The tissue fragments are removed, and

medium is replaced without disturbing the cells. The cultures are left to grow for 1–2 wk.

6. Colonies of tumor cells become visible to the eye at 1–4 wk by examining the bottom of the plate. The cells are trypsinized with a minimal volume of trypsin-EDTA diluted from a 10× stock in PBS (Invitrogen), and fresh medium is added to the dish (*see* Note 1).

7. After expanding the tumor cells to ~10^7 in a fresh flasks, they are trypsinized, and aliquots are cryopreserved in DMEM, 10% DMSO, 20% FBS in liquid nitrogen.

8. Tumor cells are characterized for expression of epithelial carcinoma markers by staining with antibodies to E-cadherin, whereas vimentin and smooth muscle α-actin positive cells indicate a myofibroblast phenotype.

Peritoneal Macrophage

1. Mice are injected in the peritoneal cavity with a sublethal dose of 25 μg lipopolysaccharide (LPS) (Sigma) in 0.5 ml of PBS to elicit activated macrophage.
2. Mice are sacrificed by CO_2 asphyxiation 5 d after the injection, and macrophage is recovered by peritoneal lavage with 10 ml of ice-cold PBS.
3. Cells are washed 1× and plated onto tissue culture dishes in macrophage serum free media (Invitrogen Cat# 12065–074). After 3 h in a 37° tissue culture incubator, the monolayers are vigorously washed with PBS to selectively recover the adherent macrophages.
4. Fresh macrophage medium is added, and the cells are maintained at 37°, 5% CO_2, and used for signaling experiments within 24 h of their isolation.

Nuclear Translocation Assay

1. Cells are plated in 96-well tissue plates (Corning #3595) at a density of 5000 cells/well in 100 μl of DMEM, 10% FBS (*see* Note 2).

2. The medium is replaced with 100 μl of DMEM for 18–24 h.

3. Cells are treated with 50 μl DMEM containing TGF-β1 or EGF at various concentrations and incubated for the indicated times (*see* Note 3).

4. Cells are then fixed with 3.7 % formaldehyde in PBS for 15 min and washed three times in PBS.

5. Cells are permeabilized with 100% methanol for 2 min, then washed three times with PBS (*see* Note 4).

6. To measure Smad2/3 nuclear translocation, mouse anti-Smad2/3 antibody is diluted 1:1000 in PBS plus 10% FBS and added at 50 μl/well. Plates can be incubated for 1 h at 20° or overnight at 4°.

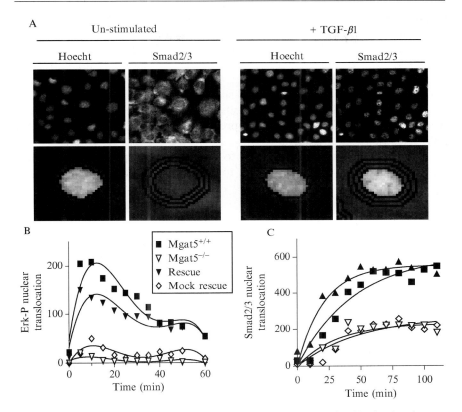

FIG. 1. Accumulation of Smad2/3 in the nuclei of control and TGF-β1–stimulated tumor cells. (A) Images of Mgat5$^{+/+}$ tumor cells stained with Hoecht 33258 and AlexaFluor 488 antibody staining for Smad2/3, either untreated or treatment with 5 ng/ml of TGF-β1 for 55 min. In the high-magnification images below, the inner black ring marks the nuclear mask of a single cell, and the area between the two outer rings is the cytoplasmic area. (B) Erk-P nuclear translocation in Mgat5$^{+/+}$ and Mgat5$^{-/-}$ tumor cells stimulated with 100 ng/ml of EGF. Mgat5$^{-/-}$ cells were infected with retroviral vectors for expression of either Mgat5 (rescued) or a mutant form of Mgat5 (L188R). (C) Smad2/3 nuclear translocation after the addition 5 ng/ml of TGF-β. Data is the mean ($n = 200$), and SE of the mean are <4%. (Reproduced from reference 1 with permission from *Science*).

7. To measure Erk phosphorylation and nuclear translocation, mouse anti-Erk-P antibodies are added at 1/1000 in PBS plus 10% FBS for 1 h at 20°.

8. The wells are washed three times with PBS.

9. AlexaFluor 488 conjugated anti-mouse antibody is diluted 1:1000 in PBS plus 10% FBS, plus 1:2000 dilution of Hoechst 33258, and added at 50 μl/well for 1 h at 20°.

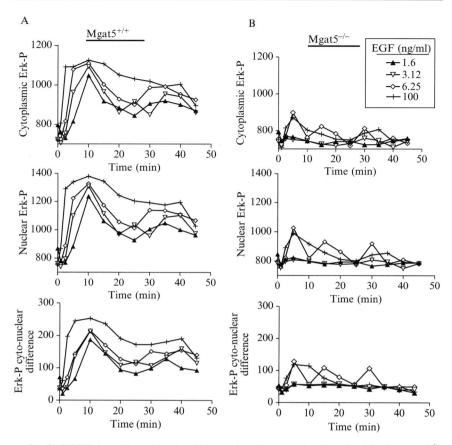

FIG. 2. EGFR-dependent phosphorylation and nuclear translocation of Erk1/2 in Mgat5$^{+/+}$ and Mgat5$^{-/-}$ tumor cells. At the indicated times after stimulation with 100 ng/ml of EGF, cells were fixed and stained for Erk-P. AlexaFluor 488 fluorescence was quantified by Scan Array in the cytoplasm (top), the nucleus (middle), and nuclear-cytoplasmic difference (bottom). The SE of the mean was <4% ($n = 100$) per assay point. (Reproduced with permission from *Science*).

10. Wells are washed three times with PBS, and a final aliquot of 100 μl/well PBS is added (*see* Note 5).

11. Cells are imaged on the Array Scan (Cellomics, Pittsburgh, PA) using the 10× objective, with channel 1 set to read Hoechst 33258 at 350 nM with an exposure of 0.5 sec, and channel 2 set for AlexaFluor 488 with an exposure of 10 sec. The cytoplasmic-nuclear translocation

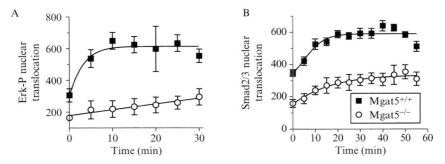

FIG. 3. Erk1/2 and Smad2/3 activation in primary macrophage. (A) Erk-P and (B) Smad2/3 nuclear translocation in LPS-elicited peritoneal macrophages from Mgat5$^{+/+}$ and Mgat5$^{-/-}$ mice stimulated 2 ng/ml of TGF-β1 and 5% FCS, respectively. The results are the mean ± SE of the well means at each time point for five mice per genotype (Reproduced with permission from *Science*).

program is used with the cytoplasmic ring width set at 2 μM, and the separation of nuclear and cytoplasmic region is set at 1.0 μM. The minimum nuclear area is set at 1.0 μm^2 and the maximum at 20 μm^2. Imaging proceeds until the program identifies 200 objects in channel 1 with the specified criteria or 20 fields are scanned before moved to the next well (*see* Note 6).

12. The difference in nuclear-cytoplasmic staining intensity is determined for 200 cells per well, and the mean ± SE is generated (*see* Note 7).

Notes

1. If tumor cell growth is robust, colonies can be isolated by trypsinization in silicone-sealed cloning rings and then transferred to new flasks.

2. Assay volumes can be adjusted for use with 12-, 24-, and 384-well tissue culture plates.

3. An 8- or 12-barrel multiwell hand pipette is used to transfer cells and reagents into the plates. Time courses are done by adding fixative to a row or column of well at each time point.

4. Plates can be covered and stored overnight at 4° with the addition of 150 μl PBS containing 10% FBS.

5. Plates can be sealed and stored for up to a week at 4° before imaging.

6. Commercial instruments are available that can quantify subcellular fluorescence of GFP-tagged proteins in live cells. Cells in 96- or 384-well plates are imaged with inverted 5×, 10×, 20×, or 40× objectives.

The stored image files can be reanalyzed with new computational constraints or with different programs.

7. Other primary cell types can be studied by automated and quantitative fluorescence imaging. We have used mouse muscle satellite cells and mouse embryonic fibroblasts. Other cytokines that activate Erk-P nuclear translocation can also be studied, and we have used basic fibroblast growth factor (b-FGF), insulin-like growth factor-1 (IGF-1), and platelet-derived growth factor (PDGF). Other proteins that translocate into the nucleus on stimulation such as STATs can be quantified using the same system.

Acknowledgments

This research was supported by operating grants from the Canadian Institute for Health Research, and J. W. D. holds a Canada Research Chair. Emily Partridge and Pam Cheung, PhD, students in my laboratory, developed and validated the methods.

References

Chen, L., Zhang, W., Fregien, N., and Pierce, M. (1998). The her-2/neu oncogene stimulates the transcription of N-acetylglucosaminyltransferase V and expression of its cell surface oligosaccharide products. *Oncogene* **17,** 2087–2093.

Chen, R. H., Sarnecki, C., and Blenis, J. (1992). Nuclear localization and regulation of erk- and rsk-encoded protein kinases. *Mol. Cell Biol.* **12,** 915–927.

Demetriou, M., Granovsky, M., Quaggin, S., and Dennis, J. W. (2001). Negative regulation of T-cell activation and autoimmunity by Mgat5 N-glycosylation. *Nature* **409,** 733–739.

Freeze, H. H. (2002). Human disorders in N-glycosylation and animal models. *Biochim. Biophys. Acta.* **1573,** 388–393.

Granovsky, M., Fata, J., Pawling, J., Muller, W. J., Khokha, R., and Dennis, J. W. (2000). Suppression of tumor growth and metastasis in Mgat5-deficient mice. *Nat. Med.* **6,** 306–312.

Haltiwanger, R. S., and Lowe, J. B. (2004). Role of glycosylation in development. *Annu. Rev. Biochem.* **73,** 491–537.

Kang, R., Saito, H., Ihara, Y., Miyoshi, E., Koyama, N., Sheng, Y., and Taniguchi, N. (1996). Transcriptional regulation of the N-acetylglucosaminyltransferase V gene in human bile duct carcinoma cells (HuCC-T1) is mediated by Ets-1. *J. Biol. Chem.* **271,** 26706–26712.

Lowe, J. B., and Marth, J. D. (2003). A genetic approach to Mammalian glycan function. *Annu. Rev. Biochem.* **72,** 643–691.

Macias-Silva, M., Abdollah, S., Hoodless, P. A., Pirone, R., Attisano, L., and Wrana, J. L. (1996). MADR2 is a substrate of the TGF-beta receptor and its phosphorylation is required for nuclear accumulation and signaling. *Cell* **87,** 1215–1224.

Oft, M., Akhurst, R. J., and Balmain, A. (2002). Metastasis is driven by sequential elevation of H-ras and Smad2 levels. *Nat. Cell Biol.* **4,** 487–494.

Partridge, E. A., Le Roy, C., Di Guglielmo, G. M., Pawling, J., Cheung, P., Granovsky, M., Nabi, I. R., Wrana, J. L., and Dennis, J. W. (2004). Regulation of cytokine receptors by Golgi N-glycan processing and endocytosis. *Science* **306**, 120–124.
Webster, M. A., Hutchinson, J. N., Rauh, M. J., Muthuswamy, S. K., Anton, M., Tortorice, C. G., Cardiff, R. D., Graham, F. L., Hassell, J. A., and Muller, W. J. (1998). Requirement for both Shc and phosphatidylinositol 3′ kinase signaling pathways in polyomavirus middle T-mediated mammary tumorigenesis. *Mol. Cell. Biol.* **18**, 2344–2359.

[2] Phenotype Changes of Fut8 Knockout Mouse: Core Fucosylation Is Crucial for the Function of Growth Factor Receptor(s)

By XIANGCHUN WANG, JIANGUO GU, EIJI MIYOSHI, KOICHI HONKE, and NAOYUKI TANIGUCHI

Abstract

α1,6-Fucosyltransferase (Fut8) catalyzes the transfer of a fucose residue to N-linked oligosaccharides on glycoproteins by means of an α1,6-linkage to form core fucosylation in mammals. In mice, disruption of Fut8 induces severe growth retardation, early death during postnatal development, and emphysema-like changes in the lung. A marked dysregulation of TGF-β1 receptor activation and signaling in Fut8-null mice lung results in over-expression of matrix metalloproteinases (MMPs), such as MMP12 and MMP13, and a down-regulation of extracellular matrix (ECM) proteins such as elastin, which contributes to the destructive emphysema-like phenotype observed in Fut8-null mice. Furthermore, therapeutic administration of exogenous TGF-β1 rescued the null mice from the emphysema-like phenotype. On the other hand, absence of Fut8 on EGF or PDGF receptor results in down-regulation of the receptor-mediated signaling, which is a plausible factor that may be responsible for the growth retardation. Reintroduction of the Fut8 gene to Fut8-null cells potentially rescued these receptor-mediated signaling impaired in null cells. Collectively, these results suggest that core fucosylation is crucial for growth factor receptors such as TGF-β1 and EGF receptor–mediated biological functions.

Overview

The remodeling of cell surface growth factor receptors by modification of their oligosaccharide structures is associated with certain functions and biological events (Akiyama *et al.*, 1989; Gregoriou, 1993; Hakomori Si, 2002;

Taniguchi *et al.*, 2001; Zheng *et al.*, 1994). Certain *N*-glycan structures of a number of glycoproteins seem to contribute to the folding, stability, and sorting of glycoproteins (Dwek, 1995; Wyss *et al.*, 1995). They have a core structure, and their branching patterns are determined by glycosyltransferases (Dennis *et al.*, 1999; Schachter, 1986; Taniguchi *et al.*, 1999).

GDP-L-Fuc:*N*-acetyl-β-D-glucosaminide α1,6-fucosyltransferase (Fut8, E.C.2.4.1.152) catalyzes the transfer of a fucose residue from GDP-fucose to position 6 of the innermost GlcNAc residue of hybrid and complex types of *N*-linked oligosaccharides on glycoproteins to form core fucosylation in mammals, as shown in Fig.1A. Fut8 is the only core FucT in mammals, but there are core α1,3-Fuc residues in plants, insects, and probably other species. The *Fut8* gene is expressed in most rat organs with a relatively high level expression in brain and small intestine (Miyoshi *et al.*, 1997). α1,6-Fucosylated glycoproteins are widely distributed in mammalian tissues and are altered under some pathological conditions. For example, the level of core fucosylation is elevated in both liver and serum during the process of hepatocarcinogenesis (Hutchinson *et al.*, 1991). The presence of core fucosylation of α-fetoprotein, a well-known tumor marker for hepatocellular carcinoma (HCC), is known to distinguish patients with HCC from those with chronic hepatitis and liver cirrhosis (Sato *et al.*, 1993; Taketa *et al.*, 1993). It has recently been reported that the deletion of the core fucose from the IgG1 molecule enhances antibody-dependent cellular cytotoxicity activity by up to 50–100 fold (Shields *et al.*, 2002; Shinkawa *et al.*, 2003) and, therefore, is thought to have considerable potential for use in antibody therapy against cancer. These findings strongly suggested that core fucosylation of *N*-glycans modifies the function of the glycoproteins.

To define the physiological roles of Fut8, Fut8-null mice were generated by gene targeting technology. A targeted disruption of *Fut8* was generated through homologous recombination in embryonic stem cells. The targeting vector was constructed by replacing exon 2 of *Fut8*, which contains the translation initiation site, with an IRES-*LacZ*-Neo-pA cassette (Fig. 1B). Disruption of Fut8 induces severe growth retardation, early death during postnatal development, and emphysema-like changes in the lung (Wang *et al.*, 2005). *Fut8*$^{-/-}$ mice were born apparently healthy with almost the expected Mendelian inheritance: Of 277 pups, there were 59 (21.3%) *Fut8*$^{-/-}$, 147 (53.1%) *Fut8*$^{+/-}$, and 71 (25.6%) *Fut8*$^{+/+}$ mice. The appearance of *Fut8*$^{-/-}$ mice could not be distinguished from *Fut8*$^{+/-}$ and *Fut8*$^{+/+}$ mice within 3 days of age, but approximately 70% of them died during this period (Fig. 2A). Most of the survivors manifested severe growth retardation (Fig. 2B). In fact, we found that dysregulation of TGF-β1 receptor activation leads to abnormal lung development and emphysema-like phenotype in Fu8-null mice (Wang *et al.*, 2005) and

A

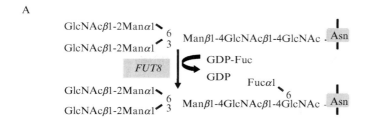

B

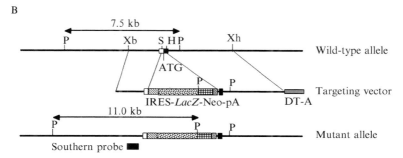

FIG. 1. Reaction pathway of core fucose synthesis and targeted disruption of Fut8 locus. (A) Man, mannose; Fuc, fucose; GDP-Fuc, guanosinediphosphofucopyranoside; Asn, asparagine. (B) The *Fut8* gene (wild-type allele; top), the targeting vector (middle), and the disrupted Fut8 locus (mutant allele; bottom). The box in the *Fut8* gene represents exon 2, which includes the translation–initiation site (ATG). A 184-bp deletion of exon 2 containing the translation–initiation site was replaced with an IRES-*LacZ*-Neo-pA cassette. The expected size of the Pst I digestion products of the gene, hybridized with the indicated probe, is shown for the wild-type allele (7.5 kb) and for the mutant allele (11.0 kb). Restriction enzyme sites: P, Pst I; Xb, Xba I; S, *Sac* I; H, *Hind* III; Xh, *Xho* I.

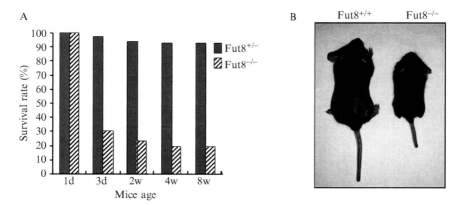

FIG. 2. Semilethality and growth retardation in Fut8$^{-/-}$ mice. (A) Survival ratio of Fut8$^{-/-}$ (–/–, striped bar), Fut8$^{+/-}$ (+/–, gray bar) mice after birth. (B) A 16-day-old Fut8$^{-/-}$ pup (–/–) with a Fut8$^{+/+}$ litter mate (+/+).

down-regulation of EGF receptor, as well as PDGF receptor activation, are plausible factors that may be responsible for the growth retardation (Wang *et al.*, 2006). Here, roles of core fucosylation on TGF-β1 and EGF receptors are described.

Lacking Core Fucosylation on TGF-β Receptor Type II Leads to Emphysema-Like Changes in Fut8-null Mice Lung

The lungs of $Fut8^{-/-}$ mice apparently displayed generalized air space enlargement and dilated alveolar ducts compared with those of $Fut8^{+/+}$ mice (Fig. 3). By calculation of mean linear intercept (MLI), diameters of the pulmonary alveoli of $Fut8^{-/-}$ mice were increased significantly from postnatal day 7 (Wang *et al.*, 2005).

Pulmonary emphysema is believed to result from decreased structural integrity of connective tissues because of a defect in their formation or to an abnormal proteolysis. Elastin and fibrillar collagen are major components of the extracellular matrix (ECM), which sustains the normal lung architecture. On the other hand, matrix metalloproteinases (MMPs) are a

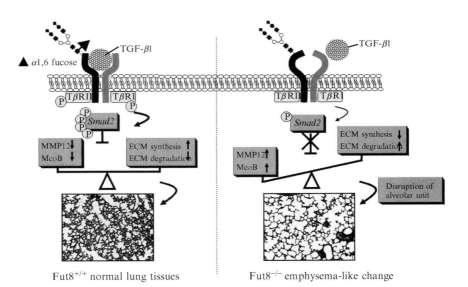

Fut8$^{+/+}$ normal lung tissues   Fut8$^{-/-}$ emphysema-like change

FIG. 3. Lacking core fucosylation on TGF-β receptor type II leads to emphysema-like changes in Fut8-null mice lung. In mice lacking core fucosylation of TGF-β type II receptor, the reduced binding ability of TGF-β1 to its receptor results in down-regulation of *Samd2* activation. This disturbed the homeostasis of ECM and MMPs. The increase in proteolytic potential leads to destruction of lung tissue and emphysema in Fut8-null mice.

group of zinc- and calcium-dependent proteinases that have an important role in the normal turnover of ECM components. Abnormal production of MMPs is implicated in the induction of emphysema. RT-PCR analysis showed that expression levels of *McolB* (a mouse ortholog of human MMP-1), MMP-12, and MMP-13 were greatly enhance in lung tissue from $Fut8^{-/-}$. Conversely, elastin expression was down-regulated in lungs from Fut8-deficient mice. On the other hand, fragmentation and a significantly reduced number of elastic fibers were observed by elastin staining in $Fut8^{-/-}$ mice. These results suggest that overexpression of a set of MMPs might be causally linked to the development of emphysema in $Fut8^{-/-}$ mice.

The TGF-β1 receptor-mediated signaling pathway is a key pathway for regulating expression of ECM proteins, including suppression of MMPs to produce a "synthetic" phenotype (Massague *et al.*, 2000). The enhancement of *McolB* and MMP12 were block by TGF-β1 treatment in $Fut8^{+/+}$ cells but not in $Fut8^{-/-}$ *cells*, indicating that the deletion of *Fut8* diminishes TGF-β1 mediated signaling. Actually, abolishment of core fucosylation on TGF-β type II receptor in $Fut8^{-/-}$ cells results in reduced binding ability of TGF-β1 to its receptor compared with $Fut8^{+/+}$ cells. Furthermore, TGF-β1 receptor-mediated signaling was suppressed in $Fut8^{-/-}$ cells and lung by carrying out *Smad*2 phosphorylation analysis. The TGF-β1 signaling deficiency was restored by re-introduction of $Fut8^{-/-}$ cells with wild-type *Fut8* gene. Importantly, administration of exogenous TGF-β1 resulted in a significant rescue of the emphysema-like phenotype, stimulated the formation of elastin fiber, and, concomitantly, reduced MMP-12 expression in Fut8$^{-/-}$ lung. In additional, using antibodies specific for surfactant protein C (SP-C, a marker of differentiated type-II alveolar epithelial cells), the expression levels of SP-C protein at each stage were slightly weaker in $Fut8^{-/-}$ lungs than in $Fut8^{+/+}$ lungs, suggesting that lung development was also disturbed by the loss of core fucosylation.

Taken together, we propose that the lack of core fucosylation of TGF-β1 receptor is crucial for developmental and progressive/destructive emphysema, suggesting that perturbation of this function could underlie certain cases of human emphysema (Fig. 3).

Core Fucosylation Regulates EGF Receptor–Mediated Intracellular Signaling

Epidermal growth factor receptor (EGFR)–mediated cellular responses to EGF and transforming growth factor-α stimulation regulate several biological functions, including cell growth and cell differentiation. The binding of these ligands to the extracellular domain of EGFR induces activation of its intrinsic tyrosine kinase activity, leading to the receptor autophosphorylation and the phosphorylation of tyrosine residues in various cellular substrates, many of which serve as intracellular signal molecules (Carpenter and Cohen, 1990; Schlessinger, 1988; Ullrich and Schlessinger, 1990). The extracellular domain of EGFR contains 12 potential N-glycosylation sites (Ullrich *et al.*, 1984), and the remodeling of N-glycans on EGFR can modulate EGFR-mediating functions (Gu *et al.*, 2004; Hazan *et al.*, 1995; Rebbaa *et al.*, 1996, 1997; Soderquist and Carpenter, 1984; Zeng *et al.*, 1995). It has been reported that the binding of EGF to EGFR was significantly reduced by treatment with some N-glycosylation inhibitors (Soderquist and Carpenter, 1984). In addition, EGF binding, as well as its tyrosine kinase activity, was reduced by addition of certain lectins (Hazan *et al.*, 1995; Rebbaa *et al.*, 1996; Zeng *et al.*, 1995), indicating that N-glycans are required for ligand binding. On the other hand, the overexpression of N-acetylglucosaminyltransferase III (GnT-III), a pivotal glycosyltransferase that plays a major role in the biosynthesis of hybrid and complex types of N-linked oligosaccharides (Nishikawa *et al.*, 1992), significantly reduces the ability of EGF to bind to its receptor, EGFR autophosphorylation, and subsequently blocks EGFR-mediated ERK phosphorylation in U373 MG glioma cells (Rebbaa *et al.*, 1997) or PC12 cells (Gu *et al.*, 2004). Recently it was also reported that N-glycans of EGFR, as well as other cytokine receptors modified by GnT-V, which catalyzes the formation of GlcNAcβ1,6 branches, play an important role in the endocytosis of EGFR to regulate its expression levels on the cell surface (Partridge *et al.*, 2004). Thus, N-linked oligosaccharides on EGFR seem to be important factors for receptor function. However, to date, the roles of core fucosylation in EGFR-mediating functions have not been identified yet.

The epidermal growth factor (EGF)–induced phosphorylation levels of the EGF receptor (EGFR) were substantially blocked in Fut8$^{-/-}$ cells compared with Fut8$^{+/+}$ cells, whereas there are no significant changes in the total activities of tyrosine phosphatase for phosphorylated EGFR between two cells. The inhibition of EGFR phosphorylation was completely restored by re-introduction of the Fut8 gene to Fut8$^{-/-}$ cells. Moreover, the tyrosine-phosphorylation levels of the EGF receptor in *Fut8*-null embryos were lower than that in wild-type embryos (Wang *et al.*, 2006). Consistent with this, EGFR-mediated JNK or ERK activation was significantly suppressed in Fut8$^{-/-}$ cells. The down-regulation of JNK and ERK activation in Fu8$^{-/-}$ cells was rescued in the restored cells. Furthermore, these differences in responsiveness to EGF stimulation between Fut8$^{+/+}$ and Fut8$^{-/-}$ cells were much more obvious at low concentrations of EGF stimulation ($\sim$0.05 ng/ml) rather than higher concentrations (0.1$\sim$ ng/ml), indicating the high binding affinity of EGF to its receptor is mainly down-regulated by a lack of core fucosylation. In fact, it has been reported that EGFR kinase activation occurs exclusively through the high-affinity subclass (Bellot *et al.*, 1990). It is noteworthy that down-regulation of phosphorylation of ERK induced by PDGF was also observed in Fu8$^{-/-}$ cells. Although there are no significant changes in FGF-mediated signaling, the possibility that other growth factor receptor-mediated signaling may also affect cell growth cannot be excluded.

The binding of ^{125}I-EGF to EGFR was reduced in Fut8$^{-/-}$ cells compared with Fut8$^{+/+}$ or the restored cells at low doses, whereas similar levels of binding were found at relatively high concentrations (Wang *et al.*, 2006). A Scatchard analysis revealed that both low- and high-affinity binding of EGFR were present in Fut8$^{+/+}$ and the restored cells, but only low affinity EGFR was detected in Fut8$^{-/-}$ cells. Thus, these results suggest that the modulation of N-glycans by core fucosylation on EGFR may regulate the high-affinity binding EGF to EGFR, which is required and sufficient for EGF-induced responses (Gregorou and Rees, 1984; Defize *et al.*, 1988, 1989), but not for the low affinity of EGFR.

Collectively, these results strongly suggest that core fucosylation is essential for EGFR-mediated biological functions. Down-regulation of EGFR-mediated signaling because of lack of core fucosylation, in part, may attribute to growth retardation in Fut8$^{-/-}$ mice (Fig. 4).

Materials and Methods

Gene Targeting

A part of the mouse *Fut8* gene spanning 13.9 kb, which includes the exon containing the translation-initiation site, was isolated by screening a mouse 129SvJ λ genomic library (Stratagene, La Jolla, CA) using a *Sac*I-*Sac*I fragment of porcine *Fut8* cDNA (nt-39 to 373) (Uozumi *et al.*, 1996; Yanagidani *et al.*, 1997) as a probe. A targeting vector was constructed by replacing the 184-bp *Sac*I–*Hind* III fragment containing the translation-initiation site with a 4.9-kb *Sac*I–*Sal*I fragment of the plasmid pGT1.8IresBgeo (Mountford *et al.*, 1994) that contains an internal ribosome entry site (*IRES*)-*LacZ-Neo'*- polyadenylation signal (*pA*) cassette, flanked with a 1.5-kb *Xho*I-*Not*I fragment of the plasmid pMC1DTpA (Yanagawa *et al.*, 1999), which encodes diphtheria toxin A chain (*DT-A*) for negative screening (Fig. 1B). The targeting vector was transfected into D3 embryonic stem cells, and clones were selected with G418. Southern blot analysis of selected clones with 5' (*A*) and 3' (*B*) probes revealed that 1.2% (4 of 343) of the embryonic stem clones had undergone correct homologous recombination. Targeted cell clones were then injected into

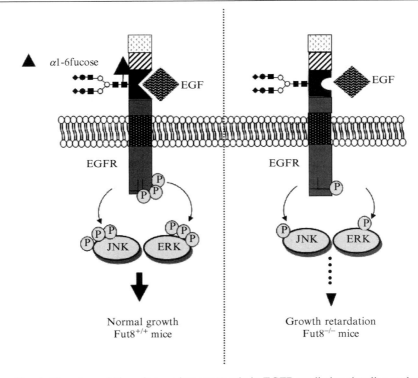

FIG. 4. Core fucosylation plays an important role in EGFR-mediating signaling pathway Absence of core fucosylation on EGFR results in down-regulation of EGFR phosphorylation. EGFR-mediated downstream signaling such as phosphorylation of JNK or ERK was also decreased in Fut8$^{-/-}$ cells. Down-regulation of EGFR-mediated signaling caused by lack of core fucosylation, in part, may attribute to growth retardation in Fut8$^{-/-}$ mice.

blastocysts from B6C3F1 mice, which are F1 mice resulting from the intercross of female C57BL/6 and male C3H mice. Germ-line transmission of the mutant allele was achieved from male chimeras derived from two independent embryonic stem cell clones.

Establishment of Embryonic Fibroblasts

For preparation of embryonic fibroblasts, a whole mouse embryo at 18.5 days after coitus was dissected, and the head and all internal organs were removed. The carcasses were minced, incubated in PBS (–) containing 0.05% trypsin, 0.53 mM EDTA, and 40 μg/ml DNase at 37° for 30 min with stirring three times, and then cells were plated on a 100-mm dish in

Dulbecco's modified Eagle's medium (DMEM) supplemented with 10% fetal calf serum and incubated at 37° in humidified air containing 5% CO_2. To obtain immortal cells, Zeocine-resistant vector (pcDNA3.1) containing the SV40 gene was introduced to these primary embryonic fibroblasts. Transfectants were screened in the presence of 400 μg/ml Zeocine, and SW and SK immortal cells were established from $Fut8^{+/+}$ and $Fut8^{-/-}$ primary fibroblasts, respectively.

Fut8 Activity Assay

The specific activity of Fut8 was determined using a synthetic substrate, 4-(2-pyridylamio)-butyl-amine (PABA)–labeled oligosaccharide as a substrate. Cells grown to subconfluence were washed with PBS (–) once, and the cell pellet was suspended in 200 μl lysis buffer containing 10 mM Tris-HCl, pH 7.4, 150 mM NaCl, and 1% Triton X-100. The cell lysate was then assayed for Fut8 activity as described before (Uozumi *et al.*, 1996).

Lectin Blotting Analysis

Whole cell lysate or immunoprecipitated EGFR was subjected to 10% or 7.5% SDS-PAGE and transferred to PVDF membranes, the membranes were blocked with 5% BSA in TBST overnight at 4°, and then incubated with 0.5 μg/ml biotinylated aleuria aurantia lectin (AAL) (Seikagaku Corp., Japan), which preferentially recognizes Fucα1,6GlcNAc structure, in TBST for 1 h at room temperature. After washing with TBST four times, lectin reactive proteins were detected using a Vectastain ABC kit (Vector laboratories, Burlingame, CA) and ECL kit.

TGF-β1 Binding Assay

The cells (1.5×10^5/well) were cultured on 24-well plates, and washed twice with 500 μl of PBS containing 0.1% BSA, and then incubated with 200 μl of PBS containing different amounts of ^{125}I-TGF-β1 in a concentration range of 0.1–1.0 ng and 10 ng of unlabeled TGF-β1. Nonspecific binding was determined by adding 100 ng of unlabeled TGF-β1. After incubation for 2 h at 4° with shaking, the cells were washed three times with ice-cold PBS containing 0.1% BSA, and then solubilized in 500 μl of 1 N NaOH. The radioactivity of the cell lysates was counted with a γ-counter (Wang *et al.*, 2005).

EGF Binding Assay

The cells were seeded at a density of 1×10^5 cell/well in 24-well plates and incubated overnight. The medium was then replaced with DMEM, which contained 0.1%BSA (M-BSA) and incubated for 20 min at 37°. After replacing the medium with ice-cold M-BSA, ^{125}I-EGF and EGF mixture was added, followed by incubation for 2 h at 4°. The cells were then washed with ice-cold PBS and hydrolyzed on 0.5 ml 1 M NaOH. The radioactivity was counted with a γ counter (Wang et al., 2006).

References

Akiyama, S. K., Yamada, S. S., and Yamada, K. M. (1989). Analysis of the role of glycosylation of the human fibronectin receptor. J. Biol. Chem. **264**, 18011–18018.

Bellot, F., Moolenaar, W., Kris, R., Mirakhur, B., Verlaan, I., Ullrich, A., Schlessinger, J., and Felder, S. (1990). High-affinity epidermal growth factor binding is specifically reduced by a monoclonal antibody, and appears necessary for early responses. J. Cell. Biol. **110**, 491–502.

Carpenter, G., and Cohen, S. (1990). Epidermal growth factor. J. Biol. Chem. **265**, 7709–7712.

Defize, L. H., Arndt-Jovin, D. J., Jovin, T. M., Boonstra, J., Meisenhelder, J., Hunter, T., de Hey, H. T., and de Laat, S. W. (1988). A431 cell variants lacking the blood group A antigen display increased high affinity epidermal growth factor-receptor number, protein-tyrosine kinase activity, and receptor turnover. J. Cell. Biol. **107**, 939–949.

Defize, L. H., Boonstra, J., Meisenhelder, J., Kruijer, W., Tertoolen, L. G., Tilly, B. C., Hunter, T., van Bergen en Henegouwen, P. M., Moolenaar, W. H., and de Laat, S. W. (1989). Signal transduction by epidermal growth factor occurs through the subclass of high affinity receptors. J. Cell. Biol. **109**, 2495–2507.

Dennis, J. W., Granovsky, M., and Warren, C. E. (1999). Glycoprotein glycosylation and cancer progression. Biochim. Biophys. Acta **1473**, 21–34.

Dwek, R. A. (1995). Glycobiology: More functions for oligosaccharides. Science **269**, 1234–1235.

Gregoriou, M. (1993). Purification of the EGF receptor for oligosaccharide studies. Methods Mol. Biol. **14**, 189–198.

Gregoriou, M., and Rees, A. R. (1984). Properties of a monoclonal antibody to epidermal growth factor receptor with implications for the mechanism of action of EGF. EMBO J. **3**, 929–937.

Gu, J., Zhao, Y., Isaji, T., Shibukawa, Y., Ihara, H., Takahashi, M., Ikeda, Y., Miyoshi, E., Honke, K., and Taniguchi, N. (2004). Beta1,4-N-Acetylglucosaminyltransferase III down-regulates neurite outgrowth induced by costimulation of epidermal growth factor and integrins through the Ras/ERK signaling pathway in PC12 cells. Glycobiology **14**, 177–186.

Hakomori Si, S. I. (2002). Inaugural article: The glycosynapse. Proc. Natl. Acad. Sci. USA **99**, 225–232.

Hazan, R., Krushel, L., and Crossin, K. L. (1995). EGF receptor-mediated signals are differentially modulated by concanavalin A. J. Cell. Physiol. **162**, 74–85.

Hutchinson, W. L., Du, M. Q., Johnson, P. J., and Williams, R. (1991). Fucosyltransferases: Differential plasma and tissue alterations in hepatocellular carcinoma and cirrhosis. Hepatology **13**, 683–688.

Massague, J., Blain, S. W., and Lo, R. S. (2000). TGFbeta signaling in growth control, cancer, and heritable disorders. Cell **103**, 295–309.

Miyoshi, E., Uozumi, N., Noda, K., Hayashi, N., Hori, M., and Taniguchi, N. (1997). Expression of alpha1-6 fucosyltransferase in rat tissues and human cancer cell lines. Int. J. Cancer **72**, 1117–1121.

Mountford, P., Zevnik, B., Duwel, A., Nichols, J., Li, M., Dani, C., Robertson, M., Chambers, I., and Smith, A. (1994). Dicistronic targeting constructs: Reporters and modifiers of mammalian gene expression. Proc. Natl. Acad. Sci. USA **91**, 4303–4307.

Nishikawa, A., Ihara, Y., Hatakeyama, M., Kangawa, K., and Taniguchi, N. (1992). Purification, cDNA cloning, and expression of UDP-N-acetylglucosamine: Beta-D-mannoside beta-1,4 N-acetylglucosaminyltransferase III from rat kidney. *J. Biol. Chem.* **267,** 18199–18204.

Partridge, E. A., Le Roy, C., Di Guglielmo, G. M., Pawling, J., Cheung, P., Granovsky, M., Nabi, I. R., Wrana, J. L., and Dennis, J. W. (2004). Regulation of cytokine receptors by Golgi N-glycan processing and endocytosis. *Science* **306,** 120–124.

Rebbaa, A., Yamamoto, H., Moskal, J. R., and Bremer, E. G. (1996). Binding of erythroagglutinating phytohemagglutinin lectin from Phaseolus vulgaris to the epidermal growth factor receptor inhibits receptor function in the human glioma cell line, U373 MG. *J. Neurochem.* **67,** 2265–2272.

Rebbaa, A., Yamamoto, H., Saito, T., Meuillet, E., Kim, P., Kersey, D. S., Bremer, E. G., Taniguchi, N., and Moskal, J. R. (1997). Gene transfection-mediated overexpression of beta 1,4-N-acetylglucosamine bisecting oligosaccharides in glioma cell line U373 MG inhibits epidermal growth factor receptor function binding of erythroagglutinating phytohemagglutinin lectin from *Phaseolus vulgaris* to the epidermal growth factor receptor inhibits receptor function in the human glioma cell line, U373 MG. *J. Biol. Chem.* **272,** 9275–9279.

Sato, Y., Nakata, K., Kato, Y., Shima, M., Ishii, N., Koji, T., Taketa, K., Endo, Y., and Nagataki, S. (1993). Early recognition of hepatocellular carcinoma based on altered profiles of alpha-fetoprotein. *N. Engl. J. Med.* **328,** 1802–1806.

Schachter, H. (1986). Biosynthetic controls that determine the branching and microheterogeneity of protein-bound oligosaccharides. *Biochem. Cell. Biol.* **64,** 163–181.

Schlessinger, J. (1988). Signal transduction by allosteric receptor oligomerization. *Trends Biochem. Sci.* **13,** 443–447.

Shields, R. L., Lai, J., Keck, R., O'Connell, L. Y., Hong, K., Meng, Y. G., Weikert, S. H., and Presta, L. G. (2002). Lack of fucose on human IgG1 N-linked oligosaccharide improves binding to human Fcgamma RIII and antibody-dependent cellular toxicity. *J. Biol. Chem.* **277,** 26733–26740.

Shinkawa, T., Nakamura, K., Yamane, N., Shoji-Hosaka, E., Kanda, Y., Sakurada, M., Uchida, K., Anazawa, H., Satoh, M., Yamasaki, M., Hanai, N., and Shitara, K. (2003). The absence of fucose but not the presence of galactose or bisecting N-acetylglucosamine of human IgG1 complex-type oligosaccharides shows the critical role of enhancing antibody-dependent cellular cytotoxicity. *J. Biol. Chem.* **278,** 3466–3473.

Soderquist, A. M., and Carpenter, G. (1984). Glycosylation of the epidermal growth factor receptor in A-431 cells. The contribution of carbohydrate to receptor function. *J. Biol. Chem.* **259,** 12586–12594.

Taketa, K., Endo, Y., Sekiya, C., Tanikawa, K., Koji, T., Taga, H., Satomura, S., Matsuura, S., Kawai, T., and Hirai, H. (1993). A collaborative study for the evaluation of lectin-reactive alpha-fetoproteins in early detection of hepatocellular carcinoma. *Cancer Res.* **53,** 5419–5423.

Taniguchi, N., Ekuni, A., Ko, J. H., Miyoshi, E., Ikeda, Y., Ihara, Y., Nishikawa, A., Honke, K., and Takahashi, M. (2001). A glycomic approach to the identification and characterization of glycoprotein function in cells transfected with glycosyltransferase genes. *Proteomics.* **1,** 239–247.

Taniguchi, N., Miyoshi, E., Ko, J. H., Ikeda, Y., and Ihara, Y. (1999). Implication of N-acetylglucosaminyltransferases III and V in cancer: Gene regulation and signaling mechanism. *Biochim. Biophys. Acta* **1455,** 287–300.

Ullrich, A., Coussens, L., Hayflick, J. S., Dull, T. J., Gray, A., Tam, A. W., Lee, J., Yarden, Y., Libermann, T. A., and Schlessinger, J. (1984). Human epidermal growth factor receptor cDNA sequence and aberrant expression of the amplified gene in A431 epidermoid carcinoma cells. *Nature* **309,** 418–425.

Ullrich, A., and Schlessinger, J. (1990). Signal transduction by receptors with tyrosine kinase activity. *Cell* **61,** 203–212.

Uozumi, N., Yanagidani, S., Miyoshi, E., Ihara, Y., Sakuma, T., Gao, C. X., Teshima, T., Fujii, S., Shiba, T., and Taniguchi, N. (1996). Purification and cDNA cloning of porcine brain GDP-L-Fuc:N-acetyl-beta-D-glucosaminide alpha1-6fucosyltransferase. *J. Biol. Chem.* **271,** 27810–27817.

Wang, X., Gu, J., Ihara, H., Miyoshi, E., Honke, K., and Taniguchi, N. (2006). Core fucosylation regulates epidermal growth factor receptor-mediated intracellular signaling. *J. Biol. Chem.* **281,** 2572–2577.

Wang, X., Inoue, S., Gu, J., Miyoshi, E., Noda, K., Li, W., Mizuno-Horikawa, Y., Nakano, M., Asahi, M., Takahashi, M., Uozumi, N., Ihara, S., Lee, S. H., Ikeda, Y., Yamaguchi, Y., Aze, Y., Tomiyama, Y., Fujii, J., Suzuki, K., Kondo, A., Shapiro, S. D., Lopez-Otin, C., Kuwaki, T., Okabe, M., Honke, K., and Taniguchi, N. (2005). Dysregulation of TGF-beta1 receptor activation leads to abnormal lung development and emphysema-like phenotype in core fucose-deficient mice. *Proc. Natl. Acad. Sci. USA* **102,** 15791–15796.

Wyss, D. F., Choi, J. S., Li, J., Knoppers, M. H., Willis, K. J., Arulanandam, A. R., Smolyar, A., Reinherz, E. L., and Wagner, G. (1995). Conformation and function of the N-linked glycan in the adhesion domain of human CD2. *Science* **269,** 1273–1278.

Yanagawa, Y., Kobayashi, T., Ohnishi, M., Tamura, S., Tsuzuki, T., Sanbo, M., Yagi, T., Tashiro, F., and Miyazaki, J. (1999). Enrichment and efficient screening of ES cells containing a targeted mutation: The use of DT-A gene with the polyadenylation signal as a negative selection maker. *Transgenic Res.* **8,** 215–221.

Yanagidani, S., Uozumi, N., Ihara, Y., Miyoshi, E., Yamaguchi, N., and Taniguchi, N. (1997). Purification and cDNA cloning of GDP-L-Fuc:N-acetyl-beta-D-glucosaminide:alpha1-6 fucosyltransferase (alpha1-6 FucT) from human gastric cancer MKN45 cells. *J. Biochem. (Tokyo)* **121,** 626–632.

Zeng, F. Y., Benguria, A., Kafert, S., Andre, S., Gabius, H. J., and Villalobo, A. (1995). Differential response of the epidermal growth factor receptor tyrosine kinase activity to several plant and mammalian lectins. *Mol. Cell. Biochem.* **142,** 117–124.

Zheng, M., Fang, H., and Hakomori, S. (1994). Functional role of N-glycosylation in alpha 5 beta 1 integrin receptor. De-N-glycosylation induces dissociation or altered association of alpha 5 and beta 1 subunits and concomitant loss of fibronectin binding activity. *J. Biol. Chem.* **269,** 12325–12331.

Section II

Neural Cell Function

[3] Cellular and Molecular Analysis of Neural Development of Glycosyltransferase Gene Knockout Mice

By KIYOHIKO ANGATA, WENJAU LEE, JUNYA MITOMA, JAMEY D. MARTH, and MINORU FUKUDA

Abstract

Recent studies demonstrate that carbohydrates synthesized by specific glycosyltransferases play important roles in the development of the central nervous system. Among these carbohydrates, polysialic acid is a unique glycan that modulates functions of the neural cell adhesion molecule (NCAM) by attenuating NCAM-mediated interaction between neural cells. During brain development, polysialic acid is synthesized in a specific spatio-temporal pattern by two polysialyltransferases, ST8SiaII and ST8SiaIV. To study *in vivo* the roles of polysialic acid synthesized by each respective enzyme, we generated ST8SiaII and ST8SiaIV knockout mice. Single knock-out ST8SiaII or ST8SiaIV mice show polysialic acid expression patterns differing from wild type, and those patterns indicate different roles of each gene during neural development. In this chapter, we discuss methods used to analyze polysialyltransferase knockout mice using immunohistochemistry of brain and primary cultures of neurons.

Overview

Recent findings demonstrate that various carbohydrate structures, in-cluding the neural cell-specific and core glycans, have important functions during development of the nervous system (Breen *et al.*, 1998; Jessell, 1990; Kleene and Schachner, 2004). In humans, mutations in specific glycosyl-transferases result in expression of unusual glycans, neurological conditions such as retardation, and morphological abnormalities in the brain (Aebi and Hennet, 2001; Endo, 2004; Jaeken and Matthijs, 2001; Olson and Walsh, 2002), and mice mutants in glycosyltransferases display similar phenotypes (Lowe and Marth, 2003). Analyzing phenotypes seen in glyco-syltransferase knockout mice is critical for evaluation of physiological roles of carbohydrates in neural development and should encourage develop-ment of treatments for neurological disease through use of these animal models to test potential therapeutics.

Polysialic acid, a linear homopolymer of $\alpha 2,8$-linked sialic acid, is mainly attached to the neural cell adhesion molecule (NCAM) in vertebrate brains

METHODS IN ENZYMOLOGY, VOL. 417   0076-6879/06 $35.00
DOI: 10.1016/S0076-6879(06)17003-2

(Edelman and Crossin, 1991; Rutishauser and Landmesser, 1996; Walsh and Doherty, 1997). Two polysialyltransferases, ST8SiaII and ST8SiaIV, are developmentally expressed and together synthesize polysialic acid on N-glycosylation sites in the fifth immunoglobulin domain of NCAM (see references in Angata and Fukuda, 2003). Given its anionic nature and long, linear and helical structure, polysialic acid can attenuate both NCAM–NCAM interaction and NCAM interactions with other molecules in the same membrane (*cis*-interactions) or in membranes of neighboring cells (*trans*-interactions) (Rutishauser and Landmesser, 1996). The biological function of polysialic acid has been explained by analyses using polysialic acid specific antibodies, endoneuraminidases that eliminate polysialic acid *in vitro* and *in vivo*, and NCAM-deficient mice (Bruses and Rutishauser, 2001; Cremer *et al.*, 2000). Phenotypes of NCAM-deficient mice, which by definition lack polysialylated NCAM, suggest that polysialylated NCAM is required for cell migration, neuronal path finding, and synaptic plasticity necessary for memory formation (Chazal *et al.*, 2000; Cremer *et al.*, 1994; 1998; Hu *et al.*, 1996; Muller *et al.*, 1996; Ono *et al.*, 1994; Tomasiewicz *et al.*, 1993). To explain the specific role of polysialic acid modifications in neural development, polysialyltransferase-deficient mice were generated (Angata *et al.*, 2004; Eckhardt *et al.*, 2000; Weinhold *et al.*, 2005). We describe methods used to analyze the role of glycosyltransferases and their products in neural development using these mouse mutants. Behavioral and electrophysiological analyses of these mice are described in Chapters 5–7.

Generation of Knockout Mice

To generate mice deficient in ST8SiaII or ST8SiaIV-deficient mice, the respective genes were disrupted by homologous recombination in ES cells, and cells were injected into blastocysts from C57Bl/6 mice. In ST8SiaII-deficient mice, exon 4 including sialyl motif L was disrupted, whereas ST8SiaIV-deficient mice lack part of exon 1 encoding the start methionine and transmembrane domain (Fig. 1A) (Eckhardt *et al.*, 2000; Angata *et al.*, 2004 and in preparation). As a control to evaluate the function of polysialic acid, NCAM-deficient mice (Cremer *et al.*, 1994) can be obtained from Jackson Laboratory. ST8SiaII or ST8SiaIV heterozygotes should be maintained on the same genetic background, in this case C57Bl/6, by repeatedly breeding with wild-type mice. All experiments using animals must be in accordance with NIH guidelines and approved in advance.

RT-PCR Analysis

RT-PCR is necessary not only to confirm gene disruption and loss of mRNA but also to analyze alterations in expression of murine ST8SiaII (mX), ST8SiaIV (mP), NCAM (mNC), or GAPDH (mGA) genes in the knockout mice. Total RNAs from the brains at various stages are prepared by

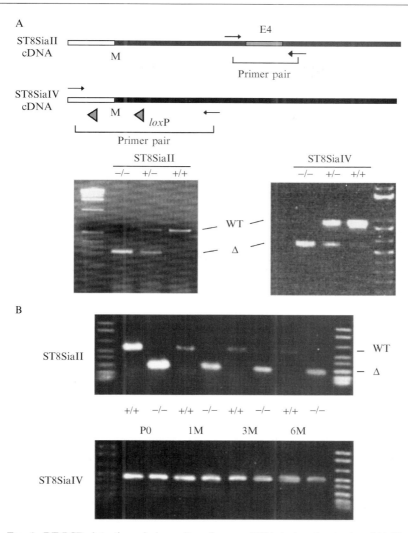

FIG. 1. RT-PCR detection of glycosyltransferase mRNA in knockout mice. (A) Upper schematic representation shows primer positions in cDNAs for ST8SiaII and ST8Sia IV. In ST8SiaII-deficient mice, shorter cDNA (△) is synthesized because of missing exon 4 (gray part in ST8SiaII cDNA). ST8SiaIV-deficient mice generate short mRNA lacking an area between two *lox*P sites. Wild-type (+/+) mice have a cDNA single band of the predicted size (WT), whereas homozygotes (−/−) show deleted cDNA (△), and heterozygotes (+/−) have both bands. (B) Developmental changes in polysialyltransferase expression in ST8SiaII-deficient mice. Total RNA was prepared from newborn pups (P0), 1-month-old (1M), 3-month-old (3M), and 6-month-old (6M) mice. Expression of ST8SiaII mRNA decreases during development. (Partly adapted from Angata *et al.*, 2004.)

using Trizol reagent (Invitrogen) and homogenized according to the manufacturer's protocol. For storage, samples homogenized in Trizol can be kept at −80° before adding chloroform. If necessary, to eliminate contaminating genomic DNA, RNase-free DNase (Roche Applied Science) should be added before cDNA synthesis. First-strand cDNA is synthesized using a reverse transcriptase, Superscript II (Invitrogen), and an oligo(dT) primer (Invitrogen) in the presence of RNase inhibitor (Promega). Single-stranded cDNAs are mixed with 10 pmol of 5′- and 3′-primers and Taq DNA polymerase (Perkin Elmer) in a 50-μl reaction. PCR is carried out for 35 cycles at 94° for 1 min, 60° for 1 min, and 72° for 2 min. RT-PCR primers are: mX-5: 5′-GAATTCTGGAGGCAGAGGTACAATCAGATC-3′ (nucleotides 99–128, nucleotides 1–3 encode the initiation methionine), mX-3: 5′-AAGGTCCTCAAAGGCCCGCTGGATGACAGA-3′ (nucleotides 651–622), mP-5: 5′-CCGCCACCTCCAATGCACAAGGTGTCACA T-3′ (nucleotides −297 to −268 or −297 − −268), mP-3: 5′-TTTCTCTGTCA CTCTCATTCCGAAAGCCTC-3′ (nucleotides 628–599), mNC-5: 5′-GCCAAGGAGAAATCAGCGTTGGAGAGTC-3′ (nucleotides 81–107), mNC-3: 5′-ATGCTCTTCAGGGTCAAGGAGGACACAC-3′ (nucleotides 1127–1100), mGA-5: 5′-CAGCAATGCATCCTGCACCACCAACTGC-3′ (nucleotides 435–462) and mGA-3: 5′-TTACTCCTTGGAGGCCATGT AGGCCATG-3′ (nucleotides 1002–975). PCR products shown in Fig. 1A show that mRNA encoding ST8SiaII or ST8SiaIV is deleted in each respective knockout mouse, suggesting no functional expression of each enzyme. RT-PCR analysis for ST8SiaII indicates that loss of functional ST8SiaII mRNA does not alter expression of ST8SiaIV mRNA (Fig. 1B).

Western Blot Analysis

Western blotting of cell lysates can assess quantitatively the extent to which loss of each polysialyltransferase affects polysialic acid synthesis, although unlike HPLC, Western blotting cannot determine the length of polysialic acid. (See later for difference of antibodies; for HPLC analysis, see Inoue and Inoue, 2003; Nakata *et al.*, 2005).

To prepare lysates, brains are removed, and various regions such as olfactory bulb, hippocampus, cerebral cortex, hypothalamus, and cerebellum are dissected and homogenized in RIPA buffer (150 mM NaCl; 50 mM Tris-HCl, pH 7.4; 1% NP40; 0.1% SDS; 5 mM EDTA) containing a proteinase inhibitor cocktail (Roche Applied Science). After centrifugation, protein concentration of supernatants is measured using a BCA kit (Pierce). Equivalent amounts of protein are loaded onto a 5% SDS-PAGE gels and transferred to PVDF membranes (Millipore). Some extracts are incubated with endoneuraminidase (Troy *et al.*, 1987) for 1 h at 37° before loading to remove polysialic acid. Membranes are blocked with 10% dry milk in 20 mM Tris-buffered saline, pH 7.6, containing 0.1% Tween-20 (TBST) and incubated with either mouse anti-polysialic acid 5A5 (Developmental Studies Hybridoma Bank, University of Iowa (http://www.uiowa.edu/~dshbwww/)) diluted 1:1000 in TBST or rat anti-NCAM H28 (Immunotech) diluted 1:200 followed by peroxidase-conjugated anti-mouse IgM (1:4000) or anti-rat IgG (1:3000), respectively. Signals are detected using an ECL kit (Amersham Biosciences).

Immunohistochemical Analysis

Specific morphological changes or changes in expression of specific markers in knockout mice can be detected by tissue staining. For immuno-histochemical analysis, the most important point is selecting primary anti-bodies and proper fixation conditions. For example, to stain sections with a polysialic acid antibody, mice are perfused intracardially with phosphate-buffered saline (PBS) followed by 4% paraformaldehyde before preparing frozen sections. On the other hand, fixation with Carnoy's fixative (60% ethanol, 30% chloroform, 10% acetic acid) followed by paraffin embedding is suitable for detecting some other antigens (see next section for an example). Several antibodies have been used to detect polysialic acid in brain sections. Some, such as 2-2B (mouse IgM, Chemicon), 5A5 (mouse IgM) and 12F8 (rat IgM, BD Biosciences), are commercially available and can recognize different lengths of polysialic acid chains (Sato *et al.*, 2000). For instance, 2-2B recognizes polysialic acid longer than a tetra-mer, whereas 5A5 can recognize chains longer than a tri-mer.

Method for Polysialic Acid Staining of Frozen Sections

1. Wild-type or glycosyltransferase knockout litter mates are deeply anesthetized with Avertin (0.015 ml/g body weight) before perfusion. Anesthesia is verified by a lack of reaction to a toe pinch. Tails are cut to prepare genomic DNA to confirm genotypes by PCR or Southern blotting.

2. PBS (10–25 ml/adult mouse) is delivered to the left atrium through a needle connected to a peristaltic pump. Tissue is then fixed with freshly prepared 4% paraformaldehyde in 0.1 M phosphate buffer, pH 7.4, at the same speed by the peristaltic pump.

3. After the brain is post-fixed overnight at 4° in the same fixative, it is cryoprotected in 30% sucrose and frozen in OCT compound (Sakura Finetek) placed in a 2-methylbutane/dry ice bath. Sagittal or coronal sections 30-μm thick cut on a cryostat are collected into PBS.

4. Floating sections are transferred to 24-well plates by forceps or a brush, treated with 0.3% H_2O_2 for 5 min to inactivate intrinsic peroxidases, and blocked with PBS containing 1% normal goat serum and 0.25% Triton-X 100 for 1 h.

5. Sections are incubated with the polysialic acid antibody (12F8) diluted in the blocking solution for at least 1 h. After three washes in PBS for 5 min, sections are treated with biotinylated secondary antibodies (anti-rat IgM, Invitrogen) for 1 h.

6. After three washes with PBS for 5 min each, sections are treated with the ABC reagent and DAB substrate (Vector Laboratories), transferred onto slides, and counterstained with cresyl violet.

Figure 2 shows polysialic acid staining in the hippocampus of wild-type and polysialyltransferase knockout mice using the preceding method (Fig. 2A) and also using an fluorescence-labeled secondary antibody (Fig. 2B). For immunofluorescence, tissues are incubated with primary antibodies in 1% normal goat serum followed by Alexa Fluor 594-conjugated goat anti-rat IgM (Invitrogen). Sections are then reacted with Hoechst 33342 (Sigma) in PBS before mounting. This method is suitable for double labeling when two different antibodies are of different isotypes or species and can tolerate the same fixation conditions. If an antibody requires a stronger fixation or an antigen retrieval procedure, use paraffin embedding instead of cryosectioning as described in the following.

Using mouse antibody to analyze mouse tissues, some antibodies need to block immunoglobulins in sections to reduce non-specific background, since secondary antibodies against mouse immunoglobulins also react with endogenous immunoglobins. Blocking before application of primary antibody (step 5) using M.O.M.^{TM} (vector) or HistoMouse^{TM} (Invitrogen) is ideal for this purpose. Alternatively, labeling the first antibody using Zenon® (Invitrogen) is useful to avoid the second antibody step. The latter method employs fluorophore-, biotin- or enzyme-labeled Fab fragments of Fc-specific anti mouse IgG antibody.

5-Bromo-2'-Deoxyuridine (BrdU)-Labeling

Polysialic acid is often expressed on neural precursors and migrating cells, particularly in the dentate gyrus of the hippocampus and the subventricular zone, which are proliferative areas of adult brain (Gage, 2000; Lois et al., 1996; Seki and Arai, 1993). BrdU-labeling is a suitable method to evaluate the role of polysialic acid in both proliferation and migration, because only dividing cells are labeled, and their localization in the brain is readily apparent using an anti-BrdU antibody. BrdU (20 mg/ml in 0.007 N NaOH and 0.9% NaCl) is intraperitoneally injected into either polysialyltransferase-deficient or wild-type mice (50 mg/kg body weight) for five times at 2-h intervals over an 8-h period.

To measure distribution of embryonic neural stem cells in brain, BrdU is administered to pregnant heterozygous mice (100 mg/kg at embryonic day 14–17) that have been mated with heterozygous male mice. Brains from neonatal and 10-day postnatal mice are analyzed.

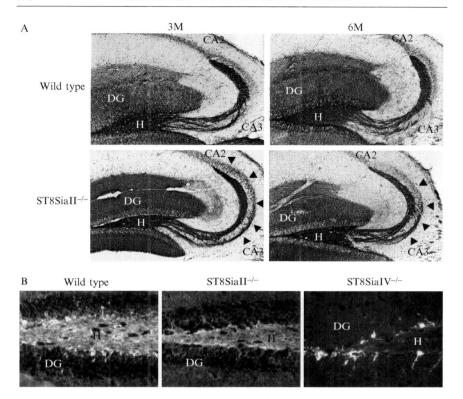

FIG. 2. Polysialic acid expression in hippocampus. (A) Immunohistochemical detection of polysialic acid in wild-type and ST8SiaII-deficient mice. Cryosections of hippocampus from 3-month-old (3M) and 6-month-old (6M) mice. Note that ST8SiaII-deficient mice have enlarged infrapyramidal mossy fibers in hippocampus, which express polysialic acid (black arrowheads) (B) Polysialic acid in the dentate gyrus is detected with an antipolysialic acid antibody, 12F8, followed by Alexa Fluor 594-conjugated secondary antibody (red). In wild-type mice, polysialic acid is expressed in the innermost granule cells and mossy fibers in the hilus. ST8SiaII-deficient mice express polysialic acid in the hilus but not in granule cells, whereas only granule cells express polysialic acid in the ST8SiaIV-deficient mice. DG, dentate gyrus; H, hilus. (Partly adapted from Angata *et al.*, 2004.) (See color insert.)

Method for BrdU Detection in Paraffin Sections

1. BrdU-injected mice are deeply anesthetized with Avertin as described previously.

2. After perfusion with PBS, brains are removed and fixed with Carnoy's fixative (60% ethanol, 30% chloroform, 10% acetic acid) overnight.

3. The fixative is replaced with ethanol, and brains are embedded in paraffin to cut sagittal or coronal sections of 10-μm thickness. Sections are collected onto glass slides.

4. Paraffin is removed with xylene, and sections are rehydrated by graded ethanol solutions. Sections are then treated with TBST (see the section for Western blot analysis) and blocked with PBS containing 1% normal goat serum and 0.25% Triton-X 100 for 1 h.

5. Sections are incubated with an anti-BrdU monoclonal antibody (Roche Applied Science) diluted in blocking solution for at least 1 h. After three washes in PBS for 5 min each, sections are treated with Alexa Fluor 488 goat anti-mouse IgG1 for 1 h.

6. Sections are counterstained with Hoechst 33342 and washed three times with PBS for 5 min. To prevent rapid quenching of fluorescence, a mounting solution containing an anti-fading reagent is recommended.

When BrdU detection is required in sections fixed in paraformaldehyde, double-stranded DNA should be denatured by HCl or DNase before treatment with the primary antibody.

Single knockout mice of ST8SiaII gene do not show a significant difference from wild-type mice in terms of neural cell growth and migration (Fig. 3). Because mice lacking either ST8SiaII or ST8SiaIV have a normal rostral migratory stream and olfactory bulb in contrast to those of NCAM-deficient mice, it is possible that both enzymes are functionally overlapped in polysialic acid synthesis required for cell migration of neural precursors.

Primary Cell Culture and Neurite Outgrowth Assay

We previously showed that polysialic acid expressed on substratum cells promotes neurite outgrowth of chicken dorsal root ganglion cells, indicating that polysialylated NCAM is a better substrate for axon extension and branching than NCAM alone (Angata *et al.*, 1997; Nakayama *et al.*, 1995). Thus, it is important to determine whether polysialic acid expressed on neurons is required for neurite outgrowth. To do so we undertook neurite outgrowth assays using primary hippocampal neurons from wild-type and knockout mice.

1. A pregnant mouse at E16 is euthanized by CO_2 gas, and embryos are removed from the uterus. The brain is removed and put in cold L-15 medium (Invitrogen). The hippocampus is isolated from brain tissue under a dissecting microscope.

2. Dissociate hippocampal neurons in 10 U Papain, 0.2 mg cysteine, 0.2 mg BSA, 5 mg glucose, and 0.01% DNase per 1 ml PBS at 37° for 15 min. The reaction is stopped by adding serum and centrifuged to precipitate hippocampal cells.

3. After resuspension in Neurobasal medium (Invitrogen) supplemented with B27 (Invitrogen), 0.5 mM glutamine, and penicillin-streptomycin, cells are further dissociated by pipetting.

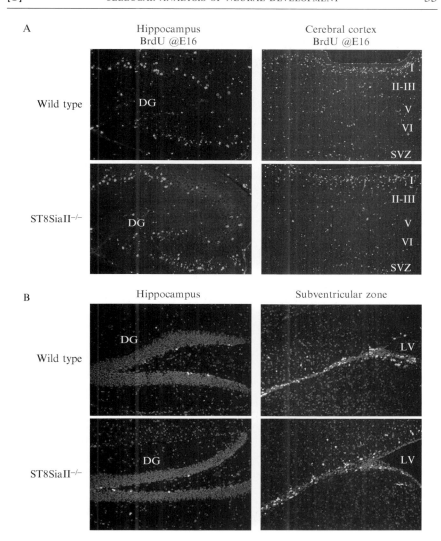

FIG. 3. BrdU-labeling and detection in mouse brain. To label mitotic cells in embryonic and adult brains, BrdU was injected into a pregnant mouse at embryonic day 16 (A) or into adult 2- to 3-month-old mice (B). BrdU-labeled cells were visualized by anti-BrdU antibody (green), and nuclei were stained with Hoechst dye (blue shown in B). (A) Migrating BrdU-labeled cells are detected in the hippocampus (left) and cerebral cortex (right). Cortical layers I–VI are shown in right panels. (B) BrdU-positive cells in the adult dentate gyrus (left) and subventricular zone (right). DG, dentate gyrus; LV, lateral ventricle; SVZ, subventricular zone. (Partly adapted from Angata *et al.*, 2004.) (See color insert.)

4. To measure neurite outgrowth, 1.5×10^5 cells per well are seeded onto coverglasses coated with 0.01% polyethylenimine (SIGMA) and 5 μg/ml laminin (SIGMA) in 6-well plates for 36 h. If HeLa or NIH3T3 cells are used as substratum cells, they should be confluent at the time of plating hippocampal neurons.

5. Neurons are stained with NCAM (rabbit polyclonal, Chemicon), 12F8, and the neuronal marker β-III tubulin (TuJ1, mouse IgG2a, BAbCO) antibodies to measure length of neurites (Fig. 4A).

When hippocampal neurons are plated on substratum cells expressing NCAM (HeLa + NCAM) or cells expressing polysialic acid (HeLa + NCAM +PSA), polysialic acid promotes neurite expression better than NCAM alone (Fig. 4B) as we have previously shown using sensory neurons prepared from chick dorsal root ganglia (Angata *et al.*, 1997; Nakayama *et al.*, 1995).

Analysis of glycosyltransferase knockout mice enables us to analyze roles of carbohydrates in the central nervous system. Evaluation of phenotypes

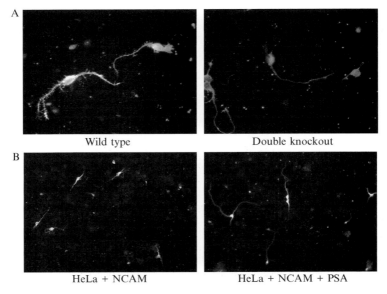

FIG. 4. Preparation of hippocampal neurons and *in vitro* neurite outgrowth assay. (A) Hippocampal neurons were prepared from wild-type (left) and double knockout mice of ST8SiaII and ST8SiaIV (right). Triple-staining with β-III tubulin (blue), NCAM (green), and polysialic acid (red, PSA) indicates that there is no polysialic acid expression on neurons from the double knockout mouse. (B) Primary neurons from the hippocampus of embryonic brain were used for an *in vitro* neurite growth assay. Prepared hippocampal neurons are plated on substratum cells expressing either NCAM alone (HeLa + NCAM) or polysialic acid (HeLa + NCAM + PSA). The neurons were stained with an antibody for β-III tubulin. (See color insert.)

exhibited by polysialyltransferase knockout mice shows that ST8SiaII and ST8SiaIV each plays a different role in polysialic acid synthesis and neural plasticity (Angata *et al.*, 2004; Eckhardt *et al.*, 2000). The specific loss of polysialic acid results in morphological alteration of neuronal projections, as visualized by immunohistochemistry. Further studies using slice cultures and culture of primary neurons and glial cells should provide further information as to the function of carbohydrates on the cellular and molecular level.

Acknowledgment

This work was supported by NIH grants CA33895 (M. F.) and DK48247 (J. D. M.).

References

Aebi, M., and Hennet, T. (2001). Congenital disorders of glycosylation: Genetic model systems lead the way. *Trends Cell. Biol.* **11,** 136–141.

Angata, K., Nakayama, J., Fredette, B., Chong, K., Ranscht, B., and Fukuda, M. (1997). Human STX polysialyltransferase forms the embryonic form of the neural cell adhesion molecule. Tissue-specific expression, neurite outgrowth, and chromosomal localization in comparison with another polysialyltransferase, PST. *J. Biol. Chem.* **272,** 7182–7190.

Angata, K., and Fukuda, M. (2003). Polysialyltransferases: Major players in polysialic acid synthesis on the neural cell adhesion molecule. *Biochimie.* **85,** 195–206.

Angata, K., Long, J. M., Bukalo, O., Lee, W., Dityatev, A., Wynshaw-Boris, A., Schachner, M., Fukuda, M., and Marth, J. D. (2004). Sialyltransferase ST8Sia-II assembles a subset of polysialic acid that directs hippocampal axonal targeting and promotes fear behavior. *J. Biol. Chem.* **279,** 32603–32613.

Breen, K. C., Coughlan, C. M., and Hayes, F. D. (1998). The role of glycoproteins in neural development function, and disease. *Mol. Neurobiol.* **16,** 163–220.

Bruses, J. L., and Rutishauser, U. (2001). Roles, regulation, and mechanism of polysialic acid function during neural development. *Biochimie.* **83,** 635–643.

Chazal, G., Durbec, P., Jankovski, A., Rougon, G., and Cremer, H. (2000). Consequences of neural cell adhesion molecule deficiency on cell migration in the rostral migratory stream of the mouse. *J. Neurosci.* **20,** 1446–1457.

Cremer, H., Lange, R., Christoph, A., Plomann, M., Vopper, G., Roes, J., Brown, R., Baldwin, S., Kraemer, P., Scheff, S., Barthels, D., Rajewsky, K., and Wille, W. (1994). Inactivation of the N-CAM gene in mice results in size reduction of the olfactory bulb and deficits in spatial learning. *Nature.* **367,** 455–459.

Cremer, H., Chazal, G., Carleton, A., Goridis, C., Vincent, J. D., and Lledo, P. M. (1998). Long-term but not short-term plasticity at mossy fiber synapses is impaired in neural cell adhesion molecule-deficient mice. *Proc. Natl. Acad. Sci. USA* **95,** 13242–13247.

Cremer, H., Chazal, G., Lledo, P. M., Rougon, G., Montaron, M. F., Mayo, W., Le Moal, M., and Abrous, D. N. (2000). PSA-NCAM: An important regulator of hippocampal plasticity. *Int. J. Dev. Neurosci.* **18,** 213–220.

Eckhardt, M., Bukalo, O., Chazal, G., Wang, L., Goridis, C., Schachner, M., Gerardy-Schahn, R., Cremer, H., and Dityatev, A. (2000). Mice deficient in the polysialyltransferase

ST8SiaIV/PST-1 allow discrimination of the roles of neural cell adhesion molecule protein and polysialic acid in neural development and synaptic plasticity. *J. Neurosci.* **20**, 5234–5244.

Edelman, G. M., and Crossin, K. L. (1991). Cell adhesion molecules: Implications for a molecular histology. *Ann. Rev. Biochem.* **60**, 155–190.

Endo, T. (2004). Structure, function and pathology of O-mannosyl glycans. *Glycoconj. J.* **21**, 3–7.

Gage, F. H. (2000). Mammalian neural stem cells. *Science* **287**, 1433–1438.

Hu, H., Tomasiewicz, H., Magnuson, T., and Rutishauser, U. (1996). The role of polysialic acid in migration of olfactory bulb interneuron precursors in the subventricular zone. *Neuron* **16**, 735–743.

Inoue, S., and Inoue, Y. (2003). Ultrasensitive analysis of sialic acids and oligo/polysialic acids by fluorometric high-performance liquid chromatography. *Methods Enzymol.* **362**, 543–560.

Jaeken, J., and Matthijs, G. (2001). Congenital disorders of glycosylation. *Annu. Rev. Genomics Hum. Genet.* **2**, 129–151.

Jessell, T. M., Hynes, M. A., and Dodd, J. (1990). Carbohydrates and carbohydrate-binding proteins in the nervous system. *Ann. Rev. Neurosci.* **13**, 227–255.

Kleene, R., and Schachner, M. (2004). Glycans and neural cell interactions. *Nat. Rev. Neurosci.* **5**, 195–208.

Lois, C., Garcia-Verdugo, J. M., and Alvarez-Buylla, A. A. (1996). Chain migration of neuronal precursors. *Science* **271**, 978–981.

Lowe, J. B., and Marth, J. D. (2003). A genetic approach to Mammalian glycan function. *Annu. Rev. Biochem.* **72**, 643–691.

Muller, D., Wang, C., Skibo, G., Toni, N., Cremer, H., Calaora, V., Rougon, G., and Kiss, J. Z. (1996). PSA-NCAM is required for activity-induced synaptic plasticity. *Neuron* **17**, 413–422.

Nakata, D., and Troy, F. A., 2nd. (2005). Degree of polymerization (DP) of polysialic acid (polySia) on neural cell adhesion molecules (N-CAMS): Development and application of a new strategy to accurately determine the DP of polySia chains on N-CAMS. *J. Biol. Chem.* **280**, 38305–38316.

Nakayama, J., Fukuda, M. N., Fredette, B., Ranscht, B., and Fukuda, M. (1995). Expression cloning of a human polysialyltransferase that forms the polysialylated neural cell adhesion molecule present in embryonic brain. *Proc. Natl. Acad. Sci. USA* **92**, 7031–7035.

Olson, E. C., and Walsh, C. A. (2002). Smooth, rough and upside-down neocortical development. *Curr. Opin. Genet. Dev.* **12**, 320–327.

Ono, K., Tomasiewicz, H., Magnuson, T., and Rutishauser, U. (1994). N-CAM mutation inhibits tangential neuronal migration and is phenocopied by enzymatic removal of polysialic acid. *Neuron* **13**, 595–609.

Rutishauser, U., and Landmesser, L. (1996). Polysialic acid in the vertebrate nervous system: A promoter of plasticity in cell-cell interactions. *Trends Neurosci.* **19**, 422–427.

Sato, C., Fukuoka, H., Ohta, K., Matsuda, T., Koshino, R., Kobayashi, K., Troy, F. A., 2nd, and Kitajima, K. (2000). Frequent occurrence of pre-existing alpha 2->8-linked disialic and oligosialic acids with chain lengths up to 7 Sia residues in mammalian brain glycoproteins. Prevalence revealed by highly sensitive chemical methods and anti-di-, oligo-, and poly-Sia antibodies specific for defined chain lengths. *J. Biol. Chem.* **275**, 15422–15431.

Seki, T., and Arai, Y. (1993). Distribution and possible roles of the highly polysialylated neural cell adhesion molecule (NCAM-H) in the developing and adult central nervous system. *Neurosci. Res.* **17**, 265–290.

Tomasiewicz, H., Ono, K., Yee, D., Thompson, C., Goridis, C., Rutishauser, U., and Magnuson, T. (1993). Genetic deletion of a neural cell adhesion molecule variant (N-CAM-180) produces distinct defects in the central nervous system. *Neuron* **11,** 1163–1174.

Troy, F. A., 2nd., Hallenbeck, P. C., McCoy, R. D., and Vimr, E. R. (1987). Detection of polysialosyl-containing glycoproteins in brain using prokaryotic-derived probes. *Methods Enzymol.* **138,** 169–185.

Walsh, F. S., and Doherty, P. (1997). Neural cell adhesion molecules of the immunoglobulin superfamily: Role in axon growth and guidance. *Annu. Rev. Cell Dev. Biol.* **13,** 425–456.

Weinhold, B., Seiden Faden, R., Rockle, I., Muhlenhoff, M., Schertzinge, F., Conzelmann, S., Marth, J. D., Gerardy-Schahn, R., and Hildebrandt, M. (2005). Genetic ablation of polysialic acid causes severe neurodevelopmental defects rescued by deletion of the neural cell adhesion molecule. *J. Biol. Chem.* **280,** 42971–42977.

[4] Roles of Glycolipids in the Development and Maintenance of Nervous Tissues

By Koichi Furukawa, Tetsuya Okuda, and Keiko Furukawa

Abstract

Glycoshingolipids are involved in a wide variety of biological events, including cell proliferation, differentiation, development, regeneration, and apoptosis in vertebrates. Expression profiles of glycolipids during the development and cell differentiation or transformation suggest that glycolipids are largely implicated in the determination of cell fates by directly transducing biosignals as receptors and/or modulating receptors' function. Despite of a number of efforts to clarify the molecular functions of glycolipids, no unambiguous results have been obtained until genetic modification of glycolipids became possible. Recent progress in the isolation of cDNAs of glycosphingolipid synthase genes has enabled us to examine roles of glycosphingolipids and strongly promoted further understanding of significances of glycosphingolipids. In particular, knock-out mice of glycosyltransferases showed quite novel aspects of glycolipid function and also redundancy among similar enzymes and glycolipid structures. Here, we summarize analytical methods with which roles of glycolipids in the development and maintenance of nervous tissues, including techniques to establish transgenic mice and gene knock-out mice, to survey fundamental behavior abnormalities, and to examine fine morphological changes lying under abnormal phenotypes of the glycolipids-modified cells and glycolipid-lacking mutant mice.

METHODS IN ENZYMOLOGY, VOL. 417   0076-6879/06 $35.00
DOI: 10.1016/S0076-6879(06)17004-4

Overview

Glycoshingolipids are amphipathic molecules and are expressed mainly on the cell surface membrane. They are involved in a wide variety of biological events, including cell proliferation, differentiation, development, regeneration, and apoptosis in vertebrates (Wiegandt, 1985). Expression profiles of glycolipids during the development and cell differentiation or transformation have been widely investigated, suggesting that carbohydrate structures in glycolipids are largely implicated in the determination of cell fates on the basis of various mechanisms such as direct transduction of biosignals as receptors and modulation of receptors' function (Hakomori et al., 1998).

Although a number of investigations aimed at clarifying the molecular functions of glycolipids have been performed, no unambiguous results have been obtained until genetic modification of glycosylation in glycolipids became possible. Recent progress in the isolation of cDNAs of glycosyltransferases responsible for the synthesis of glycosphingolipids has enabled us to examine roles of newly generated glycosphingolipids after transfection of glycosyltransferase cDNAs. Of course, genetic disruption of glycosyltransferases in experimental animals and knock-down of those enzymes with small interference RNAs or antisense DNAs in cultured cells have also strongly promoted further understanding of implications of glycosphingolipids in in vivo and in vitro systems.

Since the first example of molecular cloning as an enzyme for glycolipid synthesis in 1992 (Nagata et al., 1992), a number of glycolipid-related genes have been isolated, and genetic disruption of some of them has been achieved (Furukawa, 2001; Lowe and Marth, 2003). Results of carbohydrate remodeling with genetic manipulation of these genes have brought about tremendously huge information and quite novel findings. However, molecular mechanisms for the individual resulting phenotypes need to be clarified with precise and more sophisticated approaches. Furthermore, in many cases, knock-out mice of glycosyltransferases showed redundancy among similar enzymes and/or similar glycolipid structures, leading to requirement of further functional studies.

In this chapter, we summarize analytical methods with which roles of glycolipids in the development and maintenance of nervous tissues with focus on the techniques to establish gene knock-out mice, to survey fundamental behavior abnormalities, and to examine fine morphological changes lying under abnormal phenotypes of the glycolipid-modified cells and glycolipid-lacking mutant mice.

Genetic Approaches

Overexpression of Glycolipids in Transgenic Mice

Transgenic technique is a well-established method to generate mouse lines with exogenous gene expression. Although there are a number of modifications to regulate spatiotemporal expression patterns of introduced genes, just a standard protocol is presented in this chapter.

1. An example of the construction of a transgene (Fig. 1) (Fukumoto *et al.*, 1997). Insert cDNA, mouse β1,4GalNAc-T (Takamiya *et al.*, 1995) in pCDM8 is cleaved with *Xho*I, and the fragment containing the coding region is inserted in a *Eco*RI site of the plasmid pCAGGS (Niwa *et al.*, 1991).

2. The enzyme function of the transgene, pCAGGS/M2Tm, is confirmed by the transfection into a mouse melanoma B16 subline, KF3027 (Nagata *et al.*, 1992). The transfectant cells express high levels of GM2.

3. The injection of the transgene into fertilized eggs was performed by the standard techniques, and Fo mice are obtained.

4. Fo mice are mated with each other, and tail DNAs of F1 mice are served for PCR screening to detect the β1,4GalNAc-T transgene. PCR amplification is performed with 0.1 μg of genomic DNA. PCR buffer (50 mM KCl/Tris, pH 8.3/2.5 mM MgCl$_2$/0.01% gelatin), 0.1 μg each of primer (antisense primer sequence:440–457 in pTm3-5: 5'-AGAGCACTGATGTTGTACTC-3'/sense primer sequence:923-942: 5'-TTGACGCTGAGGAGCTGA-3'). 1 mM dNTP,

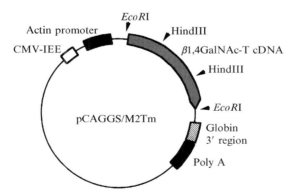

FIG. 1. Construction of a transgene. Insert cDNA of mouse β1,4GalNAc-T in pCDM8 was cleaved with *Xho* I, and the fragment containing the coding region was inserted in a *Eco*RI site of the plasmid pCAGGS (Fukumoto *et al.*, 1997).

and 0.6 unit TaqDNA polymerase (Wako, Tokyo). Amplification is carried out by thermal cycler (Astec, Fkuoka, Japan) for 3 min at 94° and 30 cycles as follows: 1 min 55°, and 2 min 72°.

5. Analyses of the transgene expression and of phenotypes. Usually, expression of the transgene is analyzed by Northern blotting and/or RT-PCR. Direct detection of the protein products is also useful when antibodies are available.

If the transcripts from the transgene contain the following properties, we can distinguish the transcripts from the transgene from that of the endogenous gene.

1. Different mRNA size
2. Different additional sequences in the transcripts
3. Tag-fused products for detection with tag-specific antibodies

An example of Northern blotting is shown in Fig. 2.

Gene Disruption by Homologous Recombination

Despite various limitations, gene knock-out of glycosyltransferases and modification enzymes is one of the most powerful approaches to access the significance of their products in tissues and organs. Depending on the steps at which the targeted gene product is located, the effects of gene knock-out varied, and approaches required are widely diverse. Here, we describe representative analyses we have performed for some mutant mice.

Knock-Out Strategy

For the construction of targeting gene, we need to know the deduced enzyme structure and entire domains such as catalytic domain. To completely destroy the enzyme activity, it is solid to delete or replace the majority of exons including the catalytic domain. When the catalytic domain is coded with multiple exons, it seems safe to delete or to replace the relatively 5'-side in the catalytic domain-coding region. It may be possible to disrupt the exon containing ATG and render the remaining coding region inactive. An example of a targeting vector for β1,4-GalNAc-T (GM2/GD2 synthase) gene is shown in Fig. 3 (Takamiya et al., 1996).

1. The targeting plasmid was constructed containing a neomycin-resistant gene (with PGK promoter) inserted into the exon 4 as shown in Fig. 3. The homologous gene spanned approximately 7.5 kbp, and the diphtheria toxin-A (DT-A) gene is attached.

2. Establishment of homologous recombinants of an ES cell line: The targeting vector (24 nM) is linearized with Not I and is mixed with ES cell

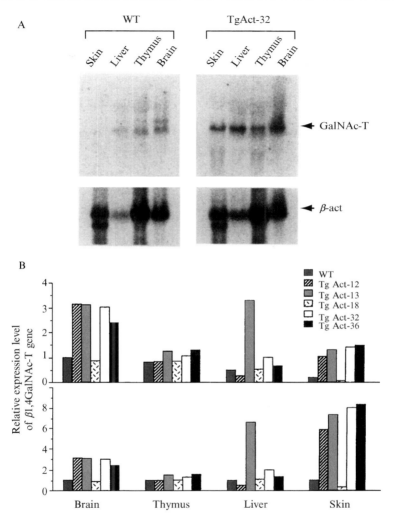

FIG. 2. Northern blotting of β1,4-GalNAc-T transgenic mice. mRNA expression levels of the gene were compared with those in the wild-type mice (Fukumoto *et al.*, 1997).

(TT2) suspension (1×10^7), then electroporated at 0.25 kV, 960 μF, using a Bio-Rad Genepulser. Forty-eight hours after electroporation, G418 is added to the medium at the concentration of 150 μg/ml. After 7–8 days, the G418-resistant clones were isolated and subjected to screening for homologous recombination by PCR. The sense primer is

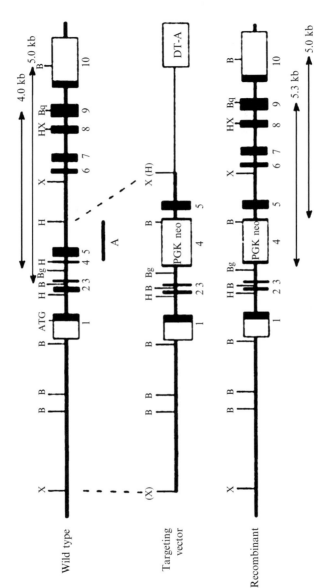

Fig. 3. An example of a targeting vector. A targeting vector for β1,4-GalNAc-T (GM2/GD2 synthase) gene is shown (Takamiya *et al.*, 1996).

5′-TCGTGCTTTACGGTATCGCCGCTCCCGATT-3′ in 3′ terminus of PGK neo, and the antisense primer was 5′-GGGTGTGGCGGCATA-CATCT-3′ in the intron of the β1.4-GalNAc-T. The reaction was started as one cycle of 95° (2 min), 55° (1 min), 74° (5 min), thereafter 35 cycles of 94° (1 min), 60° (30 sec), and 74° (1.5 min) are used. Homologous recombinant clones give a 1.1-kb fragment.

3. Another example of a targeting construct is shown in Fig. 4, which was generated for knock-out of Gb3/CD77 gene (Okuda *et al.*, 2006). Here, the neo[r] gene (Neo) was inserted between two *Ban*I sites, and the diphtheria toxin-A gene (DTA) is inserted at both sides as indicated. For screening homologous recombinant clones, Southern blotting is performed, where genomic DNA of neoresistant clones is digested with *EcoRV* or *Bgl*II and electrophoresed before being hybridized with probe-1, 10.5 kb, wild-type allele; 5.8 kb, recombinant allele (*EcoRV*); 6.2 kb, wild-type allele; and 7.3 kb, recombinant allele (*Bgl*II)(Fig. 5).

4. Northern blotting of β1,4Gal-T (Fig. 6). mRNA (2.4 kb) is detected with β1,4Gal-T cDNA probe (upper). Ribosomal RNAs detected with ethidium bromide are shown as the control (bottom). +/+, wild type; +/−, heterozygote; −/−, homozygote. Note complete loss of the transcripts.

5. The products of the enzyme assay (Gb3 synthase) using UDP[^{14}C] galactose and LacCer are analyzed by TLC/ autoradiography (left). TLC of

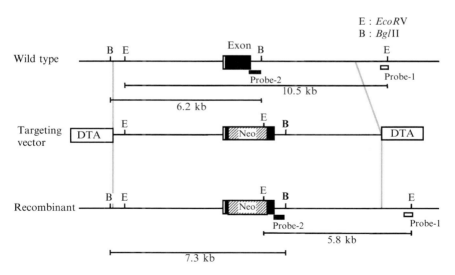

FIG. 4. Another example of a targeting construct. A targeting vector for Gb3/CD77 gene (Okuda *et al.*, 2006).

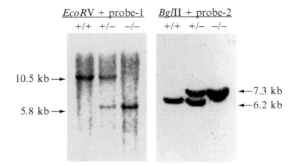

FIG. 5. Southern blotting to confirm homologous recombination of Gb3/CD77 synthase gene. Genomic DNA of neoresistant clones was digested with *EcoRV* or *BglII* and electrophoresed before hybridization with probe-1, 10.5 kb, wild-type allele; 5.8 kb, recombinant allele (*EcoRV*); 6.2 kb, wild-type allele; and 7.3 kb, recombinant allele (*BglII*) (see Fig. 4)(Okuda *et al.*, 2006).

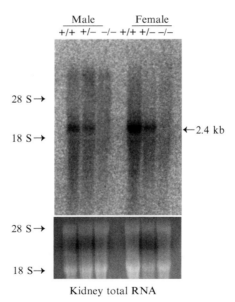

Kidney total RNA

FIG. 6. Northern blotting of α1,4Gal-T mRNA. A 2.4-kb transcript is detected with α1,4Gal-T cDNA probe (upper). Ribosomal RNAs detected with ethidium bromide are shown as the control (bottom). +/+, wild type; +/−, heterozygote; −/−, homozygote. Note complete loss of the transcripts (Okuda *et al.*, 2006).

neutral glycolipids from kidney tissues with a solvent system chloroform/methanol/water (60:35:8) is visualized by orcinol-H_2SO_4. Neutral glycolipids from B erythrocytes are used as a standard (St).

Fundamental Behavior Analysis

Because many glycolipids, in particular gangliosides, are enriched in the nervous tissues, it has been expected that glycolipid deficiency causes physical and mental disorders in null mutant mice. Therefore, various examinations concerning neurological functions, such as behavior analysis, electrophysiological analysis, responses to neuropsychological drugs, molecular biological analyses, and biochemical analyses are needed to understand entire features of effects of the gene knock-out. In this chapter, just fundamental behavior analyses for an initial examination are described.

von Frey Test

Sensory nerve is briefly tested by von Frey test and hot plate test. Mechanical nociception is assessed by applying von Frey filaments (Touch-test Sensory Evaluator, North Coast Medical Inc., San Jose, CA) ranging from 0.008–1 g to the plantar surface of the hind paw with sufficient force to cause slight bending. The glabrous skin of the hind limbs of the mice on wire grids is stimulated with several sizes of filaments to determine the threshold for the withdrawal. Testing always begins with the 0.008-g filament and proceeds from small to large monofilaments. Three trials are performed on the hind paws of each mouse, and the sources for each hind paw are averaged to yield a mean withdrawal latency.

Hot Plate Test

The hot plate test is the other brief way to examine the disorder of sensory system. Sensitivity to noxious heat stimuli is measured with a hot plate test. The animals are placed on a heated plate that is maintained at $52.5 \pm 0.5°$ by an electronic thermoregulated system. The hot plate is surrounded by a plastic cylinder (height, 20 cm; diameter, 14 cm). The heat threshold is estimated by measuring the interval between the placement of mice onto the hot plate and the initial movement of jumping. Each mouse undergoes three trials, and the average score of these three trials is used for statistical analysis. Because this test is for the examination of responses to temperature stimuli, there is often discrepancy from the results of the von Frey test as shown in Fig. 7 (Inoue et al., 2002).

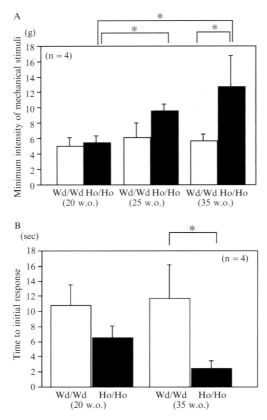

FIG. 7. Abnormal sensory function in double knock-out (Ho/Ho) mice. Sensitivity of sensory nerves was examined using two approaches (Inoue *et al.*, 2002). (A) Sensitivity to mechanical pain measured with the von Frey test. The minimum intensity of mechanical stimuli (*g*, gram) that could induce mouse reaction was determined. (B) Sensitivity to hot stimuli measured with hot plates. The results from four mice each are presented as mean ± S.D. *, $p < 0.05$. Note that sensitivity to the mechanical stimuli was reduced in 25-week-old Ho/Ho mutants.

Rota-Rod Test

Motor coordination and balance are examined by the rota-rod test. Mice are put on a rotating cylinder, approximately 3 cm diameter, and they need to continuously walk forward to keep from falling off the rotating cylinder. Usually fixed-speed rota-rods are used. After being placed on the rotating cylinder, latency to fall off is measured. Most mice are easily able to maintain balance and stay on the rota-rod for several minutes at a standard speed (i.e., 5 rpm). A 1-min cutoff maximum per trial is often used. Recently, accelerating rota-rods are in common use. Mice are placed on the rota-rod at a slow speed, for example, 4 rpm. Rotational speed gradually increases according to a predetermined program, up to a maximum rotational speed, such as 40 rpm. Mice with deficits in motor coordination or balance fall off the rota-rod before the end of the 5-min trial. Repeated performance of the rota-rod task over days is quantitated as a measure of motor learning. We need to keep in mind that cerebellar defects result in performance deficits in the rota-rod test.

Foot Printing

Gait disturbance can be briefly examined by foot printing. This test is performed by applying Chromacryl (Chroma Acrylics Inc., Lititz, PA) to the feet. Usually, a red color is used on the forelimb and a blue color is used on the hind limb. After mice walk in a narrow path with a paper bottom, shapes, strides, and width of footprints are measured as shown in Fig. 8 (Sugiura et al., 2005).

Formalin Test

The paw formalin test is used to determine the response to a prolonged nociceptive stimulus generated by localized inflammation (Wheeler-Aceto et al., 1990). Mice are gently held and injected subcutaneously with 50 μl of 1.25% formalin in 0.1 M phosphate-buffered saline into the dorsal surface of the left hind paw. After formalin injection, mice are observed for licking behavior of their injected hind paw in translucent plastic observation chamber (27 × 16 × 13 cm). The amount of time spent licking the injected paws is timed continuously every 5 min, starting immediately after the formalin injection up to 60 min after the injection (Handa et al., 2005).

Surgical Approaches for Repair Potential

Hypoglossal Nerve Resection System

To analyze the potential of repair for damaged nervous tissues, various devices have been developed in the central nervous system or peripheral nerves such as the sciatic nerve. We have developed an efficient system to estimate nerve repair potential primarily with rat hypoglossal nerve (Itoh et al., 1999).

1. Cleavage of mouse hypoglossal nerve and glycolipid injection: Mice are anesthetized by an intraperitoneal injection of 20–30 mg/kg sodium pentobarbital, and the right hypoglossal nerve (RHN) is cleaved.

+/+   −/−

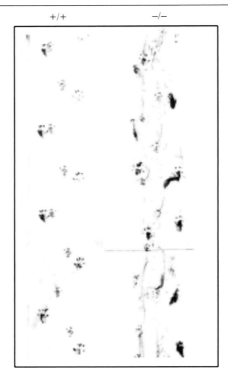

FIG. 8. Foot printing for the examination of gait disturbance. This test was performed by applying Chromacryl (Chroma Acrylics Inc) to the feet. Red color is on the forelimb and blue color is on the hind limb. Null mutant mice of β1,4-GalNAc-T gene showed shuffling traces and irregular strides (Sugiura *et al.*, 2005).

In the glycolipid-injected group, various amounts of glycolipid mixture (G-mix) (2–0.002 μg) dissolved in phosphate-buffered saline (PBS) was injected into the nerve stump site.

2. HRP injection: Ten weeks after these treatments, 20 μl of 30% horseradish peroxidase (HRP) (Toyobo, Osaka, Japan) in sterile saline is injected into various parts of the tongue of the mice as described previously (Streit and Reubi, 1977).

3. Counting of neurons at the hypoglossal nerve nucleus: After 24 h, the animals are anesthetized deeply and perfused intracardially with 0.9% saline containing heparin-Na, and fixed with 10% formalin in 0.1 M phosphate buffer. The lower brain stem is dissected, and 50-μm serial cross-sections are prepared on a freezing microtome. The sections are then incubated with a mixture of 3,3′-diaminobenzidine and hydrogen peroxide

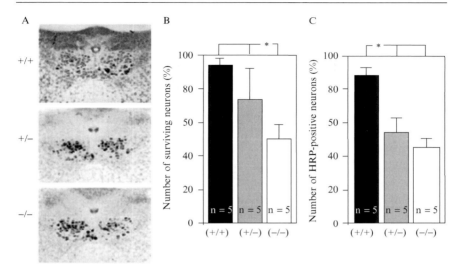

FIG. 9. Reduced regeneration of the axotomized hypoglossal nerves in the mutant mice of GD3 synthase gene. (A) HRP-stained patterns of the hypoglossal nerve nuclei in the wild-type (+/+), heterozygote (+/−), and homozygote mice (−/−) 10 weeks after the nerve resection. Regeneration of the axotomized nerves was examined by staining the sections with 3,3′-diaminobenzidine. (B) Numbers of surviving neurons 10 weeks after the operation were counted and plotted. (C) Numbers of HRP-positive neurons were counted and plotted. B and C show the mean ± S.D. (n is as indicated). * represents $p < 0.005$ (Okada et al., 2002).

at room temperature for 40 min, mounted on 3-aminopropyl-trienthox-ysilane (Aldrich, Milwaukee, Wl)–coated glass slides, and counterstained with 1% cresyl violet (Chroma, Kongen, Germany). Only cells containing a clearly visible HRP vesicle in the cytoplasm are counted in every fifth section, as described previously (Taniuchi et al., 1986). The same counting procedure is performed for the untreated left hypoglossal nerve (LHN), and its result is used to obtain the percent ratio (RHN/LHN). To follow the time course of neuronal death and nerve regeneration, mice are sacrificed at 2, 4, 6, 8, 10, and 20 weeks after the operation, and saved for the histological analyses.

An example of nerve repair experiments using GD3 synthase null mutants is presented in Fig. 9 (Okada et al., 2002).

Morphological Approaches

To evaluate the effects of gene knock-out, morphological changes in mice should be the most straightforward point and the easiest approach to be taken. For the detailed analyses of morphological changes, electron

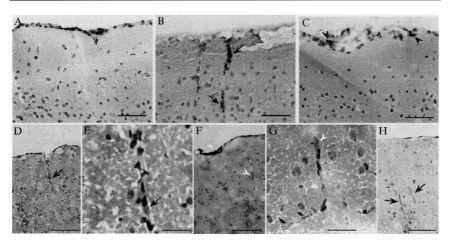

FIG. 10. Gb3 localization in brain cortex. Cryosections from mice brain were stained with mAb 38.13 (A–E and H) or an anti-CD31 (F, G) using DAB as a substrate and counterstained with hematoxylin (Okuda *et al.*, 2006). (A) α1,4Gal-T$^{-/-}$ mouse tissues. (B–H) Wild-type mouse tissues; (E and G) high-magnification images of (D and F), respectively. Gb3 was found in the cortex (B, D, E, arrows) and the pia mater (C, arrowheads). These expression patterns correspond to the staining of CD31 (F, G, white arrowheads), which was used as an endothelial cell marker.

microscopy techniques seem indispensable as we reported (Inoue *et al.*, 2002), but observation of conventional tissue sections is a fundamental step to get an initial feeling and find a clue for further analyses. When null mutant mice are available, they are very useful not only as an objective of the investigation but also as a negative control for carbohydrate epitopes that are considered to be synthesized by means of the action of the disrupted gene product. In particular, tissues from the null mutants are quite valuable, when the expression levels of the glycolipid products are minimal. Here, methods with which we detected faint but solid staining of Gb3 on endothelial cells in brain using Gb3 synthase knock-out mice as controls.

Detection of Low-Level Glycolipids: Histology and Immunohistochemistry

1. Preparation of tissue sections: tissues from 10–15-week-old mice before and after treatment with VT-2 are fixed with 3.7% formalin in PBS and embedded in paraffin. The sections were stained with hematoxylin–eosin. For immunohistochemical analysis, tissues were frozen in liquid nitrogen, and 7-μm sections were prepared on a cryostat (Leica) and fixed with ice-cold acetone for 15 min at -20°.

2. Immunohistostaining with DAB: After blocking with 0.05% H_2O_2 and 10% normal goat serum, cryosections are incubated with mAb-38.13 (1:50) or an anti-CD31 mAb-390 (1:50) (eBioscience, Kobe, Japan), and the antibody bindings are detected with Histofine Simple Stain Mouse MAX PO (Rat) (Nichirei, Tokyo) and 3,3'-diaminobenzidine-tetrahydrochloride (DAB) (DOJIN, Kumamoto, Japan) as a substrate. Nuclei were stained with hematoxylin. For negative controls of antibodies, immunohisto-chemistry is carried out using nonrelevant antibodies with the same isotypes. An example is presented in Fig. 10.

3. To determine the expression sites of Gb3 in tubules, we carry out immunohistochemistry using fluorescence-conjugated second antibodies and performed our analysis with a Fluoview FV500 confocal laser microscope (Olympus, Tokyo). Antibody binding to Gb3 and CD31 is detected with anti-rat-IgM-FITC (ICN Pharmaceuticals, Aurora, OH) and anti-rat-IgG-Alexa-488 (Molecular Probe, Invitrogen), respectively. GM1 expression in tissues is detected with CTB-Alexa-555 (1:100) (Molecular Probe, Invitrogen).

References

Fukumoto, S., Yamamoto, A., Hasegawa, T., Abe, K., Takamiya, K., Okada, M., Min, Z. J., Furukawa, K., Miyazaki, H., Tsuji, Y., Goto, G., Suzuki, M., Shiku, H., and Furukawa, K. (1997). Genetic remodeling of gangliosides resulted in the enhanced reactions to the foreign substances in skin. *Glycobiology* **8**, 1111–1120.

Handa, Y., Ozaki, N., Honda, T., Furukawa, K., Tomita, Y., Inoue, M., Furukawa, K., Okada, M., and Sugiura, Y. (2005). GD3 synthase gene knockout mice exhibit thermal hyperalgesia and mechanical allodynia but decreased response to formalin-induced prolonged noxious stimulation. *Pain* **117**, 271–279.

Inoue, M., Fujii, Y., Furukawa, K., Okada, M., Okumura, K., Hayakawa, T., Furukawa, K., and Sugiura, Y. (2002). Refractory skin injury in complex knock-out mice expressing only the GM3 ganglioside. *J. Biol. Chem.* **277**, 29881–29888.

Itoh, M., Fukumoto, S., Baba, N., Kuga, Y., Mizuno, A., and Furukawa, K. (1999). Prevention of the death of the rat axotomized hypoglossal nerve and promotion of its regeneration by bovine brain gangliosides. *Glycobiology* **9**, 1247–1252.

Lowe, J. B., and Marth, J. D. (2003). A genetic approach to Mammalian glycan function. *Annu. Rev. Biochem.* **72**, 643–691.

Nagata, Y., Yamashiro, S., Yodoi, J., Lloyd, K. O., Shiku, H., and Furukawa, K. (1992). Expression cloning of beta 1,4 N-acetylgalactosaminyltransferase cDNAs that determine the expression of GM2 and GD2 gangliosides. *J. Biol. Chem.* **267**, 12082–12089.

Niwa, H., Yamamura, K., and Miyazaki, J. (1991). Efficient selection for high-expression transfectants with a novel eukaryotic vector. *Gene* **108**, 193–199.

Okada, M., Itoh, M. M., Haraguchi, M., Okajima, T., Inoue, M., Oishi, H., Matsuda, Y, Iwamoto, T., Kawano, T., Fukumoto, S., Miyazaki, H., Furukawa, K., Aizawa, S., and Furukawa, K. (2002). b-series ganglioside deficiency exhibits no definite changes in the neurogenesis and the sensitivity to Fas-mediated apoptosis but impairs regeneration of the lesioned hypoglossal nerve. *J. Biol. Chem.* **277**, 1633–1636.

Okuda, T., Tokuda, N., Numata, S., Ito, M., Ohta, M., Kawamura, K., Wiels, J., Urano, T.,
   Tajima, O., Furukawa, K., and Furukawa, K. (2006). Targeted disruption of Gb3/CD77
   synthase gene resulted in the complete deletion of globo-series glycosphingolipids and loss
   of sensitivity to verotoxins. *J. Biol. Chem.* **281**, 10230–10235.
Streit, P., and Reubi, J. C. (1977). A new and sensitive staining method for axonally transported
   horseradish peroxidase (HRP) in the pigeon visual system. *Brain Res.* **126**, 530–537.
Sugiura, Y., Furukawa, K., Tajima, O., Mii, S., Honda, T., and Furukawa, K. (2005). Sensory
   nerve-dominant nerve degeneration and remodeling in the mutant mice lacking complex
   gangliosides. *Neuroscience* **135**, 1167–1178.
Takamiya, K., Yamamoto, A., Yamashiro, S., Furukawa, K., Haraguchi, M., Okada, M., Ikeda,
   T., Shiku, H., and Furukawa, K. (1995). T cell receptor-mediated stimulation of mouse
   thymocytes induces up-regulation of the GM2/GD2 synthase gene. *FEBS Lett.* **358**, 79–83.
Takamiya, K., Yamamoto, A., Furukawa, K., Yamashiro, S., Shin, M., Okada, M., Fukumoto,
   S., Haraguchi, M., Takeda, N., Fujimura, K., Sakae, M., Kishikawa, M., Shiku, H,
   Furukawa, K., and Aizawa, S. (1996). Mice with disrupted GM2/GD2 synthase gene lack
   complex gangliosides but exhibit only subtle defects in their nervous system. *Proc. Natl.
   Acad. Sci. USA* **93**, 10662–10667.
Taniuchi, M., Clark, H. B., and Johnson, E. M., Jr. (1986). Induction of nerve growth factor
   receptor in Schwann cells after axotomy. *Proc. Natl. Acad. Sci. USA* **83**, 4094–4098.
Wheeler-Aceto, H., Porreca, F., and Cowan, A. (1990). The rat paw formalin test: Comparison of
   noxious agents. *Pain.* **40**, 229–238.
Wiegandt, H. (1985). Glycolipids. pp. 199–260. Elsevier, Amsterdam.

Further Reading

Furukawa, K., Takamiya, K., Okada, M., Inoue, M., Fukumoto, S., and Furukawa, K. (2001).
   Novel functions of complex carbohydrates elucidated by the mutant mice of glycosyltransferase genes. *Biochim. Biophys. Acta* **1525**, 1–12.
Hakomori, S., Yamamura, S., and Handa, A. K. (1998). Signal transduction through glyco
   (sphingo)lipids. Introduction and recent studies on glyco(sphingo)lipid-enriched microdomains. *Ann. N.Y. Acad. Sci.* **845**, 1–10.

[5]   Analysis of Neural Cell Functions in Gene Knockout Mice: Electrophysiological Investigation of Synaptic Plasticity in Acute Hippocampal Slices

By OLENA BUKALO and ALEXANDER DITYATEV

Abstract

      Several knockout mice deficient in transferases, required for glycosyla-
tion of cell adhesion and extracellular matrix molecules, have recently been
produced. Extracellular recordings of field excitatory postsynaptic potentials in acute hippocampal slices prepared from these mutant mice proved
to be a highly sensitive method to reveal the roles of transferases and

METHODS IN ENZYMOLOGY, VOL. 417   0076-6879/06 $35.00
                                                DOI: 10.1016/S0076-6879(06)17005-6

related carbohydrates in synaptic transmission and plasticity. Although most available data have been collected for synaptic connections between CA3 and CA1 pyramidal cells, several other synapses are assessable for extracellular recording in the hippocampus, including connections between mossy fibers and CA3 pyramidal cells. Analysis of distinct forms of short- and long-term plasticity in these connections may be instrumental for dissection of mechanisms by which carbohydrates affect synaptic functions.

Overview

The efficacy of synaptic transmission may vary in an activity-dependent manner. This phenomenon of synaptic plasticity appears in many forms and is believed to underlie learning and memory. Analysis of synaptic plasticity has been greatly enhanced by the use of acute brain slices that provide a convenient preparation for precise placement of stimulating and recording electrodes and experimental manipulations with cell environment. Particularly, the highly laminated structure of the hippocampus is ideal for electrophysiological recordings, because lamina contains stereotyped portions of the hippocampal neuronal circuitry (e.g., CA3–CA1 synapses in the *stratum radiatum* of the CA1 region; Fig. 1A), making it possible to record interpretable extracellular field potentials (field extracellular excitatory postsynaptic potentials or population spikes) from defined

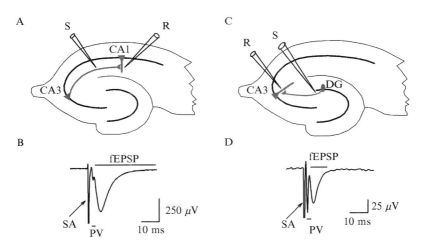

FIG. 1. Positioning of electrodes (A, C) and examples of fEPSPs (B, D) recorded in CA3-CA1 (A, B) and mossy fiber-CA3 (C, D) synapses. S, Stimulating electrode; R, recording electrode; SA, stimulus artefact; PV, presynaptic volleys; DG, dentate gyrus. Horizontal bars indicate duration of fEPSP and PV.

synapses. Also remarkable ability to induce synaptic plasticity in the hippocampal slices makes this preparation so attractive.

Persistent enhancement or reduction of synaptic strength of stimulated synapses is known as long-term potentiation and depression (LTP and LTD). The molecular mechanisms of hippocampal LTP and LTD are complex, and many molecules are involved in synaptic changes, including cell adhesion and extracellular matrix molecules and associated carbohydrates (Bliss and Collingridge, 1993; Dityatev and Schachner, 2003; Sanes and Lichtman, 1999). The involvement of carbohydrates was initially demonstrated by experiments involving enzymatic removal of carbohydrates such as polysialic acid, heparan sulfates, and chondroitin sulfates. Enzymatic removal of polysialic acid or heparan sulfates results in impaired LTP and abolished formation of perforated synapses in the CA1 area of the hippocampus (Becker *et al.*, 1996; Dityatev *et al.*, 2004; Lauri *et al.*, 1999; Muller *et al.*, 1996). Also, digestion of chondroitin sulfates reduces CA1 LTP and LTD (Bukalo *et al.*, 2001). In contrast, application of L-fucose or antibodies to the HNK-1 (first described on *h*uman *n*atural *k*iller cells) carbohydrate promotes induction of CA1 LTP (Matthies *et al.*, 1996; Saghatelyan *et al.*, 2000). Identification of two polysialyltransferases, ST8SiaII and ST8SiaIV, required for synthesis of polysialic acid, made it possible to perform a genetic analysis of polysialic acid functions. Electrophysiological recordings in mice deficient in these enzymes show that ablation of ST8SiaIV, but not ST8SiaII, leads to deficits in LTP and LTD in the CA3-CA1 connections (Angata *et al.*, 2004; Eckhardt *et al.*, 2000). Remarkably, two mouse mutants deficient either in glucuronyltransferase or HNK-1 sulfotransferase, which are the final pathway enzymes in the synthesis of the HNK-1 carbohydrate, exhibit a similar reduction in CA1 LTP (Senn *et al.*, 2002; Yamamoto *et al.*, 2002). In summary, electrophysiological analysis of mutant mice deficient in transferases provided important genetic evidence for the roles of polysialic acid and HNK-1 in hippocampal synaptic plasticity.

Here, we describe methods used in the studies previously cited to prepare hippocampal slices, to extracellularly record and analyze field excitatory postsynaptic potentials (fEPSPs), and to induce short- and long-term synaptic plasticity in the CA3–CA1 and mossy fiber–CA3 connections. These methods can be effectively used to assess the roles of carbohydrates in synaptic functions. Analyses of the stimulus-response curve, paired-pulse facilitation, and frequency facilitation (described in this chapter) provide a way to evaluate basal synaptic transmission and probability of release in hippocampal synapses (Bortolotto *et al.*, 2001). Because LTP in CA3–CA1 synapses depends on signaling by means of NMDA receptors, whereas LTP in mossy fiber-CA3 synapses is NMDA receptor independent (cyclic AMP dependent), recordings of these two forms of LTP are of help to reveal specificity of synaptic abnormalities in mutant mice and, thus, may be a guide in explanation of underlying mechanisms.

Animal Considerations

All protocols using live animals must first be reviewed and approved by an Institutional Animal Care and Use Committee and must follow officially approved procedures for the care and use of laboratory animals. Information about age, gender, and genetic line should be carefully collected for all animals used.

Level of LTP depends on the genetic background of mice. Therefore, it is highly recommended to use litter mate wild-type controls, particularly if the background of genetically modified mice is mixed and mice were not inbreed for >six generations. If mice are originating from different parents, it is important that they are raised under similar conditions. If the animal facility is located in another building than where the slices are prepared, mice should be brought to the place of slice preparation at least 1 day before slice preparation and kept in a quiet place (to minimize well-known effects of stress on synaptic plasticity).

Often mice of both sexes are used for LTP recordings. Whether a genetic manipulation leads to gender-specific effects may be statistically evaluated using the two-way ANOVA, considering gene and gender as two factors underlying variability of LTP. LTP can be easily induced in slices derived from mice of different ages; however, age should be within a well-defined interval, because some forms of LTP show age dependence, and there may be a strong gene-dependent interaction between age and LTP in genetically modified mice (Eckhardt *et al.*, 2000). Two ages are particularly attractive: 3-week-old (if extracellular recordings are planned to complemented by patch-clamp study) and 3-month-old mice (if electrophysiological analysis is planned to be complemented by behavioral analysis of cognitive functions).

Materials for Slice Preparation

Genetically modified (e.g., knockout) and wild-type mice
Compressed CO_2 tank with regulator
Transparent vessel of sufficient size to place a mouse
Vibratome (e.g., VT 1000M from Leica, Nussloch, Germany)
Razor blade (a new blade is used for every slice preparation)
Acetone (all chemical are from Sigma, St. Louis, MO, unless otherwise stated)
Compressed gas mixture (95% O_2/5% CO_2) tank with regulator

Aquarium air stone, tubing of an appropriate diameter to connect regulator and air stone

Gelatin blocks (20% dissolved in water, approx. 1 cm × 1 cm × 1 cm)

Gelatin solution (10% dissolved in water and kept at 60° in a water bath)

Water bath (e.g., Inkubationsbad from GFL, Burgwedel, Germany)

A chamber for incubation of brain slices

Rapid (super) glue

Small weighing spatula, scalpel, forceps, large scissors, and a small sharp-nosed dissecting scissors

Soft small paintbrush for transfer of slices

Solutions for Slice Preparation

Dissection buffer for recordings in CA3–CA1 synapses: 250 mM sucrose, 25 mM $NaHCO_3$, 25 mM glucose, 2.5 mM KCl, 1.25 mM NaH_2PO_4, 2 mM $CaCl_2$, and 1.5 mM $MgCl_2$, pH 7.4.

Dissection buffer for recordings in mossy fiber–CA3 synapses: 250 mM sucrose, 25 mM $NaHCO_3$, 25 mM glucose, 2.5 mM KCl, 1.25 mM NaH_2PO_4, 0.5 mM $CaCl_2$, and 6 mM $MgCl_2$, pH 7.4.

The solutions should be kept ice cold and gassed with 95% O_2/5% CO_2 at least 20 min before the use.

Method for Slice Preparation

1. A mouse is kept in a closed glass vessel into which CO_2 is slowly released through tubing until the mouse reaches an anesthetic state, after that it is quickly decapitated using large scissors.

2. The skin and skull are cut with sharp-nose scissors, and brain is quickly removed using a small spatula and placed into a petri dish with ice-cold dissection buffer. It is important to minimize time required for this step to 1–1.5 min.

3. The cerebellum is dissected away, and the brain is cut into two hemispheres by scalpel. One of hemispheres is placed to the Vibratome chamber filled with ice-cold dissection buffer bubbled with 95% O_2, 5% CO_2 (to be sliced later on). The other half is glued upright on the object-mounting platform with the rapid glue next to the block of 20% gelatin, which is glued before the dissection of brain. For recordings in CA1, the cut surface is glued to the brain-mounting platform. For CA3 recordings, the ventral surface of each hemisphere is cut with the angle of 20 degrees, and this surface is glued to the platform (Claiborne *et al.*, 1993). Thereafter 10% gelatin is applied to embed the hemisphere and fill a gap between it and the block. The platform with the glued hemisphere is immediately placed in the Vibratome chamber.

4. The attached cerebral hemisphere is then cut with a Vibratome into slices. The blade should be cleaned in acetone for several minutes and then washed in water. The blade should be fixed in the blade holder to ensure positioning of the blade parallel to the tissue. The thickness of slices depends on age. We use 400-μm-thick coronal slices for mice younger than 1 month, 350 μm for 1- to 3-month-old mice, 300 μm form mice older than 3 months.

5. The Vibratome chamber and brain mounting platform used for physiological recordings should be different from those used for sectioning of formaldehyde fixed brains to avoid influence of formaldehyde on live slices.

6. The slices are separated from gelatin with the scalpel and transferred to the slice incubation chamber filled with ACSF and maintained at room temperature for 1.5–2 h before the beginning of electrophysiological recordings.

7. A chamber for incubation of brain slices can be made from a plastic vessel (10 cm in diameter and height) inside which a plastic cylinder (5 cm in diameter, 3 cm in height) is glued at the middle of the wall. The slices are placed at the bottom of the cylinder, which can be made of nylon stockings glued to the edges of the cylinder. The vessel is filled with ACSF to the top of cylinder. The aquarium stone is placed at the bottom of the vessel, outside of cylinder, and gas pressure is adjusted to saturate solution but not strong enough to make the slices move or float.

8. At the end of preparation, a 2–3-mm long piece of mouse tail is cut with scissors (cleaned with ethanol), placed into a 1.5-ml tube, labeled with a code of the animal, and stored in a freezer for regenotyping (particularly necessary in cases in which an unexplainable variability in results is observed).

Materials for Extracellular Recordings of fEPSPs

Glass microelectrode puller (e.g., DMZ Universal Puller, Zeitz, Munich, Germany)

Glass pipettes for extracellular recording (2–4 MΩ) and stimulation (1 MΩ). The pipettes are filled with artificial cerebrospinal fluid (ACSF)

Upright microscope (BX50WI from Olympus, Hamburg, Germany) or stereomicroscope (e.g., Stemi 2000 from Zeiss, Jena, Germany)

Two micromanipulators (e.g., LN Mini 25 from Luigs & Neumann, Ratingen, Germany)

Stimulus isolator unit (e.g., A360, World Precision Instruments, USA)

Amplifier (voltage amplifier or patch clamp amplifier, e.g., ECP9 from HEKA Lambrecht/Pfalz, Germany or Axoclamp 2A, Axon Instruments, Molecular Devices Corporation, Union City, USA)

Recording chamber, preferably with a possibility to heat ACSF (e.g., slice mini Badkammer I with Controller type V, from Luigs and Neuman)

Small pieces of nylon mesh (7 × 10 mm; made from insets NY80, Millipore, Billerica, MA) with a hole (2 mm in diameter) in the middle and two small metal loads to fix the mesh in the recording chamber

Perfusion system consisting of a container of ACSF and polyethylene tubing that circulates ACSF through a peristaltic pump (e.g., Minipuls3 from ADInstruments GmbH, Spechbach, Germany) and the recording chamber

Computer (e.g., IBM PC)

Acquisition board (e.g., CIO-DAS08/JR-AO from Measurement Computing, Middleboro, MA)

Software for data acquisition and measurement of fEPSPs (e.g., Pulse from HEKA or "LTP Program" freely available at the http://www. ltp-program.com and is described by Anderson and Collingridge, 2001).

Solutions for Extracellular Recordings of fEPSPs

Artificial cerebrospinal fluid (ACSF) for recordings in CA3–CA1 synapses: 124 mM NaCl, 25 mM NaHCO$_3$, 25 mM glucose, 2.5 mM KCl, 1.25 mM NaH$_2$PO$_4$, 2 mM CaCl$_2$, and 1.5 mM MgCl$_2$, pH 7.4.

Artificial cerebrospinal fluid (ACSF) for recordings in mossy fiber–CA3 synapses: 124 mM NaCl, 25 mM NaHCO$_3$, 25 mM glucose, 2.5 mM KCl, 1.25 mM NaH$_2$PO$_4$, 2.5 mM CaCl$_2$, and 1.5 mM MgCl$_2$, pH 7.4.

The solutions should be gassed with 95% O$_2$/5% CO$_2$ at least 30 min before the use.

Method for Extracellular Recordings of fEPSPs

1. A slice with a clearly visible hippocampus is transferred to the recording chamber using a paintbrush and spatula. The recording chamber with slice is continuously superfused at a rate of 2–3 ml/min with ACSF bubbled with 95%O$_2$, 5%CO$_2$.

2. A mesh is put on the top of slice in such a way that a hole in the mesh is located above the hippocampus. The mesh is fixed by two small loads. Slices of good quality normally do not stick to the paintbrush and have clearly visible (light) layers of pyramidal and granule cells.

3. *Recordings in CA3-CA1 synapses:* Stimulating and recording electrodes are placed at the slice surface approximately 400 μm apart from each other in the *stratum radiatum,* approximately at the same

distance from the *stratum pyramidale* of the CA1 region as shown in Fig. 1A. Stimulation (50–70 μA, 0.2 ms) is applied to CA3 axons every 20 sec, and the recording and stimulation electrodes are slowly advanced toward the slice until maximal amplitude of fEPSPs is attained (Fig. 1B).

4. *Recordings in mossy fiber–CA3 synapses:* The stimulating electrode is placed close to the granule cell layer. The recording pipette is placed in the *stratum lucidum* of the CA3 region (Fig. 1C). Mossy fiber fEPSPs have characteristic fast rise-time and decay (total duration of fEPSP <10 ms, rise time <3.5 ms), large paired-pulse facilitation (>170%), and prominent frequency facilitation (>200%; Figs. 1D and 3A). Those selected for analysis responses should not have hallmarks of polysynaptic activation, such as jagged decay phase or multiple peaks, or variable latencies of fEPSPs. To obtain such responses, one needs to probe at different positions for both stimulating and recording electrodes. To avoid contamination of mossy fiber responses by polysynaptic activity, current pulses of small intensity (0.2 ms, 20–40 μA) are used, which elicit responses of 40–60 μV. Furthermore, to confirm that the fEPSPs recorded were evoked by the stimulation of mossy fibers and not by the associational/commissural pathway, an agonist of type II metabotropic glutamate receptors (L-CCG1, 10 μM or DCG-IV, 2 μM; Tocris), which is known to reduce synaptic transmission in CA3 mossy fiber synapses, should be applied at the end of each experiment (Fig. 3B). Only experimental data in which responses are reduced by at least 70% are selected for analysis.

Analysis of Synaptic Transmission and Plasticity

Calculation of Stimulus-Response Curve in CA3–CA1 Synapses. fEPSPs are collected in response to stimulation pulses (0.2 ms, every 20 sec) of increasing intensity (e.g., from 10 μA by step of 20 μA; Fig. 2A) until a population spike (upward going inflection) appears on the decaying phase of fEPSPs. In wild-type mice, the amplitude of supramaximal fEPSPs (i.e., without population spike) is expected to be >1 mV, and the amplitude of presynaptic volleys (Fig. 1C) is expected to be several times smaller than fEPSPs.

Paired-Pulse Stimulation. Pairs of stimuli with different intervals (e.g., 15, 25, 50, 100, and 200 ms) are delivered. Five responses are collected and averaged for each interval (Fig. 2B).

Frequency Facilitation. A train of 40 stimuli is applied at 0.33 Hz. This protocol should produce a strong increase (>200%) of mossy-fiber responses (Fig. 3A).

Baseline Recording. fEPSPs are elicited every 20 sec for at least 10 min with the stimulation strength that provides fEPSPs with a slope of 50% of

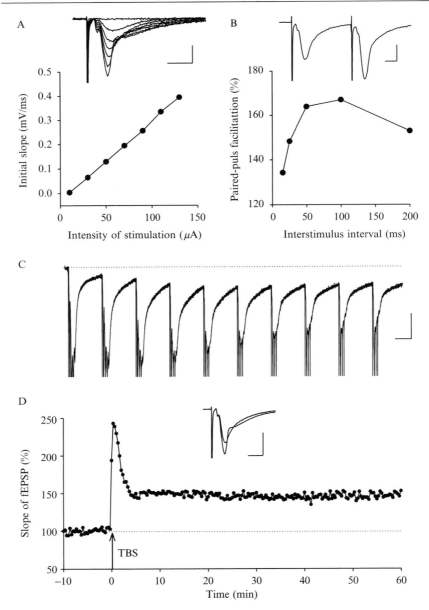

FIG. 2. Basal synaptic transmission, paired-pulse facilitation, short- and long-term potentiation in CA3-CA1 synapses. (A) Stimulus-response curve. Inset shows traces representing fEPSPs evoked with different stimulus strengths (from 10 μA by steps of 20 μA) in CA3–CA1 synapses. Bars, 10 ms and 0.5 mV. (B) Analysis of paired-pulse facilitation.

the subthreshold maximum. It is very important to obtain measurements of fEPSPs before induction of LTP to verify that the baseline is stable (i.e., there is no progressive increase or decrease in the amplitude/slope of responses during recording) (Fig. 2D). If systematic increase/decrease of fEPSPs is observed, the recording of baseline should be continued until it reaches a stable level. Thereafter, the stimulus-response curve should be determined again to select a stimulation intensity that provides fEPSPs with a slope of 50% of the subthreshold maximum.

Induction of LTP by Theta-Burst Stimulation (TBS). Ten bursts are delivered at 5 Hz, each burst consists of four pulses (0.2 ms) delivered at 100 Hz. Five TBSs are applied every 20 sec to induce LTP. The stimulation strength is the same as for the baseline recording. Examples of fEPSP elicited by TBS and TBS-induced LTP in CA3–CA1 synapses are shown in Fig. 2 C, D. This protocol (5 × TBS) has been shown to reliably induce LTP in CA3-CA1 synapses. Induction of this form of LTP is dependent on activity of the NMDA receptor and the L-type of Ca^{2+} channels (Evers *et al.*, 2002). The contribution of L-type Ca^{2+} channels may be pharmacologically prevented using blockers of these channels (e.g., nifedipine) (Evers *et al.*, 2002). Alternately, application of 1–2 × TBS results in the L-type Ca^{2+} channel independent form of LTP.

Induction of LTP by High-Frequency Stimulation. Four stimulus trains, each having a frequency of 100 Hz and 1 sec duration, with an interval of 20 sec are applied. This protocol produces strong LTP in mossy fiber synapses. To evoke LTP exclusively in mossy fiber synapses, which are known to undergo LTP in a NMDA receptor–independent manner, the NMDA receptor antagonist AP-5 (50 μM; Tocris, Bristol, UK) is applied 15 min before and during HFS. Mossy fiber responses should follow presynaptic stimulation at 100 Hz and show no changes in the shape of responses after induction of LTP (Fig. 3C). An example of LTP recording in CA3–CA1 synapse is shown in Fig. 3D.

Field EPSPs elicited by TBS or HFS normally (i.e., in healthy slices from wild-mice) should not decline to the baseline during stimulation (Figs. 2C and 3C). This slow component partially reflects depolarization

---

A ratio between the slopes of fEPSPs evoked by the second and first pulses is plotted as a function of interstimulus interval. An example of recordings is given in the inset for the interval of 50 ms. Bars, 10 ms and 0.25 mV. (C) Response elicited by TBS. Bars, 100 ms and 0.25 mV. (D) A profile of TBS-induced short- and long-term potentiation. Traces on the top provide fEPSPs (average of 30 sweeps) collected before and 50–60 min after TBS administration. The mean slope of fEPSPs recorded 0–10 min before TBS is taken as 100%. Bars, 10 ms and 0.5 mV.

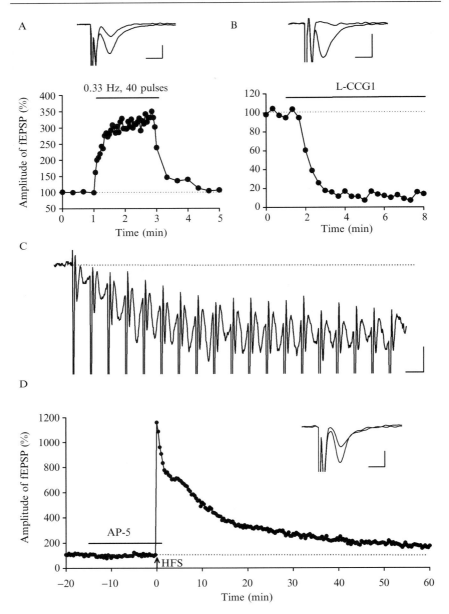

FIG. 3. Frequency facilitation, post-tetanic, and long-term potentiation in mossy fiber-CA3 synapses. (A) Frequency facilitation. Stimulation of mossy fibers with a frequency of 0.33 Hz increases the amplitudes of fEPSPs. Inset shows fEPSPs collected before and during frequency facilitation. Bars, 5 ms and 50 μV. (B) Application of the type II metabotropic

of neuronal membrane because of summation of long-lasting synaptic responses mediated by NMDA receptors. However, depolarization of electrodes may also result in a similar pattern. Therefore, it is useful to collect (after the end of LTP recording) TBS- or HFS-induced stimulus artefact by taking the stimulating electrode out of the slice. It can be digitally subtracted from TBS-induced fEPSP, thus, providing "true" synaptic response.

Induction of LTD by Low-Frequency Stimulation (LFS). Two trains are applied at 1 Hz for 10 min with a 10-min interval between them. Stimulation strength during baseline recordings and after induction of LTD is set to 30–40% of supramaximal fEPSPs. Stimulation strength is set to 60–70% when 1 Hz trains are delivered. The amplitude of fEPSP should be not less then 1 mV at this stimulation strength. This protocol has been shown to induce input-specific NMDA receptor–dependent LTD in CA3–CA1 synapses in adult rodents (Eckhardt *et al.*, 2000). An example of LTD is shown in Fig. 4.

It is important to compare fEPSPs before and after induction of LTP and LTD, as shown in Figs. 2D, 3D, and 4. If the conditions of recordings are stable (hence, the experiment is acceptable for analysis), the stimulus artefact and the amplitude of presynaptic volley should be identical before application of high-frequency stimulation and at the end of LTP and LTD recording.

Measurements and Calculation of Parameters

It is important to select an appropriate time window for slope measurements in such a way that it would be in the phase of linear changes in fEPSPs for all stimulation intensities used to estimate the stimulus-response curve and for fEPSPs collected during paired-pulse stimulation and after induction of LTP.

For CA3–CA1 synapses, in addition to the plot of stimulus intensity, slope of fEPSP, analysis of relationship between the amplitude of presynaptic volleys, and slope of fEPSP is often performed. It provides a measure of relationship between the number of activated axons (presynaptic

---

glutamate receptor agonist LCCG1 (10 μM) reduces the amplitude of fEPSPs. Inset shows fEPSPs collected before and during LCCGI application. Bars, 5 ms and 50 μV. (C) Synaptic response evoked by high-frequency stimulation (HFS) of mossy fibers. Bars, 10 ms and 50 μV. (D) A profile of HFS-induced posttetanic and long-term potentiation. The time interval of the application of the NMDA receptor antagonist AP-5 is shown by a horizontal bar. Mean slope of fEPSPs recorded 0–10 min before HFS was taken as 100%. Inset shows averages of 10 fEPSPs recorded before and 60 min after induction of LTP. Scale bars, 5 ms and 50 μV.

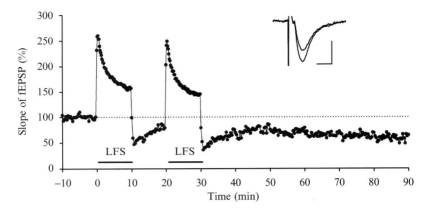

FIG. 4. Long-term depression in CA3–CA1 synapses. Two trains of low-frequency stimulation (LFS) of CA3 axons induce a persistent decrease in slopes of fEPSPs in CA3–CA1 synapses. The mean slope of fEPSPs recorded 0–10 min before LFS was taken as 100%. Horizontal bars indicate delivery of LFS. Inset shows fEPSPs recorded before and 60 min after LFS. Scale bars, 10 ms and 0.25 mV.

volleys) and resulting fEPSPs. It is important to keep recording and stimulating electrode at the defined and sufficiently long distance, because the amplitude of presynaptic volleys depends on this distance and may partially reflect generation of action potentials in directly activated CA1 neurons.

For mossy fiber fEPSPs, measurements of slope are very noisy (because of small amplitudes of responses) and, therefore, mean amplitude is measured at time window around the peak of responses. Amplitude/slope measurements from three consecutive time points may be averaged to reduce noise.

Paired-pulse facilitation is calculated as a ratio between the slopes of fEPSPs elicited by the second and first pulses. To calculate paired-pulse facilitation for short interpulse intervals (e.g, 10 or 20 ms), the fEPSP elicited by a single stimulus should be digitally subtracted from the sum of responses elicited by two pulses.

The mean magnitude of fEPSPs recorded during the baseline (at least 10 min before TBS) is taken as 100%, and changes are expressed relative to this level (Fig 3C). The transient potentiation immediately after TBS (or STP, short-term potentiation) is measured as a maximal increase in the fEPSP slope during 1 min after LTP induction. The values of LTP are calculated as increase in the mean slopes of fEPSPs measured at 50–60 min after induction of LTP.

Acknowledgments

The authors thank A. Artola, H. Beck, T. Katafuchi, P. Castillo, A. Saghatelyan, and B. Salmen for fruitful discussion of slice preparation and LTP recordings; Kodeeswaran Parameshwaran and Vishnu Suppiramaniam for their comments on the manuscript; and M. Schachner for constant support. This work was supported by Deutsche Forschungsgemeinschaft (Di 702/4–1 and –2 to A.D.).

References

Anderson, W. W., and Collingridge, G. L. (2001). The LTP Program: A data acquisition program for on-line analysis of long-term potentiation and other synaptic events. *J. Neurosci. Methods* **108**, 71–83.

Angata, K., Long, J. M., Bukalo, O., Lee, W., Dityatev, A., Wynshaw-Boris, A., Schachner, M., Fukuda, M., and Marth, J. D. (2004). Sialyltransferase ST8Sia-II assembles a distinct repertoire of polysialic acid that alters hippocampal axonal targeting and promotes fear behavior. *J. Biol. Chem.* **279**, 32603–32613.

Becker, C. G., Artola, A., Gerardy-Schahn, R., Becker, T., Welzl, H., and Schachner, M. (1996). The polysialic acid modification of the neural cell adhesion molecule is involved in spatial learning and hippocampal long-term potentiation. *J. Neurosci. Res.* **45**, 143–152.

Bliss, T. V., and Collingridge, G. L. (1993). A synaptic model of memory: Long-term potentiation in the hippocampus. *Nature* **361**, 31–39.

Bortolotto, Z. A., Anderson, W. W., Isaac, J. T. R., and Collingridge, G. L. (2001). "Synaptic Plasticity in the Hippocampal Slice Preparation" *In* Current Protocols in Neuroscience, (J. Crawley, C. Gerfen, M. Rogawski, D. Sibley, P. Skolnick, S. Wray, and R. McKay, eds.), pp. 6.13.1–6.13.23. John Wiley and Sons Inc., New York.

Bukalo, O., Schachner, M., and Dityatev, A. (2001). Modification of extracellular matrix by enzymatic removal of chondroitin sulfate and by lack of tenascin-R differentially affects several forms of synaptic plasticity in the hippocampus. *Neuroscience* **104**, 359–369.

Claiborne, B. J., Xiang, Z., and Brown, T. H. (1993). Hippocampal circuitry complicates analysis of long-term potentiation in mossy fiber synapses. *Hippocampus* **3**, 115–121.

Dityatev, A., and Schachner, M. (2003). Extracellular matrix molecules and synaptic plasticity. *Nat. Rev. Neurosci.* **4**, 456–468.

Dityatev, A., Dityateva, G., Sytnyk, V., Delling, M., Toni, N., Nikonenko, I., Muller, D., and Schachner, M. (2004). Polysialylated neural cell adhesion molecule PSA-NCAM promotes formation and remodeling of hippocampal synapses. *J. Neurosci.* **24**, 9372–9382.

Eckhardt, M., Bukalo, O., Chazal, G., Wang, L., Goridis, C., Schachner, M., Gerardy-Schahn, R., Cremer, H., and Dityatev, A. (2000). Mice deficient in the polysialyltransferase ST8SiaIV allow discrimination of the roles of neural cell adhesion molecule protein and polysialic acid in neural development and synaptic plasticity. *J. Neurosci.* **20**, 5234–5244.

Evers, M. R., Salmen, B., Bukalo, O., Rollenhagen, A., Bösl, M. R., Morellini, F., Bartsch, U., Dityatev, A., and Schachner, M. (2002). Impairment of L-type Ca^{2+} channel-dependent forms of hippocampal synaptic plasticity in mice deficient in the extracellular matrix glycoprotein tenascin-C. *J. Neurosci.* **22**, 7177–7194.

Lauri, S. E., Kaukinen, S., Kinnunen, T., Ylinen, A., Imai, S., Kaila, K., Taira, T., and Rauvala, H. (1999). Regulatory role and molecular interactions of a cell-surface heparan sulfate proteoglycan (N-syndecan) in hippocampal long-term potentiation. *J. Neurosci.* **19**, 1226–1235.

Matthies, H., Staak, S., and Krug, M. (1996). Fucose and fucosyllactose enhance *in-vitro* hippocampal long-term potentiation. *Brain Res.* **725**, 276–280.

Muller, D., Wang, C., Skibo, G., Toni, G., Cremer, H., Calaora, V., Rougon, G., and Kiss, J. Z. (1996). PSA-NCAM is required for activity-induced synaptic plasticity. *Neuron.* **17,** 413–422.

Saghatelyan, A. K., Gorissen, S., Albert, M., Hertlein, B., Schachner, M., and Dityatev, A. (2000). The extracellular matrix molecule tenascin-R and its HNK-1 carbohydrate modulate perisomatic inhibition and long-term potentiation in the CA1 region of the hippocampus. *Eur. J. Neurosci.* **12,** 3331–3342.

Sanes, J. R., and Lichtman, J. W. (1999). Can molecules explain long-term potentiation? *Nat. Neurosci.* **2,** 597–604.

Senn, C., Kutsche, M., Saghatelyan, A., Bosl, M., Lohler, J., Bartsch, U., Morellini, F., and Schachner, M. (2002). Mice deficient for the HNK-1 sulfotransferase show alterations in synaptic efficacy and spatial learning and memory. *Mol. Cell. Neurosci.* **20,** 712–729.

Yamamoto, S., Oka, S., Inoue, M., Shimuta, M., Manabe, T., Takahashi, H., Miyamoto, M., Asano, M., Sakagami, J., Sudo, K., Iwakura, Y., Ono, K., and Kawasaki, T. (2002). Mice deficient in nervous system-specific carbohydrate epitope HNK-1 exhibit impaired synaptic plasticity and spatial learning. *J. Biol. Chem.* **277,** 27227–27231.

[6] Analysis of Neural Cell Function in Gene Knockout Mice: Behavior

By JEFFREY M. LONG

Abstract

A powerful tool to investigative gene function is the ability to create mice with targeted gene mutations. Analysis of the resulting phenotype is sometimes difficult, however, because individual genes have more than one function, and observed effects on complex behaviors are often a result of abnormalities of any of a number of individual processes. One way to address this issue is by examining mice in a battery of behavioral tests to assess the specificity of any observed differences among genotypes. This chapter describes a test battery used to examine metabolic and behavioral phenotypes in mice with mutations in specific glycan-binding proteins and glycosyltransferases genes. Because the potential consequences of these genetic deletions are varied, a large number of assays across a variety of domains was included in the battery. The power and usefulness of this approach is in discovering areas for more detailed investigation.

Introduction

As part of a multiarmed approach to define the functions of glycan-binding proteins (GBPs), glycosyltransferases (GTs), and GBP–ligand

METHODS IN ENZYMOLOGY, VOL. 417
Copyright 2006, Elsevier Inc. All rights reserved.

0076-6879/06 $35.00
DOI: 10.1016/S0076-6879(06)17006-8

relationships, mice were generated with specific targeted mutations in GBPs and GTs genes (Chapter 39). The consequence of these genetic deletions were then assessed in a variety of domains (Chapters 39–41), including the topic of this chapter, metabolism and behavior. Because GBPs and GTs are involved in a large array of biochemical processes (many of which are still being discovered), the mice were assessed in a large number of tests that examined a variety of parameters. The use of a test battery allows for the assessment of a larger number of behavioral domains than if each assay used a separate cohort of mice.

Behavioral test batteries have several other advantages. First, they allow the possibility to correlate phenotypes across different behaviors. Second, they give increased assurance in a specific phenotype when observed across tasks with overlapping sensory/motor/cognitive requirements. Third, they reduce the number mice needed for each study (Contet et al., 2001; Crawley, 1999; Crawley and Paylor, 1997; Hatcher et al., 2001; Rogers et al., 1999).

However, there are disadvantages to test batteries as well. For example, it may be necessary to modify the testing protocol for a specific test so the overall length of the test battery is manageable. There may also be several interesting alternate protocols for a specific assay, and only one must be chosen for inclusion in the battery. In addition, there is the possibility that running mice on prior assays influences performance on subsequent tests (McIlwain et al., 2001; Voikar et al., 2004). Furthermore, without methodological or statistical corrections, examining a large number of dependent measures within the same study will increase the probability of making a type I statistical error beyond the individual alpha level of 0.05 (Benjamini et al., 2001). That is, the probability of finding a statistically significant difference between groups of mice for any one test, when in reality there is no difference, is higher than the generally accepted value of 5%.

With these caveats in mind, the Metabolism and Phenotyping Core of the Consortium for Functional Glycomics designed a test battery to provide a gross overall assessment of mutant mice on a variety of parameters. The test battery serves as an initial examination designed to detect large differences and to stimulate additional, more focused, experiments. The test order was planned so that mice were evaluated in more invasive tests either before a 2-day rest or toward the end of the battery. The effects of test order and multiple comparisons were not directly addressed, and, as such, the results of this battery are appropriate for hypothesis formation, not hypothesis confirmation. Statistically significant findings should be replicated with a separate cohort of mice using a more focused research design. Evaluating a different cohort in only a subset of tasks will address confounds in the data because of the previous test history of the mouse, test order, and increased type I error (because of multiple comparisons) inherent in large test batteries.

The power and proper use of this battery is to discover areas for more detailed investigation.

The battery includes assessment of physical features, reflexes, metabolism, pulmonary function, blood pressure, sensorimotor gating, nociception, motor movement, muscle strength, and learning and memory. Testing took 4.2 weeks and occurred Monday–Friday with the exception of the metabolism cages. Figure 1 lists the order in which the tests were administered. This chapter takes a different approach from the previous chapters, because the test battery includes multiple and diverse behavioral methods, and space limitations do not allow for detailed protocols to be included for each task. None of the tasks in the battery are novel however, and the reader is directed to alternate sources for detailed descriptions of the methods. Two valuable resources for this information are (1) Chapter 8 ("Behavioral Neuroscience") in *Current Protocols in Neuroscience* (Crawley, 2003) and (2) the book *What's Wrong with my Mouse? Behavioral Phenotyping of Transgenic and Knockout Mice* (Crawley, 2000).

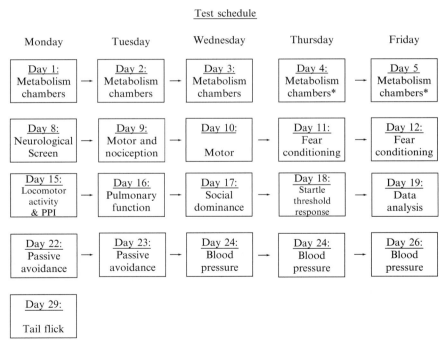

Fig. 1. Test schedule for the metabolism and behavioral test battery.

Materials

Subjects

1. Ten null and 10 litter mate control mice 3–5 months of age.*

Day 1–3 (4–6)[†] Metabolism Chambers

1. Comprehensive Lab Animal Monitoring System (CLAMS, Columbus Instruments, Columbus, OH) or equivalent.[†]
2. Mouse chow ground to a powder.
3. Drinking water.

Day 8 Neurological Screen

1. Standard shoebox mouse cage.
2. Clean bedding.
3. Scale and mouse-sized rectal thermometer.
4. Cotton tipped applicators.
5. 3-mm diameter wooden dowel covered with masking tape.
6. 20-cm diameter raised platform.
7. Notebook or data input template.[‡]

---

* Sex, estrus cycle, and age can affect the performance of many of the tasks in the battery. In this instance, the decision was made not to examine these variables. Male mice were tested past their stage of rapid development but before the onset of age-related impairments. Litter mates are the preferred control subjects, especially in non-congenic strains. Because of the inherent variability in behavior data, no less than 10 mice per groups is recommended. All automated equipment in the Metabolism and Behavior Phenotyping Core could test four mice simultaneously, thus allowing 20 mice to be tested by a single investigator within a 5–7-hour period. The investigator conducting the experiments should be blinded to experimental assignment of the mice. If mice are transported from a housing room to a testing room, they should acclimate to the testing room for 1 h. before the start of testing.

† Columbus Instruments of OH (http://www.colinst.com/brief.php?id=61) working with Jackson Laboratories http://aretha.jax.org/pub-cgi/phenome/mpdcgi?rtn=prodocs/92 developed the eight-chamber metabolism system used here. With only eight chambers it was decided to test only eight mice from each genotype and to divide the mice into two separate batches, tested on days 1–3 and 4–6, respectively. An equal number of control and experimental mice was assigned to each batch.

‡ Because all mice are individually assessed on a large number of assays that elicit a small number of set responses (e.g., yes/no, normal/abnormal), it is efficient to create paper or computer-based templates for recording these data.

Day 9 Motor and Nociception

1. Modified wire mouse cage lid.[§]
2. Rotating drum (UGO Basile, Varese, Italy) or equivalent.
3. Grip strength meter (UGO Basile, Varese, Italy) or equivalent.
4. Hot plate apparatus (Columbus Instruments, Columbus, OH) or equivalent.

Day 10 Motor

1. Taut string or wire.
2. Balance beam apparatus (10 and 6 mm diameter × 60 cm long wooden dowel 120 cm above a cushion, suspended on each end by a 10-cm × 10-cm escape platform).

Days 11–12 Memory: Fear Conditioning

1. Four fear conditioning shock chambers (26 × 22 × 18 cm high) made of clear Plexiglas placed in a 2 × 2 array (Med Associates Inc., East Fairfield, VT) or equivalent.
2. A video camera connected to a video-based system for digital recording and subsequent analysis of freezing behavior (Freeze Frame; Actimetrics, St. Evanston, Il) or equivalent.

Day 15 Locomotor Activity and Prepulse Inhibition of the Acoustic Startle Response

1. Photocell-based activity monitor (Digiscan; Accuasacn Electronics, Columbus, OH) or equivalent.
2. SR-LAB startle apparatus with digitized electronic signal output (San Diego Instruments, San Diego, CA) or equivalent.

Day 16 Pulmonary Function

1. Carbon monoxide uptake monitor (Columbus Instruments, Columbus, OH) or equivalent.
2. Acclimation apparatus.*

[§] Placing tape around the perimeter of the cage will keep for the mouse from walking over the edge.

* Placing the mice in the testing chambers increases stress and respiration rate. We have found that placing a small amount of home-cage bedding, covering the chambers with dark contact paper, and allowing the mice to acclimate for 2 h results in an asymptotic respiration rate. To save time, we made an acclimation apparatus using the same chambers and air pump that are used in the CO uptake equipment. Quick-snap connectors allow the mice to stay in the same chamber for acclimation and testing and allow several batches of mice to acclimate while previously acclimated mice undergo testing.

Day 17 Social Dominance

 1. 30-cm-long by 3.0 cm-diameter opaque plastic tube.

Day 18 Startle Threshold Response Test

 1. SR-LAB startle apparatus with digitized electronic signal output
    (San Diego Instruments, San Diego, CA) or equivalent.

Days 22–23 Memory: Passive Avoidance

Automated passive avoidance chamber (35.5 × 18 × 30.5 cm high) with
a shock floor, photobeams, and a central guillotine door attached to a
computer (Coulbourn Instruments, Allentown, PA) or equivalent.

Days 24–26 Blood Pressure

 1. Noninvasive blood pressure tail-cuff system (Columbus Instruments,
    Columbus, OH) or equivalent.

Day 29 Nociception Test: Tail Flick

 1. Tail flick analgesia meter (Columbus Instruments, Columbus, OH)
    or equivalent.

Methods

Subjects

The mice are tested in the following assays in the order presented in the
following (also see Fig. 1). Be sure to note any extraordinary features or
behaviors that occur during testing.[†]

Day 1–3 (4–6) Metabolism Chambers

To ascertain whether the genes of interest were involved in metabolic
processes, individual mice were placed in indirect, open-circuit calorimeter
chambers. These metabolic cages provide measures of oxygen consumption
and carbon dioxide production, locomotor activity, and food and water

---

[†] Separate assays testing the same general domain (i.e., muscle strength) are not necessarily
administered together in the battery, but at the completion of testing, these data generally
organized together.

intake. Data are collected every 30 min over three 12-h dark cycles and two 12-h light cycles.[‡] The most relevant metabolic measurements include the volume of carbon dioxide produced (VCO^2), the volume of oxygen consumed (VO^2), the respiration measurement ($RER = VCO^2/VO^2$), and the caloric (heat) value (($3.815 + 1.232 \times RER) \times VO^2$)).

1. Record the body weight of the mice.
2. Power the CO and CO_2 sensors for at least 90 min before following the software prompts to calibrate the sensors.
3. Place drinking water in the reservoir, and test the functionality of the lick spouts with a paper towel.
4. Fill the food hoppers and ensure the food scales are functional.
5. Place mice in the chambers and follow the software prompts to start the experiment.
6. Check daily to ensure all mice are eating and drinking; remove mice if necessary.
7. At the end of the experiment, remove mice and record their body weight.
8. Because a large amount of data is generated during the experiment, it is useful to use a computer program (e.g., Excel, Matlab) to create a software data template for summarizing the variables of interest. One basic analysis is to look at the different measures as a function of the light/dark cycle.

Day 8 Neurological Screen

This is a formalized procedure designed to note remarkable physical abnormalities.

1. Mice are tested one at a time in all assays, with clean bedding and cotton-tipped applicator used for each mouse.
2. Gross physical assessment: Individual mice are placed in a clean shoebox cage and during a 2-min observation period several behavioral (e.g., presence of wild running, excessive grooming, freezing, rearing, jumping, defecation, urination, cage exploration, hunched body posture) and physical (e.g., presence of dirty fur, ulcerated skin, bald spots, thinning fur, trimmed whiskers, labored breathing, piloerection, exophthalmos, palpebral closure;

[‡] The number of chambers in the system will dictate the time period between gas measurements for any specific cage. For symmetry, all measurements were collected every 30 min, although activity and food consumption can be recorded with shorter intervals between measurements.

condition of teeth, nails, nose, rectum area; fur color) features of the mice are noted.

3. Sensorimotor reflexes: With the mouse still in the cage, and with the use of a cotton-tipped applicator, each mouse is assessed for several sensorimotor reflexes (e.g., eye blink, ear twitch, whisker-orienting, sound orientating) and its response to an approaching object.

4. Postural reflexes: Latency to return to upright posture, ability to maintain upright balance in a rapidly moving cage, and the tail suspension response are noted.

5. Pole test: A measure of muscle strength. Place mouse at end of a 3-cm diameter pole lying in a horizontal position. The pole is gradually lifted to a vertical position with the latency to fall off the pole being the dependent measure. These values are converted to a pole test score: fell before the pole reached 45-degree or 90-degree angle = 0 or 1; fell in 0–10 seconds = 2, 11–20 sec =3, 21–30 sec=4, 31–40 sec =5, 41–50 sec = 6, 51–60 sec = 7; stayed on 60 sec and climbed halfway down the pole = 8, climbed to lower half of the pole = 9, climbed down and off in 51–60 sec =10, 41–50 sec = 11, 31–40 sec = 12, 21–30 sec = 13, 11–20 sec = 14, 1–10 sec = 15.

6. Body temperature and weight are recorded last because of the invasiveness of the temperature probe.

7. See Chapter 3 in Crawley (2000) and Crawley and Paylor (1997) for more theoretical and methodological detail of the tasks included in this neurological screen.

Day 9 Motor and Nociception Tests

1. Cage top hang test: This test of muscle strength allows the mouse to use both its fore and hind limbs to maintain its grip. The mouse is placed on a modified cage lid (duct tape is placed around the edges) and the lid inverted. The latency to fall on a cushioned surface is the dependent measure. Trial maximum is 60 sec. See Chapter 4 in Crawley (2000) for more detail.

2. Rotorod: Locomotor coordination and balance are measured by placing mice on an accelerating, 3-cm diameter, rotating drum for three trials with a minimum of 30-min interval between trials. The Rotorod starts at 4 rpm and increases to 40 rpm over a 5-min period. The mean latency to fall over the three trials is the dependent measure. See Chapter 8.12 in Crawley (2003) and Chapter 4 in Crawley (2000) for more detail.

3. Grip strength: In a separate assessment of forelimb strength, mice are suspended by the tail and lowered until it grasps the loop of a mouse grip strength meter. The mouse is then gently pulled away from the loop, and the

maximum grip force exerted by the mouse before loosing its grip is recorded. Five trials are run, with the mean of the middle three scores used for analysis.

4. Hot plate: This reflex involves both brain and spinal cord circuitry to detect painful thermal stimuli. The mouse is placed on a hot plate analgesia meter set at 55°. The latency for the mouse to jump, hind paw lick, or hind paw shake is recorded. Trail maximum is 30 sec. See Malmberg (1999) and Chapter 5 in Crawley (2000) for more detail.

Day 10 Motor Tests

1. Wire suspension: Forelimb strength is assessed by suspending mice by the tail and gently lowering them until it grasps a 1-mm diameter wire with its front paws. The mean latency to fall (maximum 60 sec) over three trials is the dependent measure.

2. Balance beam: Locomotor coordination and balance are measured in this task. Two round wooden rods (10 and 6 mm in diameter × 60-cm long) were suspended 120 cm above a cushion. At each end of the rod is a 10-cm × 10-cm escape platform. The mouse is first placed on the escape platform for 10 sec then placed in the middle of each the beam. The latency to either fall or reach the escape platform is recorded. The maximum trial length is 120 sec. See Carter *et al.* (2001) for more detail.

Days 11–12 Memory: Fear Conditioning

Conditioned fear is a form of classical conditioning in which the mouse must remember an association between a negative experience (a foot shock) and environmental cues. The amygdala is involved in processing a single cue (e.g., a tone) test, whereas both the amygdala and the hippocampal formation are involved in processing during the context (e.g., the multi-cued environment) test.

1. Follow the software instructions to create the desired testing protocol. In this example, the conditioned stimulus (CS) is an 85-dB, 2800 Hz, 20-sec tone, and the unconditioned stimulus (US) is a scrambled foot shock at 0.45 mA presented during the last 3 sec of the CS. Mice are placed in the test chamber for 3 min before the CS, and freezing behavior is recorded. Freezing thresholds were selected by means of the software and show a high correlation with human observers. Three CS/US pairings are given with 1-min spacing and freezing during the CS is recorded. Twenty-four hours later, each mouse was placed back into the shock chamber, and freezing response was recorded for 3 min (context test). Two hours later, the chambers were modified to present a different environmental context (e.g., shape, odor, color changes), and the mouse was placed in this novel environment. Freezing behavior was recorded for 3 min before and during

each of three CS presentations (cued conditioning). The time spent freezing is converted to a percent freezing value.

2. See Wehner and Radcliffe (2001) for more methodical detail and theoretical background.

3. The mice are given a 2-day rest period before the next test day.

Day 15 Locomotor Activity and Prepulse Inhibition of the Acoustic Startle Response

Both assays are automated and take approximately 30 min per mouse. Mice are tested in batches, first for locomotor activity then for prepulse inhibition (PPI). It is efficient to immediately test the mice that have just completed the locomotor activity assay in the PPI test and at the same time test the next batch of mice in the open-field activity chambers.

Locomotor Activity. The open-field measures exploratory activity in a novel environment and overall activity levels

1. Follow software instructions for preparing the open field environment and data acquisition. Most software programs will output a large number of variables.

2. Basic measurements include horizontal activity (locomotor activity), vertical activity (rearing), total distance (cm), and center distance. The center distance divided by the total distance is an indicator of anxiety-related behavior. In this instance, the mice were placed in the activity chambers for 30 min, and data were analyzed in 2-min intervals.

3. Additional background and detail for the open-field test is provided in Pierce and Kalivas (1997) and Chapter 4 in Crawley (2000).

PPI: Prepulse Inhibition Is a Measure of Sensorimotor Gating and Deficits Suggest Abnormal Processing of Incoming Sensory Stimuli (Geyer and Braff, 1987; Geyer et al., 2002)

1. Follow software instructions for preparing the test chambers and programming an experiment. In this instance, a test session began by placing the mouse in the Plexiglas cylinder where it was left undisturbed for 5 min. A test session consisted of seven trial types. One trial type was a 40 ms, 120-dB sound burst used as the startle stimulus. There were five different acoustic prepulse plus acoustic startle stimulus trials. The prepulse sound was presented 100 ms before the startle stimulus. The 20-ms prepulse sounds were 72, 74, 76, 78, or 80 dB. Finally, there were trials in which no stimulus was presented to measure baseline movement in the cylinders. Six blocks of the seven trial types were presented in pseudorandom order such that each trial type was presented once within a block of seven trials. The average

intertrial interval was 15 sec (range, 10–20 sec). The startle response was recorded for 65 ms (measuring the response every 1 ms) starting with the onset of the startle stimulus. The background noise level in each chamber was 70 dB. The maximum startle amplitude recorded during the 65-ms sampling window was used as the dependent variable. Percent prepulse inhibition of a startle response was calculated as: 100 – [(startle response on acoustic prepulse and startle stimulus trials/startle response alone trials) × 100].

2. Methodical detail and theoretical background for the prepulse inhibition task can be found in Geyer and Dulawa (2003) and Geyer and Swerdlow (1998).

Day 16 Pulmonary Function

The uptake of carbon monoxide (CO) can be used as an assessment of lung function; CO uptake decreases as lung function decreases. A carbon monoxide uptake monitor can measure the CO level in an animal chamber after exposing the mouse to a 60-sec interval of air with 0.17% carbon monoxide. The mean number of breaths per minute is also recorded.

1. Follow software instructions for setting up an experiment.
2. Test chamber for air leaks.
3. Place mice in acclimation apparatus.
4. After 2 h acclimation, connect chamber to the CO uptake measurement machine and start the computer program.

Day 17 Social Dominance

The tube test is a task designed to assess social dominance without the need for lengthy housing isolation or the possibility of injury that could possibly occur in other test of murine aggression.

1. Someone other than the investigator testing the mice must decode the grouping data.
2. Record the body weight of all mice.
3. Create a data sheet to record the result of each trial. Each mutant mouse and control mouse will experience three confrontations, each with a different opponent that is matched for body weight as closely as possible.
4. Before the start of testing, allow each mouse to move throughout the tube unopposed in both directions. Clean the tube between mice.
5. For the test, place the appropriate mice in opposite ends of the tube and record if one mouse backs the other out.
6. If no mouse backs the other out in 60 sec, the trial is over and is recorded as a tie.
7. More detail and background information is found in Chapter 9 of Crawley (2000).

Day 18 Startle Threshold Response Test

This test will ascertain the dB required to elicit a startle response from the subject. Higher than normal dB levels required to elicit a response suggest impairment either in hearing or the startle reflex. These are not mutually exclusive factors, and care must be given when interpreting data from other tasks that require one of both of these factors.

1. A separate startle threshold experiment is conducted using procedures identical to that described for PPI in "Day 15 Locomotor Activity and Prepulse Inhibition of the Acoustic Startle Response," except that startle stimulus only trials were presented at dB levels ranging from 0–48 dB above the 70-dB baseline. The different startle stimuli were presented in a random order, with five presentations at each level. This procedure provided a gross indication of the hearing ability of each mouse. Methodical detail and theoretical background for the startle threshold response test can be found in Geyer and Swerdlow (1998) and Geyer and Dulawa (2003).

2. Three rest days are provided to the mice before testing resumes.

Days 22–23 Memory: Passive Avoidance

Passive avoidance requires the mouse to recall the location of a negative event.

1. Follow software instructions for setting up the chamber and a test protocol.

2. In this instance, the guillotine door separated the dark half of the chamber from the brightly lit half. Mice were placed in the lit chamber for 10 sec, and then the door was raised. When the mouse entered the dark chamber, the computer automatically recorded the latency, lowered the door, and delivered a 0.65-mA foot shock for 2 sec. Twenty-four hours later the mouse was again placed in the brightly lit half of the chamber, and the latency to enter the dark chamber was recorded. Chapter 6 in Crawley (2000) provides methodical detail and theoretical background.

Days 24–26 Blood Pressure

1. Follow the software prompts to start an experiment.*
2. Check for leaks in the airline and sensor cuffs.

---

* The mice should be tested for several days to acclimate the mice to this stressful procedure. Five days is preferable, because a compromise of running a test battery two acclimations days and a test day was given here.

3. Place the mouse in the cylinder restraining chamber and secure the occlusion and sensor cuffs on near the base of the tail.

4. Set the equipment to warm to tail to 37° for 30 min.

5. Following the software prompt and examining each waveform, take repeated blood pressure measurements for each mouse. Careful attention must be given to the waveform to ensure that tail movements or low-amplitude heart rate readings do not introduce artifacts into the recorded values. The software is not a good as the human eye to detect artifacts.

6. Continue to record blood pressure measurement until four good readings of systolic blood pressure, heart rate, and relative changes in diastolic and mean blood pressures are obtained. The mean of these four readings is used as the dependent measure.

7. Two days of recovery are provided before testing resumes.

Day 29 Nociception Test: Tail Flick

1. The intensity of a light beam in a tail flick analgesia meter (Columbus Instruments, Columbus, OH) was adjusted to produce tail flick of 4–6 sec on a previous set of calibration mice. For testing, the mice were wrapped in a soft towel with their tail extending over the light path. Three trials were run on different areas of the tail with at least a 50min intertrial interval. Trial maximum was 10 sec.

2. See Malmberg and Bannon (1999) and Chapter 5 in Crawley (2000) for methodical detail and theoretical background.

Acknowledgments

The author thanks Drs. Anthony Wynshaw-Boris and Richard Paylor for advice in setting up the test battery. This work was supported by U54GM62116.

References

Benjamini, Y., Drai, D., Elmer, G., Kafkafi, N., and Golani, I. (2001). Controlling the false discovery rate in behavior genetics research. *Behav. Brain Res.* **125,** 279–284.

Carter, R. J., Morton, A. J., and Dunnett, S. B. (2001). Motor Coordination and Balance in Rodents. *In* "Current Protocols in Neuroscience" (J. N. Crawley, C. R. Gerfen, and M. A. Rogawski *et al.*, eds.), Vol. 3, pp. 8.12.1–8.12.14. John Wiley & Sons, New York.

Contet, C., Rawlins, J. N., and Deacon, R. M. (2001). A comparison of 129S2/SvHsd and C57BL/6JOlaHsd mice on a test battery assessing sensorimotor, affective and cognitive behaviours: Implications for the study of genetically modified mice. *Behav. Brain Res.* **124,** 33–46.

Crawley, J. N. (1999). Behavioral phenotyping of transgenic and knockout mice: Experimental design and evaluation of general health, sensory functions, motor abilities, and specific behavioral tests. *Brain Res.* **835,** 18–26.

Crawley, J. N. (2000). What's Wrong with my Mouse? Behavioral Phenotyping of Transgenic and Knockout Mice Wiley-Liss, New York.

Crawley, J. N. (2003). Behavioral Neuroscience. In "Current Protocols in Neuroscience" (J. N. Crawley, C. R. Gerfen, and M. A. Rogawski et al., eds.), Vol. 3, pp. 8.0.1–8.0.2. John Wiley & Sons, New York.

Crawley, J. N., and Paylor, R. (1997). A proposed test battery and constellations of specific behavioral paradigms to investigate the behavioral phenotypes of transgenic and knockout mice. Horm. Behav. 31, 197–211.

Geyer, M. A., and Braff, D. L. (1987). Startle habituation and sensorimotor gating in schizophrenia and related animal models. Schizophr. Bull. 13, 643–668.

Geyer, M. A., and Dulawa, S. C. (2003). Assessment of Murine Startle Reactivity, Prepulse Inhibition, and Habituation. In "Current Protocols in Neuroscience" (J. N. Crawley, C. R. Gerfen, and M. A. Rogawski et al., eds.), Vol. 3, pp. 8.17.1–8.17.15. John Wiley & Sons, New York.

Geyer, M. A., McIlwain, K. L., and Paylor, R. (2002). Mouse genetic models for prepulse inhibition: An early review. Mol. Psychiatry 7, 1039–1053.

Geyer, M. A., and Swerdlow, N. R. (1998). Measurement of Startle Response, Prepulse Inhibition and Habituation. In "Current Protocols in Neuroscience" (J. N. Crawley, C. R. Gerfen, and M. A. Rogawski et al., eds.), Vol. 3, pp. 8.7.1–8.7.15. John Wiley & Sons, New York.

Hatcher, J. P., Jones, D. N., Rogers, D. C., Hatcher, P. D., Reavill, C., Hagan, J. J., and Hunter, A. J. (2001). Development of SHIRPA to characterise the phenotype of gene-targeted mice. Behav. Brain Res. 125, 43–47.

Malmberg, A. B., and Bannon, A. W. (1999). Models of Nociception: Hot-Plate, Tail-Flick, and Formalin Tests in Rodents. In "Current Protocols in Neuroscience" (J. N. Crawley, C. R. Gerfen, and M. A. Rogawski et al., eds.), Vol. 3, pp. 8.9.1–8.9.15. John Wiley & Sons, New York.

McIlwain, K. L., Merriweather, M. Y., Yuva-Paylor, L. A., and Paylor, R. (2001). The use of behavioral test batteries: Effects of training history. Physiol. Behav. 73, 705–717.

Pierce, R. C., and Kalivas, P. (1997). Locomotor Behavior. In "Current Protocols in Neuroscience" (J. N. Crawley, C. R. Gerfen, and M. A. Rogawski et al., eds.), Vol. 3, pp. 8.1.1–8.1.8. John Wiley & Sons, New York.

Rogers, D. C., Jones, D. N., Nelson, P. R., Jones, C. M., Quilter, C. A., Robinson, T. L., and Hagan, J. J. (1999). Use of SHIRPA and discriminant analysis to characterise marked differences in the behavioural phenotype of six inbred mouse strains. Behav. Brain Res. 105, 207–217.

Voikar, V., Vasar, E., and Rauvala, H. (2004). Behavioral alterations induced by repeated testing in C57BL/6J and 129S2/Sv mice: Implications for phenotyping screens. Genes Brain Behav. 3, 27–38.

Wehner, J. M., and Radcliffe, R. A. (2001). Cued and Contextual Fear Conditioning in Mice. In "Current Protocols in Neuroscience" (J. N. Crawley, C. R. Gerfen, and M. A. Rogawski et al., eds.), Vol. 3, pp. 8.5C.1–8.5C.12. John Wiley & Sons, New York.

[7] Electrophysiological Analysis of Interactions Between Carbohydrates and Transmitter Receptors Reconstituted in Lipid Bilayers

By Vishnu Suppiramaniam, Thirumalini Vaithianathan, and Kodeeswaran Parameshwaran

Abstract

The investigation of functional interactions between carbohydrates and neurotransmitter receptors is a challenging task. The presence of a wide variety of carbohydrates in the nervous system may preclude electrophysiological analysis using intact brain slice preparations or isolated neurons. The purification of transmitter receptors and their subsequent reconstitution into an artificial lipid bilayer can serve as a valuable tool to study carbohydrate and transmitter receptor interaction in a controlled environment. The "tip-dip" bilayer technique along with patch clamp electronics provides a unique means to explore carbohydrate interactions with a single transmitter receptor channel. This technique is also helpful in analyzing the interaction of carbohydrates with synaptic transmitter receptors using isolated synaptosomal preparations. Here, we illustrate the methods involved in reconstituting transmitter receptors in tip-dip bilayers and the subsequent study of carbohydrate interaction with the receptors.

Overview

Proper neuronal transmission in the nervous system depends on the appropriate functioning of neurotransmitter receptors. Various carbohydrate moieties have been shown to interact with and modulate neurotransmitter receptors (Chicoine *et al.*, 2004; Hall *et al.*, 1996; Sinnarajah *et al.*, 1999) and thereby alter neuronal physiology (Vaithianathan *et al.*, 2004). The functional interaction between specific transmitter receptors and carbohydrates can be investigated by a variety of techniques. Electrophysiological methods such as traditional patch clamp experiments using brain slice preparations or isolated neurons can be used to demonstrate the functional effects of carbohydrates on native transmitter receptors. However, the complexity of the receptor environment and the presence of endogenous carbohydrates in the nervous system limit the use of these techniques in obtaining valuable information regarding direct interaction of specific carbohydrates with transmitter

METHODS IN ENZYMOLOGY, VOL. 417   0076-6879/06 $35.00
DOI: 10.1016/S0076-6879(06)17007-X

receptors. In contrast, isolation and purification of neurotransmitter receptors from their native environment and reconstitution of these receptors in an artificial lipid bilayer is a powerful tool for investigating the direct modulation of receptors by carbohydrates in a controlled environment.

The artificial lipid bilayer technique has been used to study the functional properties of ion channels. The tip-dip bilayer technique (Coronado and Lattore, 1983; Wilmsen *et al.*, 1983) is one of the methods used to investigate single-channel properties of purified transmitter receptors (Suarez-Isla *et al.*, 1983; Suppiramaniam *et al.*, 2001; Vaithianathan *et al.*, 2004; Vodyanoy *et al.*, 1993). In addition, the tip-dip bilayer technique has also been useful in studying single-channel properties of synaptic receptors. Thus, the tip-dip method provides a unique opportunity to investigate the interaction between carbohydrates and purified receptors or receptors present in synaptic membranes (Vaithianathan *et al.*, 2004, 2005).

A key feature of this sensitive technique described here is that it provides a "clean and controlled environment" to study the direct interaction between the neurotransmitter receptors and carbohydrate molecules. This chapter provides a step-by-step protocol for successful tip-dip bilayer formation and subsequent reconstitution of isolated transmitter proteins or synaptosomal membranes for the analysis of receptor and carbohydrate interaction.

Materials for Phospholipid Preparation

    Sterile syringe filters (10-mm diameter, 0.02-μm pore size)
    pH meter
    Hexane, 95% anhydrous, 100 ml (Sigma-Aldrich)
    Nitrogen gas, compressed with tank regulator
    Hamilton syringe, 1.0 ml; gas tight with 22-guage needle
    Gas drying jar with drierite
    Millipore filter holder (250 ml)
    1,2-diphytanoyl-sn-glycero-3-phosphocholine in chloroform, 1-mg vials
        (Avanti Polar Lipids)

Solutions

    AECF: 125.0 mM NaCl, 5.0 mM KCl, 1.25 mM NaH$_2$PO$_4$, and 5.0 mM
        Tris HCl.
    AICF: 110.0 mM KCl, 4.0 mM NaCl, 2.0 mM NaHCO$_3$, 0.1 mM
        CaCl$_2$, 1.0 mM MgCl$_2$, and 2.0 mM 3-(N-Morpholino propane
        sulfonic acid).

Preparation of Artificial Extracellular and Intracellular Fluids

Reconstitution of neurotransmitter receptors in tip-dip bilayers is carried out using asymmetrical saline conditions across the lipid bilayer. Stock solutions of artificial intracellular fluid (AICF) and artificial extracellular fluid (AECF) are prepared in advance and stored in refrigerator for up to a month. The solutions are warmed slightly to remove the dissolved gases immediately before use. Once the solutions reach the room temperature, the pH of the solutions is adjusted to 7.4. Most solutions are prepared in distilled water. The ion channel blockers are prepared in accordance with product specifications. The AICF is filtered using a new 0.02-μm syringe filter into a clean dust-free centrifuge tube. The patch pipette is filled with AICF and AECF is placed into the microbeaker.

Preparation of Phospholipid

Lipids are prepared carefully, because it is easy to introduce contaminants that can affect membrane stability. Because some fungal and bacterial toxins have the ability to form ion channels, care should be taken to avoid contamination of solutions. Fresh phospholipids made once a week minimize these problems. The vial containing 0.5 mg phosphatidylcholine (PC) in chloroform (custom ordered from Avanti Polar Lipids Inc., Alabaster, AL) is opened at room temperature. The chloroform is evaporated by gently blowing a steady stream of filtered, dry nitrogen gas into the vial kept in a closed container (Fig. 1c). Dried PC is then dissolved in 0.5 ml anhydrous hexane. A three-way T connecter with valve is used to connect the container with anhydrous $CaSO_4$ (Drierite) (Fig. 1b) to a Millipore

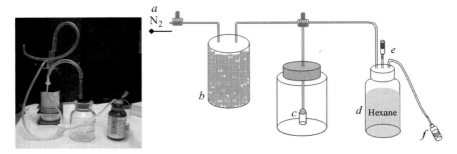

FIG. 1. Setup for PC preparation. The PC is prepared by first evaporating chloroform using dry nitrogen gas and then dissolving the dried PC in anhydrous hexane (1 mg/ml). (a) Attached to nitrogen gas; (b) drierite; (c) vial containing PC/chloroform; (d) hexane bottle; (e) air releasing valve; (f) valve for syringe attachment. (See color insert.)

filter holder and the bottle containing hexane (Fig. 1d). Dried nitrogen is forced into bottle containing hexane, and the pressure-driven hexane coming out through a vent is collected in a 1.0-ml Hamilton syringe attached to a valve (Fig. 1f). Hexane (0.5 ml) is collected in the syringe and released into the vial containing dried PC. The dried PC on the wall and bottom of the vial are completely dissolved by repeatedly swirling the bottle. Once the PC is completely dissolved, the solution is drawn into the syringe and released into a clean small glass vial equipped with an airtight cap. The cap is closed at once, and the solution is used immediately or stored at −4°. If the PC preparation absorbs moisture, bilayer formation will become difficult. Therefore, the vial containing PC is kept hermetically sealed at all times except when removing PC.

Cleaning of Microbeakers

Materials for Cleaning of Microbeakers

Borosilicate glass microbeakers (500 μl)
Polypropylene wide mouth bottles (e.g., Qorpak, Fisher Scientific, 4 oz/120 ml bottles)
Chloroform (99.9% assay, Fisher Chemicals)
Methanol (99.9% assay, Fisher Chemicals)
Acetone (99.5% assay, Sigma Chemicals)
Ethyl alcohol (Fisher Chemicals)
Petroleum ether (Fisher Chemicals)

The 500-μl microbeakers used in the experiments are cleaned by a five-step washing procedure. The following cleaning liquids are kept in separate and labeled small polypropylene stout wide mouth bottles: (1) chloroform and methanol mixture (3:1 ratio vol/vol), (2) acetone, (3) distilled and filtered water, (4) ethyl alcohol, and (5) petroleum ether (Fig. 2). The bottles containing microbeakers are sonicated in a water bath sonicator for 1 min in each of the first four solutions in the same order. After the final sonication in ethyl alcohol, the microbeakers are placed in petroleum ether for about 5 min, then lifted with a clean forceps and kept in a dust-free container for drying.

Chloriding the Electrodes

We use Ag/AgCl$_2$ reference and recording electrodes. Periodical chloriding helps in forming a thin uniform coating of silver chloride on the surface of electrodes. Both electrodes are kept immersed in a 30-ml beaker containing 6.0% sodium hypochlorite solution (Clorox bleach) for about 5 min. After this, the electrodes are inspected under a microscope for

FIG. 2. Steps involved in washing microbeakers. Microbeakers and microstir bars undergo a five-step washing procedure by sonicating them in the first four solutions in the same order as shown and finally placing them in petroleum ether for 5 min. (See color insert.)

uniformity and completeness of coating. Subsequent chloriding requires removal of existing silver chloride coating by a fine sand paper.

Electronic Assembly

The patch clamp assembly used in our laboratory is illustrated in Fig. 3. The assembly consists of Axopatch 200B amplifier. The amplified electrical signals are passed through an analog-to-digital converter to a VCR or passed through a digitizer for storage into a computer. An amplifier, digitizer, and computer are sufficient to build the assembly if storage on a VHS tape is not required. The Faraday cage containing a simple bilayer assembly is shown in Fig. 3.3. The microbeaker is placed on an air-driven stirrer to drive the magnetic bar in the beaker. A single axis manipulator is adequate to perform tip-dip method. The bilayer assembly is set on an isolation table.

Materials for Tip-Dip Bilayer Method

Patch clamp amplifier (Molecular Devices, Foster City, CA)
Digital data recorder-VR 10B (Instrutech Corp., Elmont, NY)
Oscilloscope
Analog input-digital output data acquisition system, Digidata-1200 (Molecular Device, CA)
Microelectrode puller (Sutter Instrument Inc.)
Borosilicate glass capillaries
Videocassette recorder
Grounding strap wire
BNC cables, audio and video cable and 5″ silver wire
Borosilicate glass microbeakers (500 μl)
Stir bars (7.0 mm L × 2.0 mm D; Fisher)
Syringe needle for filling patch pipettes (34G, ~67 mm long)

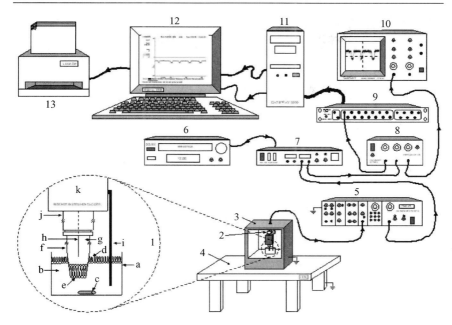

FIG. 3. Patch clamp assembly. Single-channel currents are amplified, filtered, digitized, and stored in video tapes and/or computer discs. The components of patch clamp assembly required for tip-dip bilayer technique are shown: (1a) microbeaker; (1b) artificial extracellular fluid; (1c) microstir bar; (1d) lipid monolayer; (1e) lipid bilayer; (1f) patch pipette; (1g) artificial intracellular fluid; (1h) recording electrode; (1i) reference electrode; (1j) electrode holder; (1k) head stage; (2) head stage; (3) Faraday box; (4) isolation table; (5) patch clamp amplifier; (6) video cassette recorder; (7) analog-to-digital converter; (8) low-pass filter; (9) digitizer; (10) oscilloscope; (11) computer hard drive; (12) monitor; (13) printer.

Manipulator (World Precision Instruments Inc. Sarasota, FL)
Faraday cage ($12'' \times 12'' \times 12''$ or greater dimensions)
Isolation table ($23'' \times 35''$)
Purified receptors/synaptosomes/membrane fragments
Receptor specific agonist and antagonist
Carbohydrate of interest

Tip-Dip bilayer Method

1. Borosilicate glass pipettes with about 100 MΩ resistances are pulled using a pipette puller.
2. A clean microbeaker (Fig. 4e) is placed on the stage inside the Faraday cage. The stage is positioned immediately below the manipulator.

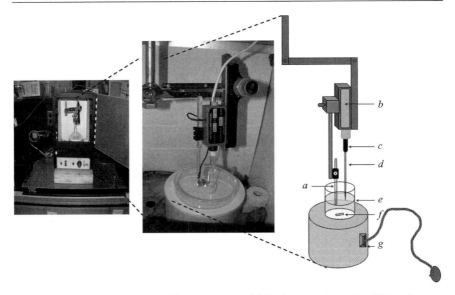

FIG. 4. Assembly for tip-tip bilayer method. (a) Reference electrode; (b) head stage; (c) electrode holder; (d) recording electrode; (e) artificial extracellular fluid in micro beaker; (f) micro stir bar; (g) magnetic stir bar mixer. (See color insert.)

The air-driven rotating steel inside the stage (Fig. 4g) facilitates the rotation of the stir bar (Fig. 4f), which is placed in the microbeaker.

3. Approximately 1.0 ml of AECF is filtered into a clean dust-free centrifuge tube. From this, 300 μl of AECF is added into the microbeaker. Next, approximately 1.0 ml of AICF is filtered through the syringe filter into a clean and dust-free centrifuge tube. The sharp tip of the recording pipette is immersed in the AICF for about 20 sec for the fluid to be taken up by capillary action. Then the pipette is backfilled with AICF using the micropipette-filling syringe needle.

4. The recording pipette filled with AICF is observed under a light microscope for the presence of air bubbles or dust particles. Any bubbles trapped inside the pipette are removed by gently tapping the pipette with a finger.

5. The patch pipette filled with AICF (Fig. 4d) is then attached to electrode holder with recording electrode and the head stage of the patch amplifier. The recording electrode should be straight and is inserted into the patch pipette carefully to avoid touching the flat end of the pipette. This would prevent silver chloride coatings of the electrode being scraped off and introduced into the pipette. The microbeaker that is filled with 300 μl of

AECF is placed directly under the electrodes, and the position of the reference electrode (Fig. 4a) is adjusted so that neither of the electrodes will touch the walls of the microbeaker, because such contacts often result in noise.

6. The head stage (Fig. 4b), which is attached to the manipulator, is lowered until the pipette tip is just below the surface of the AECF (Fig. 5.1). After the reference and recording electrode are immersed in the AECF, the pipette resistance is measured.

7. Approximately 3–4 μl of the prepared PC in hexane solution is dispensed into the AECF in the microbeaker (Fig. 5.2). After approximately 20 sec (allowing the lipid monolayer to spread), the tip of the recording pipette is taken out of AECF (Fig. 5.3) and quickly dipped back in (Fig. 5.4). This is repeated until stable lipid bilayer is formed at the tip of the recording electrode. The existence and the stability of the bilayer can be verified by measuring membrane resistance. A GΩ resistance is preferred.

8. The next step involves the incorporation of neurotransmitter receptors of interest into the lipid bilayer. In the specific example given here, approximately 2 μl of purified α-amino-3-hydroxy-5-methyl-4-isoxazolepropionic acid (AMPA)-glutamate receptor suspension is mixed with 2 μl of PC and added into the AECF (Fig. 5.4a). A gentle stirring with a microstir bar will assist incorporation of receptors or membrane fragments into the bilayer (Fig. 5.5). The purification of the AMPA receptor is carried out as described by Bahr *et al.* (1992). For recording from single synaptic receptors, synaptosomes are isolated by technique described by Taupin *et al.* (1994), and data collection is performed as per Vaithianathan *et al.* (2004).

9. If partially purified AMPA receptors or membrane preparations such as synaptosomes are used, the following cocktail of ion channel blockers are added to the AECF: 1.0 μM tetrodotoxin, 2.0 μM tetraethyl ammonium, 50 μM aminophosphovalerate, 1.0 μM methyl glutamate analog (2S, 4R)-4-methylglutamate (SYM 2081), and 100 μM picrotoxin. These antagonists will block sodium channels, potassium channels, NMDA receptors, kainite receptors, GABA$_A$, and glycine receptors, respectively.

10. The single-channel AMPA receptor activity is elicited by delivering 290 nM of AMPA to the ECF (Fig. 5.6a) at a particular holding potential. The resulting single-channel fluctuations can be viewed using an oscilloscope or computer. If channel events are not observed at various voltages, it may indicate that receptors are not incorporated into the lipid bilayer. Repeated mild stirring can facilitate the incorporation of receptors into the bilayer.

11. Once consistent AMPA receptor channel activity is observed, the single-channel currents are recorded on videotape or directly saved on

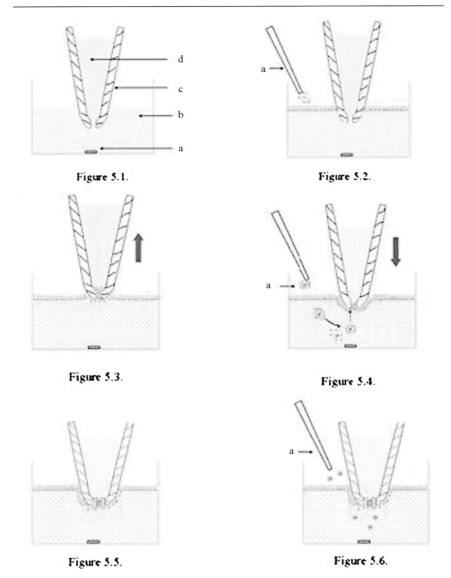

Figure 5.1.

Figure 5.2.

Figure 5.3.

Figure 5.4.

Figure 5.5.

Figure 5.6.

FIG. 5. Tip-dip method. (5.1) Recording electrode dipped in AECF: (a) micro stir bar; (b) artificial extracellular fluid; (c) patch pipette; (d) artificial intracellular fluid. (5.2) Adding $3-4\,\mu l$ of PC into AECF. (5.3) Patch pipette is moved upwards, out of AECF. (5.4) Pipette moved down into AECF and bilayer is formed. (5.4a) The receptor preparation is added, and the incorporation into artificial bilayer is aided by stirring using micro stir bar. (5.5) Bilayer with receptor incorporated. (5.6) Activation of receptor with specific agonist and modulators. (See color insert.)

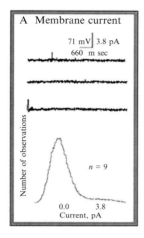 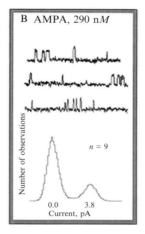 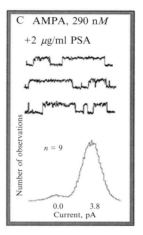

FIG. 6. Modulatory effects of PSA on reconstituted AMPA receptors. (A) Membrane current recorded in the absence of any agonist with voltage clamped at +71 mV. (B) Single-channel currents elicited by the addition of 290 nM AMPA. Channel openings are indicated by upward current deflections. (C) The lifetime of the channel open state is markedly prolonged by coapplying 2 μg/ml of polysialic acid (PSA). The amplitude histograms (lower panels) show the respective bimodal distribution with peaks corresponding to the current levels at open and closed states. The maximum unitary current is 3.8 pA. The open probability with 290 nM AMPA is 0.23 and it is increased to 0.83 with coapplication of 2 μg/ml PSA.

computer. After recording at various voltages, the carbohydrate of interest is delivered to the AECF using a micropipette. In the data shown here (Fig. 6C), 2 μg/ml polysialic acid (PSA) is added to demonstrate the modulatory effects on AMPA receptors. Single-channel currents are recorded continuously after adding the carbohydrate for a particular period to enable recording of adequate single channel events until sufficient channel events are collected. At the end of the experiment, specific AMPA receptor antagonists, CNQX, may be added to AECF to block the channel activity.

12. The data files saved digitally are analyzed off line using software pCLAMP 6.0, as described by Vaithianathan *et al.* (2004).

Acknowledgment

The authors thank Drs. Vitaly Vodyanoy and Muralikrishnan Dhanasekaran for their useful comments and Mr. Solomon Yilma for the drawings. This work was supported by National Institute of Health grant (NS 02128) and Auburn University Biogrant Program.

References

Bahr, B. A., Vodyanoy, V., Hall, R. A., Suppiramaniam, V., Kessler, M., Sumikawa, K., and Lynch, G. (1992). Functional reconstitution of alpha-amino-3-hydroxy-5-methylisoxazole-4-propionate (AMPA) receptors from rat brain. *J. Neurochem.* **59,** 1979–1982.

Chicoine, L. M., Suppiramaniam, V., Vaithianathan, T., Gianutsos, G., and Bahr, B. A. (2004). Sulfate- and size-dependent polysaccharide modulation of AMPA receptor properties. *J. Neurosci. Res.* **75,** 408–416.

Coronado, R., and Lattore, R. (1983). Phospholipid bilayers made from monolayers on patch-clamp pipettes. *Biophys. J.* **43,** 231–236.

Hall, R. A., Vodyanoy, V., Quan, A., Sinnarajah, S., Suppiramaniam, V., Kessler, M., and Bahr, B..A. (1996). Effects of heparin on the properties of solubilized and reconstituted rat brain AMPA receptors. *Neurosci. Lett.* **217,** 179–183.

Sinnarajah, S., Suppiramaniam, V., Kumar, K. P., Hall, R. A., Bahr, B. A., and Vodyanoy, V. (1999). Heparin modulates the single channel kinetics of reconstituted AMPA receptors from rat brain. *Synapse.* **31,** 203–209.

Suarez-Isla, B. A., Wan, K., Lindstrom, J., and Montal, M. (1983). Single channel recordings from purified acetylcholine receptors reconstituted in bilayers formed at the tip of patch pipette. *Biochemistry* **22,** 2319–2323.

Suppiramaniam, V., Bahr, B. A., Sinnarajah, S., Owens, K., Rogers, G., Yilma, S., and Vodyanoy, V. (2001). Member of the Ampakine class of memory enhancers prolongs the single channel open time of reconstituted AMPA receptors. *Synapse.* **40,** 154–158.

Taupin, P., Zini, S., Cesselin, F., Ben-Ari, Y., and Roisin, M. P. (1994). Subcellular fractionation on Percoll gradient of mossy fiber synaptosomes: Morphological and biochemical characterization in control and degranulated rat hippocampus. *J. Neurochem.* **62,** 1586–1595.

Vaithianathan, T., Manivannan, K., Kleene, R., Bahr, B. A., Dey, M. P., Dityatev, A., and Suppiramaniam, V. (2005). Single channel recordings from synaptosomal AMPA receptors. *Cell Biochem. Biophys.* **42,** 75–85.

Vaithianathan, T., Matthias, K., Bahr, B., Schachner, M., Suppiramaniam, V., Dityatev, A., and Steinhauser, C. (2004). Neural cell adhesion molecule-associated polysialic acid potentiates alpha-amino-3-hydroxy-5-methylisoxazole-4-propionic acid receptor currents. *J. Biol. Chem.* **279,** 47975–47984.

Vodyanoy, V., Bahr, B. A., Suppiramaniam, V., Hall, R. A., Baudry, M., and Lynch, G. (1993). Single channel recordings of reconstituted AMPA receptors reveal low and high conductance states. *Neurosci. Lett.* **150,** 80–84.

Wilmsen, U., Methfessel, C., Henke, W., and Boheim, G. (1983). "Channel Current Fluctuation Studies with Solvent Free Planar Lipid Bilayers Using Neher-Sakmann Pipettes." *In* Physical Chemistry of Transmembrane Ion Motion, (G. Spach, ed.), pp. 479–485. Elsevier, Amsterdam.

Section III

Notch Signaling and Muscular Dystrophy

[8]   Methods for Analysis of O-Linked Modifications
on Epidermal Growth Factor-Like and
Thrombospondin Type 1 Repeats

By ALEKSANDRA NITA-LAZAR and ROBERT S. HALTIWANGER

Abstract

The identification of novel forms of O-linked glycosylation on epidermal growth factor and thrombospondin type 1 repeats, and their emerging functional significance, require the development of new methods for their analysis. This chapter describes detailed methods to analyze both the structure and the site of modification of O-fucose and O-glucose glycans on proteins. These methods use both traditional biochemical methods of carbohydrate composition analysis and electrospray ionization–mass spectrometry of glycopeptides.

Overview

Novel O-linked carbohydrate modifications have been described in the context of two small, cysteine-knot protein motifs: epidermal growth factor–like (EGF) repeats, and thrombospondin type 1 repeats (TSRs) (Adams and Tucker, 2000; Campbell and Bork, 1993). EGF repeats can be modified by both O-linked fucose (O-fucose) and O-linked glucose (O-glucose) saccharides (Harris and Spellman, 1993) and TSRs by O-fucose saccharides (Gonzalez de Peredo et al., 2002; Hofsteenge et al., 2001). These modifications have attracted a great deal of interest because of the proposed role of O-fucose in Notch signaling (Haltiwanger, 2002; Haines and Irvine, 2003; Haltiwanger and Lowe, 2004; Haltiwanger and Stanley, 2002). The extracellular domain of Notch1 contains up to 36 tandem EGF repeats (Artavanis-Tsakonas et al., 1999), many of which are modified by O-fucose and O-glucose saccharides (Moloney et al., 2000b; Rampal et al., 2005a; Shao et al., 2003). Elimination of the enzyme responsible for addition of O-fucose to EGF repeats, protein O-fucosyltransferase 1 (POFUT1), causes severe, Notch-like phenotypes in both mice and D. melanogaster (Okajima and Irvine, 2002; Sasamura et al., 2003; Shi and Stanley, 2003). Furthermore, extension of the O-fucose monosaccharide by fucose-specific β1,3-N-acetylglucosaminyltransferases of the Fringe family modulates Notch activity (Bruckner et al., 2000; Moloney et al., 2000a; Rampal et al., 2005b). Recent studies in our laboratory suggest that the O-glucose modifications are equally important in Notch function

METHODS IN ENZYMOLOGY, VOL. 417   0076-6879/06 $35.00
          DOI: 10.1016/S0076-6879(06)17008-1

(Nita-Lazar, A. and Haltiwanger, R. S., manuscript in preparation). Thus, O-glycosylation of small cysteine-knot motifs seems to play an important role in the biology of the proteins it modifies.

EGF repeats are characterized by six conserved cysteine residues forming three disulfide bonds, resulting in their distinct fold (Campbell and Bork, 1993). They are found in a wide variety of cell-surface and secreted proteins and are known to function in protein–protein interactions. EGF repeats from many proteins, including Notch, contain consensus sites for O-fucosylation (C^2X_{4-5} ($\underline{S/T}$)C^3, where C^2 and C^3 are the second and third conserved cysteines, and ($\underline{S/T}$) is the modification site [Panin $et\ al.$, 2002; Shao $et\ al.$, 2003]) and O-glucosylation ($C^1X\underline{S}XPC^2$) (Harris and Spellman, 1993). The O-fucose mono- saccharide on EGF repeats can be elongated by β1,3-N-acetylglucosa-minyltransferases of the Fringe family to form a GlcNAc-β1,3-Fuc-α1-O-Ser/Thr disaccharide (Bruckner $et\ al.$, 2000; Moloney $et\ al.$, 2000a). The disaccharide can be further elongated to the tetrasaccharide Sia-α2,3/6-Gal-β1,4-GlcNAc-β1,3-Fuc-α1-O-Ser/Thr in mammalian systems, although no evidence for elongation beyond the disaccharide exists in $D.\ melanogaster$. Elongation beyond the monosaccharide occurs only on a subset of O-fucosylated EGF repeats, and recent results from our laboratory show that this specificity is mediated by the recognition of specific amino acids within individual EGF repeats by the Fringe enzymes (Rampal $et\ al.$, 2005b; Shao $et\ al.$, 2003). O-Glucose also exists in both monosaccharide and elongated forms (Hase $et\ al.$, 1990). Elongated O-glucose exists as the trisaccharide Xyl-α1,3-Xyl-α1,3-Glc-β1-O-Ser (Nishimura $et\ al.$, 1989). Although enzymatic activities capable of adding O-glucose glycans to EGF repeats have been detected (Minamida $et\ al.$, 1996; Omichi $et\ al.$, 1997; Shao $et\ al.$, 2002), the genes encoding the enzymes are not yet known.

TSRs also contain six conserved cysteine residues forming three disulfide bonds, although the disulfide-bonding pattern is distinct from that of EGF repeats (Adams and Tucker, 2000). In addition, TSRs are characterized by conserved Trp, Ser, and Arg residues. Like EGF repeats, TSRs occur in numerous secreted proteins and are known to function in binding to receptors and components of the extracellular matrix. Many of the proteins containing TSRs are predicted to be modified with O-fucose on the basis of the presence of the consensus sequence $C^1X_{2-3}(\underline{S/T})C^2X_2G$ (where C^1 and C^2 are the first and second conserved cysteines of the TSR), and ($\underline{S/T}$) is the modification site (Gonzalez de Peredo $et\ al.$, 2002; Hofsteenge $et\ al.$, 2001). The enzyme responsible for addition of O-fucose to TSRs has recently been identified as protein O-fucosyltransferase 2 (POFUT2) (Luo $et\ al.$, 2006a,b). The O-fucose on TSRs can be elongated by a fucose-specific β1,3-glucosyltransferase, leading to the formation of the disaccharide Glc-β1,3-Fuc-α1-O-Ser/Thr (Gonzalez de Peredo $et\ al.$, 2002; Hofsteenge $et\ al.$, 2001;

Luo et al., 2006a,b; Moloney and Haltiwanger, 1999). An enzymatic activity for the fucose-specific β1,3-glucosyltransferase has been identified (Moloney and Haltiwanger, 1999), but the gene encoding the enzyme has not. A recent study showed that members of the Fringe family are not able to add GlcNAc to O-fucose on TSRs, and the fucose-specific β1,3-glucosyltransferase is not able to add glucose to O-fucose on EGF repeats (Luo et al., 2006b). Thus, these two O-fucosylation pathways seem completely distinct. Although the function of O-fucose on TSRs is not known, the fact that the O-fucosylation site in thrombospondin-1 is within a putative cell-binding site suggests that the glycan may modulate interactions in a way similar to that seen with Fringe and Notch (Moloney et al., 2000a). Studies on the biological importance of O-fucose on TSRs are ongoing.

Analysis of O-Fucose and O-Glucose Glycan Structure by Metabolic Radiolabeling Coupled with Alkali-Induced β-Elimination

Traditional methods of glycan analysis have been adapted to analyze the O-fucose and O-glucose composition and structure (Hardy and Townsend, 1994; Kobata, 1994; Varki, 1994). We have used these methods successfully to determine the structure of O-fucose and O-glucose glycans on EGF repeats from Notch (Arboleda-Velasquez et al., 2005; Moloney et al., 2000a,b; Rampal et al., 2005b; Shao et al., 2003) and O-fucose glycans on TSRs (Moloney et al., 1997; Sturla et al., 2003; Luo et al., 2006a). These methods use metabolic labeling of Chinese hamster ovary (CHO) cells with radiolabeled monosaccharides. We take advantage of CHO cell lines developed in the laboratory of Dr. Pamela Stanley (Albert Einstein College of Medicine) harboring mutations in traditional glycosylation pathways to favor incorporation of the radiolabel into the less predominant O-fucose and O-glucose pathways. For instance, Lec1-CHO cells have a defect in GlcNAc transferase 1, preventing formation of complex-type N-glycans (Stanley and Siminovitch, 1977; Stanley, 1992). This mutation causes a significant reduction in incorporation of [^{3}H]fucose into N-glycans, with no effect on incorporation into O-fucose glycans (Lin et al., 1994). Similarly, Lec8-CHO cells have a defect in the Golgi UDP-galactose transporter, resulting in significant reductions in galactosylation (Deutscher and Hirschberg, 1986; Stanley, 1981). When these cells are metabolically radiolabeled with [^{3}H]galactose, some of the radiolabel is converted to [^{3}H]glucose (by the UDP-galactose 4′-epimerase), which can be transported and incorporated into O-glucose glycans (Moloney et al., 2000b). Once radiolabel is incorporated into the appropriate structures, the glycans are released by alkali-induced β-elimination and analyzed using a combination of size-exclusion chromatography, exoglycosidase digestions, and high-pH anion exchange chromatography.

Materials

    Lec1 and Lec8 mutant CHO cells (ATCC)
    α-MEM cell culture media (Invitrogen) with 10% bovine calf serum (Hyclone) and 10 mM penicillin/streptomycin
    OptiMem I cell culture media (Invitrogen)
    Radioactive (^{3}H) monosaccharides: L-[6-^{3}H]fucose (60 Ci/mmol), D-[6-^{3}H] galactose (60 Ci/mmol) (American Radiolabeled Chemicals)
    Dowex-50, H$^+$-form (Bio-Rad)
    Acetone (stored at $-20°$)
    Disposable columns (Bio-Rad)
    Sep-Pak C18 (Waters)
    1-ml and 10-ml disposable syringes and 18-gauge needles
    MilliQ water
    Superdex peptide gel filtration column (Pharmacia) connected to an HPLC capable of 0.5 ml/min flow, with refractive index detector for standardization
    Dionex DX300 HPLC system (or equivalent) equipped with pulsed amperometric detection (PAD-2 cell) and CarboPac MA-1 column
    Ni-NTA Sepharose beads (Qiagen)
    Nitrocellulose membrane (Bio-Rad)
    Antibodies for Western blot protein detection: monoclonal anti-myc epitope antibody 9E10 (kind gift from Dr. Jen-Chih Hsieh, Stony Brook University), mouse anti-His$_6$ (Santa Cruz Biotechnology)
    Antibodies for immunoprecipitation: rabbit anti-myc epitope polyclonal antibody (Abcam)
    En^{3}Hance autoradiography enhancer (Perkin Elmer). Highly corrosive!
    Exoglycosidases: N-Acetylneuraminidase I (α2,3-specific sialidase) from Glyco, Inc; β-galactosidase (*Diplococcus pneumoniae*) from Roche Molecular Biochemicals; β-hexosaminidase (jack bean) from Sigma.

Standards

    Dextran D-4133 (Sigma) after mild hydrolysis as described by Kobata (Kobata, 1994)
    Monosaccharides (Sigma)
    Alditol sugar standards (fucitol, glucitol, galactitol, galactosaminitol, glucosaminitol) prepared by reduction of the corresponding sugar according to (Haltiwanger *et al.*, 1990)

Disaccharide standards synthesized and converted to alditols as described (Moloney *et al.*, 2000b)

Stock Solutions

RIPA buffer (50 m*M* Tris-HCl, pH 8.0, 150 m*M* NaCl, 1% NP40, 0.5% DOC, 0.1% SDS)
TBS (Tris-buffered saline: 10 m*M* Tris, pH 7.4, 0.15 *M* NaCl)
0.1 *M* EDTA, pH 8.0
4 *M* acetic acid
50% slurry of Dowex-50 (H$^+$-form) in 20% methanol
1 *M* NaCl

Fresh Solutions

β-Elimination solution: 2 *M* NaBH$_4$, 100 m*M* NaOH
2 *M* trifluoroacetic acid (TFA)

Metabolic Radiolabeling

These methods can be used to analyze *O*-fucose and *O*-glucose structures on endogenous proteins in CHO cells (Moloney *et al.*, 2000b) or CHO cells transiently or stably transfected with expression plasmids encoding the protein of choice (Arboleda-Velasquez *et al.*, 2005; Moloney *et al.*, 2000a; Rampal *et al.*, 2005a; Shao *et al.*, 2003). Typically, cells are transfected with a construct encoding the protein of choice with an N-terminal secretion signal and C-terminally tagged with MYC and 6×His tags for detection and purification. Appropriate controls include cells transfected (stably or transiently) with an empty plasmid. Approximately 10^6 cells (Lec1-CHO if analyzing *O*-fucose, Lec1 or Lec8-CHO if analyzing *O*-glucose) are seeded on a 100-mm tissue culture dish in 10 ml α-MEM with 10% calf serum and grown to about 50% confluency. The medium is then replaced with 5 ml α-MEM with 10% CS containing 20 μCi/ml of the appropriate radiolabeled monosaccharide ([^{3}H]fucose for analysis of *O*-fucose, [^{3}H]galactose for *O*-glucose). The cells are incubated for 48 h, and the medium containing the radioactively labeled recombinant glycoprotein is collected for purification. Intracellular proteins can be immunopurified from the cell lysates by the same method (Luo and Haltiwanger, 2005; Moloney *et al.*, 2000b). To reduce the amount of protein present in the medium, the radiolabeling can be done using a serum-free medium such as Opti-MEM. This dramatically reduces the amount of contaminants in the purified protein (and is essential for the mass spectral analyses described later) but also results in a decrease in the efficiency of radiolabel incorporation. To use Opti-MEM, after seeding the cells and allowing them to become 50% confluent (or 24 h after transient transfection), the cells are washed three times with 10 ml of TBS, and replaced with 5 ml of α-MEM (or Opti-MEM; Invitrogen) serum-free medium, containing the appropriate radiolabeled monosaccharide. The remainder of the procedure is the same as described previously.

Protein Purification

Endogenous proteins can be purified by immunoprecipitation from medium or cell lysates using appropriate antibodies (Moloney et al., 2000b). The C-terminal His and MYC tags on the transfected proteins permit fairly straightforward purification and detection, respectively. To purify a tagged protein from the harvested medium, Ni-NTA agarose chromatography is used. The medium is centrifuged to remove any cell debris, and 100 μl of Ni–NTA-agarose (50% slurry, washed thoroughly with RIPA buffer) per 5 ml of medium is added. The mix is rotated for 1 h at 4°, and the Ni–NTA-agarose is collected by centrifugation at 5000g. The supernatant is removed and the Ni–NTA-agarose is washed by resuspension, vortexing, and centrifugation three times, using 1 ml RIPA buffer for each wash. The supernatants are removed with an 18-gauge needle attached to a 1-ml syringe to ensure complete removal of liquid and no loss of Ni–NTA-agarose. The protein is eluted from the Ni–NTA-agarose with 250 μl of 100 mM EDTA. The Ni–NTA-agarose is removed by centrifugation, and the supernatant is collected. A small aliquot (2–5%) of the eluate is counted in the liquid scintillation counter to approximate efficiency of radiolabel recovered in the protein of interest versus controls (untransfected cells for tagged proteins, control immunoprecipitation for endogenous proteins). If a tagged protein is being purified from cell lysates, immunoprecipitation with polyclonal anti-MYC antibodies is recommended rather than the Ni–NTA chromatography (see Luo and Haltiwanger [2005]) for details).

Fluorography

Fluorography and SDS-PAGE are performed using the purified radiolabeled protein to establish radiochemical purity and relative stoichiometry of glycosylation. Equivalent amounts of radioactivity (at least 2000 cpm) are loaded onto two SDS-PAGE gels. One gel is used for the Western blot probed with appropriate antibodies (e.g., anti-MYC for MYC-tagged proteins). The second gel is used for fluorography using En[3]Hance (Perkin Elmer, very corrosive!) according to the manufacturer's instructions. After development, both films are scanned (both the exposures should be in the linear range of the film), and the ratio of radioactivity (from the fluorograph) to protein (from the Western Blot) is determined. Relative stoichiometries for proteins analyzed in the same experiment can be compared using this approach (see Shao *et al.* [2003] and Arboleda-Velasquez *et al.* [2005]) for details).

Alkali-Induced β-Elimination

The purified, radiolabeled glycoprotein (10,000–25,000 cpm) is precipitated with 4–8 volumes of cold acetone for 2–16 h at −20°. The precipitate is pelleted by centrifugation in a Microfuge at maximum speed, the supernatant is removed with an 18-gauge needle, and the pellet is air-dried. Alkali-induced β-elimination is initiated by addition of 500 μl of β-elimination solution to the acetone precipitated protein pellet. If necessary, the pellet is broken up with the pipette tip and repeated vortexing. The sample is then incubated at 55–65° for 18–24 h. The sample is cooled on ice and neutralized by drop wise addition of 4 *M* acetic acid until the pH is below 6.0 (monitored with pH paper after addition of a few drops—typically about 10 drops of acid are enough). The neutralized sample is allowed to outgas for 30 min on ice. The sample is then desalted by passing through 1.5 ml of Dowex-50(H^+-form) (pre-washed with 15 ml MilliQ-water before sample loading) in a disposable Bio-Rad column with an 18-gauge needle. The Dowex-50 exchanges the Na^+ in the sample for H^+. The flow-through is collected in a 50-ml conical tube. The column is washed with 10 ml MilliQ-water, and the wash is combined with the flow-through. The combined flow-through and wash is then passed through the Sep-Pak C18 column to remove peptides and protein from the sample. The Sep-Pak column is preconditioned with 10 ml of 100% methanol and then with 10 ml MilliQ-water. After loading the sample, the Sep-Pak is washed with an additional 5 ml MilliQ-water and combined with the Sep-Pak flow-through in a 50-ml conical tube. The desalted/deproteinized sample is frozen and lyophilized. Boric acid (which was formed from neutralizing the borohydride) is subsequently removed by methanol dry-down: sufficient 100% methanol is added to the sample to resuspend it, mixed, and evaporated using a vacuum centrifuge set at medium heat (about 60°). The dry-down is repeated twice with 0.5 ml of methanol or until the volume of the white powder no longer decreases. The sample is dissolved in 100 μl of water, and the recovered radioactivity is determined by counting 5 μl of the sample in the liquid scintillation counter.

Oligosaccharide Structural Analysis by Size-Exclusion Chromatography and Exoglycosidase Digestion

Size exclusion chromatography of each sample is performed using a Superdex column (Moloney *et al.*, 1997). Approximately 10,000 cpm of each sample is injected onto a calibrated Superdex column equilibrated with MilliQ-water at a flow rate of 0.5 ml/min. Fractions (0.5 min = 250 μl) are collected, and aliquots (10–50%) are analyzed for radioactivity in a liquid scintillation counter. The elution position of the radioactive species is compared with that of partially hydrolyzed dextran standards (see Kobata, [1994] and Haltiwanger and Philipsberg [1997] for the details of how the Superdex column is calibrated). Because the elution positions of dextran standards on a Superdex column are highly reproducible, dextran standards do not need to be included with each sample.

Exoglycosidase digestions are performed to determine the terminal monosaccharide on the glycan as previously described (Jacob and Scudder, 1994). The fractions containing the radioactive peaks from size fractionation on the Superdex column are pooled and dried in the vacuum centrifuge. The dried sample is resuspended in 50 μl of sodium acetate buffer, pH 5.5, and the appropriate exoglycosidase is added to the sample. Typical exoglycosidases to try are N-acetyl neuraminidase I (α2,3-specific sialidase, 10 milliunits), β1, 4-galactosidase (5 milliunits), or β-hexosaminidase (250 milliunits). Mock-digested controls (containing no enzyme) should be prepared alongside the samples. The digests are incubated for 18 h at 37°, and the reactions are terminated by diluting the samples with 50 μl of water and heating for 5 min at 100°. The samples are then reanalyzed by size-exclusion chromatography as described previously. A shift in migration of the sample to a smaller size indicates removal of the respective sugar.

Oligosaccharide Charge Analysis

To examine the charge state of specific oligosaccharides, charge analysis is performed as previously described (Lin *et al.*, 1994; Moloney *et al.*, 2000b). The oligosaccharide fraction suspected of having a charge is treated with and without sialidase as described previously and dried in the vacuum centrifuge. The sample is resuspended in 500 μl of 2 mM Tris-base and passed through a 330-μl QAE-Sephadex column equilibrated with 2 mM Tris-base. The column is then washed with three 0.5-ml portions of 2 mM Tris-base, each 0.5-ml portion being collected as a separate fraction. The unbound material in these fractions represents the neutral species. The charged species are released from the column with increasing concentrations of NaCl: one negative charge, 20 mM NaCl; two negative charges, 70 mM NaCl; three negative charges, 100 mM NaCl; four negative charges, 140 mM NaCl; five or more negative charges, 1 M NaCl. Fractions (0.5 ml) are collected and monitored by scintillation counting. Each concentration of NaCl is prepared in 2 mM Tris-base, and each is used to elute the column with four 0.5-ml portions.

Acid Hydrolysis and HPAEC

To confirm that the radiolabeled sugar moiety is *O*-linked to the protein, the oligosaccharides released by alkali-induced β-elimination are hydrolyzed, and the resulting monosaccharides are analyzed by high pH anion exchange chromatography (HPAEC) in the presence of monosaccharide standards. This is essential for studies involving metabolic radiolabeling with monosaccharides, because the sugars can become incorporated as terminal modifications in other classes of glycans. The sugar residue directly linked to the protein forms a corresponding alditol as a result of the alkali-induced β-elimination procedure (e.g., fucitol for *O*-fucose, glucitol for *O*-glucose). HPAEC allows easy separation of aldoses from their corresponding sugar alditols. An appropriate amount of the sample to be analyzed (1000–5000 cpm) is dried in a 1.5-ml screw-cap conical tube with an O-ring. The sample is hydrolyzed in 200 μl of freshly prepared 2 *M* trifluoroacetic acid by heating to 100° for 2 h. After hydrolysis, the sample is dried in the vacuum centrifuge and resuspended in 100 μl of MilliQ-water for HPAEC. Before injection, the sample is mixed with appropriate standards: for *O*-fucose glycan analysis, 1 nmol each fucitol, fucose, and glucose; for *O*-glucose glycan analysis 1 nmol each galactosaminitol, glucose, glucitol, galactitol, and galactose. The mix is injected onto the CarboPac MA-1 column run at 0.4 ml/min, using the following gradient: 0–11 min: isocratic at 0.1 *M* NaOH; 11–21 min, 0.1–0.7 *M* NaOH; 21–40 min, 0.7 *M* NaOH. The gradient can be adjusted to optimize the standard peak resolution. The standards are followed by pulsed amperometric detection (PAD-2 cell, settings as follows: E1 = 0.00 V, T1 = 4; E2 = 0.80 V, T2 = 3; E3 = -0.25 V, T3 = 6; Range = 2; Response time = 3 sec). Fractions (0.5 min =0.2 ml) are collected and monitored for radioactivity by scintillation counting.

Glycosylation Site Mapping by Mass Spectrometry

Although sites of glycosylation can be identified by observing loss of radiolabeling after mutation of predicted glycosylation sites (Arboleda-Velasquez *et al.*, 2005; Shao *et al.*, 2003; Yan *et al.*, 2002), this approach becomes exceedingly difficult on proteins such as Notch that contain multiple sites of glycosylation. As a more robust approach to confirm without doubt whether the predicted *O*-glycosylation sites are actually modified in the context of proteins with numerous predicted sites, we have turned to mass spectrometry. Mass spectrometry is superior to site-directed mutagenesis, because it provides positive data for the presence of the modifications at specific sites, and it provides information regarding the structure of the glycan at that site. Similar approaches have been used to examine *O*-fucosylation on TSRs (Gonzalez de Peredo *et al.*, 2002; Hofsteenge *et al.*, 2001; Macek *et al.*, 2001). We have applied these methods to mapping *O*-linked glucose and fucose glycans to specific EGF repeats using a 3D ion trap mass spectrometer equipped with an ESI source. For site mapping, we perform LC-ESI-MS on in-gel proteolytic digests of protein bands separated by SDS-PAGE (10 pmol of protein is generally enough to identify most glycosylation sites). The procedures described here are an adaptation of those described elsewhere (Duelli *et al.*, 2005). We have also developed a method for quick assay for the degree of modification on a particular site using direct infusion of single EGF repeats, which were subject to *in vitro* glycosylation assay and HPLC purification, into the mass spectrometer (Rampal *et al.*, 2005b; Luo *et al.*, 2006b). If *N*-glycosylation sites are present, removal of *N*-glycans by PNGaseF digestion helps to achieve less complicated spectra.

Materials

Bond-Breaker TCEP (Tris[2-carboxyethyl] phosphine-HCl) Solution, Pierce, cat. # 77720 (0.5 *M* TCEP, pH 7.0)

Iodoacetamide (Sigma, I-1149) (20-mg aliquots stored in 1.5-m tubes at 20°)

Novex NuPAGE 4–12% Bis-Tris prepoured Gel, 1.0 mm, 9 well (Invitrogen cat. # NP0327Box)

MES electrophoresis buffer (Invitrogen)

Methanol (Optima grade, Fischer)

Acetonitrile (HPLC grade, B&D)

Formic Acid (96%, Aldrich)

Water (Sigma ACS grade or better for all solutions, MilliQ for gel staining)

Bio-Rad Zinc Staining kit (Cat. # 161-0440)

0.22-μm Steriflip and Stericup units from Millipore

Sequencing grade modified trypsin (Promega)

Zip tips, C18 (Millipore, cat. # ZTC18S960)

LC/MSD Trap XCT mass spectrometer (Agilent Technologies) coupled with capillary LC 1100 series (Agilent Technologies) and with infusion pump (kd Scientific).

Stock Solutions

Tris-HCl, pH 8.0 (sterile filtered)

1 *M* di-AP (diammonium phosphate), pH 8.0 (sterile filtered, 1-ml aliquots are stored at −20°)

250 m*M* ammonium phosphate (sterile filtered, store in 1-ml aliquots at −20°)

3× Laemmli sample buffer (SB) without reducing agents (sterile filtered, 1-mL aliquots stored at −20°)

20 m*M* di-AP, pH 8.0 (sterile filtered, stored in 1-ml aliquots at −20°)

0.1 *M* EDTA, sterile filtered

0.5 mg/ml trypsin (resuspended in buffer provided by manufacturer, stored in 6-*μ*l aliquots at −80°)

5% formic acid

0.1% formic acid

50% acetonitrile, 0.1% formic acid

95% acetonitrile, 0.1% formic acid

30% methanol, 0.5% formic acid

Fresh Solutions

50 m*M* Tris-HCl, pH 8.0 (2.5 ml 1 *M* Tris-HCl, pH 8.0, + 47.5 ml ACS-grade water, sterile filtered)

50 m*M* Tris-HCl, pH 8.0, 100 m*M* iodoacetamide (20 mg iodoacetamide + 1 ml 50 m*M* Tris-HCl, pH 8.0)

TCEP/3X SB stock: 50 parts 3× Laemmli sample buffer + 1 part 0.5 *M* TCEP

50% methanol in 20 m*M* di-AP (diammonium phosphate), pH 8.0 (1 ml 1 *M* di-AP, pH 8.0, 25 ml methanol, 24 ml ACS-grade water, sterile filtered)

20 m*M* di-AP, pH 8.0 (1 ml 1 *M* di-AP, 49 ml ACS-grade water, sterile filtered)

Trypsin solution: 6 *μ*l 0.5 mg/ml trypsin + 94 *μ*l 20 m*M* di-AP, pH 8.0

Sample Preparation

Before running the sample on SDS-PAGE, the cysteine residues in the sample protein are reduced and alkylated using TCEP and iodoacetamide. Typically, 10 pmol protein is enough for the analysis. Sensitivity could be improved 10-fold or more by using nanoflow chromatography systems. If the protein is lyophilized, it can be resuspended in 5 *μ*l 50 m*M* Tris-HCl, pH 8.0. Otherwise, 5 *μ*l of sample can be used or the protocol scaled accordingly, considering the size of the wells in the SDS-PAGE system. First, 5 *μ*l of sample is mixed with 5 *μ*l 3× SB/TCEP mix and heated for 5 min at 100°. This treatment is sufficient to break the disulfide bonds and reduce the cysteines. Efficient reduction of the disulfide bonds is essential for subsequent digestion of proteins containing either EGF repeats or TSRs. In the second step, 5 *μ*l of 50 m*M* Tris-HCl, pH 8.0, 100 m*M* iodoacetamide (made fresh) are added, the sample is vortexed and incubated at room temperature in the dark (iodoacetamide is light-sensitive) for 30 min. This causes carbamidomethylation of cysteine residues preventing the formation of disulfide bonds.

The reduced and alkylated sample is then loaded onto the NOVEX NuPage 4–12% gradient gel and subjected to the electrophoresis in the MES running buffer at 150 V, until the tracking dye gets to the bottom. The gel is stained with Bio-Rad Zinc Staining kit (sensitive to about 10 ng) following the manufacturer's instructions. This kit is recommended not only because of sensitivity but also because it is a negative stain, leaving the proteins unmodified for mass spectral analysis. Alternately, the gel can be stained overnight with colloidal Coomassie (e.g., Pierce GelCode Blue Stain Reagent, cat. # 24590).

To cut the bands of interest out of the gel, the gel is transferred to an unused 150-mm plastic petri dish. The dish should be kept covered for most of the time to reduce keratins or dust contamination of the sample. The gel and the tubes in all subsequent steps should be handled with gloved hands, and, if possible, the procedure should be performed in a sterile environment (e.g., a biosafety cabinet). After taking a picture of the gel, the bands of interest are cut out using a fresh scalpel, taking as little excess polyacrylamide as possible. The scalpel should be well rinsed with water between bands. The bands are transferred to clean, labeled 1.5-ml tubes using clean pipette tips (P10 or smaller) and cut into three or four pieces using the tip once inside the tube to fit in the smallest possible volume at the bottom of the tube. The gel pieces can be stored frozen at this stage.

In-Gel Digestions

Gel pieces are washed once by adding 500 μl of 0.1 M EDTA and vortexing for 15 min at room temperature. The supernatant is aspirated, and the gel pieces are washed at least five times (the more washes the better, the goal is to remove the SDS from the gel pieces) with 1 ml of 50% methanol, 20 mM di-AP, pH 8.0, vortexing at room temperature for 15–30 min for each wash. The next wash is performed overnight, vortexing at 4°. The next morning, three more washes with 1 ml of 50% methanol, 20 mM di-AP, pH 8.0, are performed. The gel pieces are then dehydrated by addition of 50 μl of 100% acetonitrile, vortexing for 30 sec. The gel is dried in a vacuum centrifuge (approximately 15–20 minutes; the gel pieces should look white, but should not be overdried). Trypsin solution (10 μl) is added to each sample. Other proteases (like chymotrypsin, Asp-N or Glu-C) can be used in the same digestion buffer. The choice of protease should be determined by performing *in silico* digestion (e.g., using the Peptide Mass tool from Proteomics tools at www.expasy.com). Ideally, peptides after proteolysis should be less than 4 kDa and contain as few glycosylation sites as possible (ideally one site per peptide).

The gel is allowed to absorb the trypsin solution for 15 min at 37° in an incubator. Prewarmed 20 mM di-AP, pH 8.0, is added in sufficient amount to cover the gel pieces (approximately 20 μl, depending on the size of the gel pieces). After 45–60 min, the gel pieces should be checked to make sure they are still completely covered with liquid. If not, more 20 mM di-AP, pH 8.0, should be added. The gel pieces are then incubated 4–24 h at 37° in the incubator. After the digest, the supernatant is removed from each sample and placed in a clean, labeled tube. Sufficient 5% formic acid is added to each tube with gel pieces to cover them (20–50 μl), and the tubes are placed in a bath sonicator for 20 min to extract the tryptic peptides from the gel. After sonication, the supernatant is combined with the original and the gel slices are discarded.

Zip Tip Cleanup

The sample is bound to a C-18 Zip Tip to remove salts and contaminants. A C-18 Zip Tip is placed on a P20 set at 20 μl, conditioned two times with 20 μl 95% acetonitrile, 0.1% formate and washed two times with 20 μl 0.1% formate. The sample is passed through the Zip Tip 20 μl at a time, transferring to a clean tube. From that tube, the sample is passed through a second time, transferring back to original tube. The peptides should now be bound to the Zip Tip. The Zip Tip is washed six times with 20 μl of 0.1% formate. The peptides are then eluted using 5 μl of 50% acetonitrile, 0.1% formate, pipetting up and down in the tube 10 times. The eluate is dried in the vacuum centrifuge (approximately 10 min, longer drying times reduce yields) and stored at $-20°$.

PNGase Digest of Tryptic Peptides (Optional)

The dry, Zip-tipped sample is resuspended in 20 μl of digestion buffer: 50 mM Tris-HCl, pH 8.6, with 10 mM EDTA containing 1 μl of PNGase F (0.1 U, Sigma). The reaction is incubated for 6–8 h at room temperature alongside negative controls (PNGase F only, and sample with no PNGase F). To stop the reaction, the tube is heated for 5 min at 100° to inactivate the enzyme. The sample is acidified to 0.5% formate and desalted using a Zip-Tip as described previously.

LC-MS of the Digested Sample

The sample is resuspended in 5 μl of 20% acetonitrile in 0.1% formic acid by sonicating for 20 min in a bath sonicator. After centrifugation to remove any insoluble material, the sample is injected onto Zorbax 300SB-C8 column

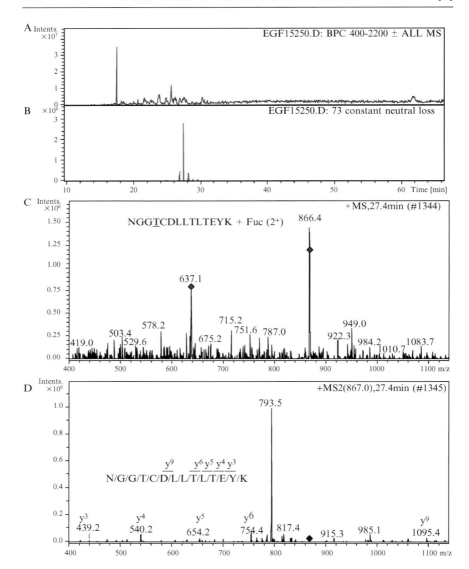

FIG. 1. LC-ESI-MS analysis of *O*-fucose glycopeptides from EGF repeats 15 of mouse Notch1. EGF repeats 1–5 from mouse Notch1 were expressed in Lec1-CHO cells and purified as described in the text (Shao *et al.*, 2003). Approximately 250 ng of protein was subjected to in gel tryptic digestion and analyzed by LC-ESI-MS analysis as described in the text. (A) Base peak chromatogram showing the most abundant ion species at each time point in the chromatogram. (B) Constant neutral loss of 73 amu. The data in A were searched for ions that lost 73 amu between MS analysis (parent ions) and MS/MS analysis (fragmentation).

(3.5 μm beads, 150 $\times$ 0.3 mm, Agilent). The column is eluted at 5 μl/min with the following gradient: 0–5 min, 5% B; 5–85 min, 5–35% B; 85–105 min, 35–95% B; 105–115 min, 95% B (where buffer A = 0.1% formic acid, and buffer B = 95% acetonitrile in 0.1% formic acid). The effluent from the column is sprayed directly into an Agilent XCT Ion Trap mass spectrometer using a capillary spray needle, with nebulizer gas (nitrogen) at 4 l/min, and drying gas (nitrogen) at 10 l/min, 300°. The scanning range is 400–2200 m/z, and MS/MS (MS2) is performed on the two most intense ions in each spectrum with exclusion after two spectra.

Ions containing *O*-fucose or glucose are identified by constant neutral loss searches of the MS/MS data. The data are queried for neutral losses of m/z corresponding to the mass of mono- or oligosaccharide in an appropriate charge state. Table I shows the masses of all known *O*-fucose and *O*-glucose glycans. Peptides from a tryptic digest can exist in number of charge states (e.g., 2+, 3+, 4+). An example of such an analysis is shown in Fig. 1, where constant neutral loss analysis of 73 amu (fucose in a 2+ state) was performed. The search revealed a major species eluting at 27.4 min that lost 73 amu on fragmentation (Fig. 1B). A parent ion of m/z 866.4 (Fig. 1C) was identified that, on fragmentation, yielded a major product ion of m/z 793.5 (Fig. 1D). This indicates a loss of 72.9 amu, consistent with the loss of fucose from a double charged peptide. The m/z of the product ion corresponds to a peptide with a singly charged mass of 1586.0. A tryptic peptide containing a predicted *O*-fucose site (NGGTCDLLTLTEYK +H$^+$) from the protein analyzed (EGF repeats 1–5 from mouse Notch1) has a calculated monoisotopic mass of 1584.8, consistent with this assignment. Confirmation that this is the correct peptide comes from a number of peptide fragments (y-ions, see Fig. 1D). Although this analysis does not identify the specific amino acid modified with the glycan, it does demonstrate that this peptide, containing an *O*-fucose consensus sequence, is modified with a monosaccharide form of *O*-fucose. Similar searches can

---

(C) MS scan at 27.4 min, identified in the neutral loss search shown in B, indicating two major ion species that were selected for fragmentation (indicated by diamonds), including ion at m/z 866.4. This ion corresponds to the mass of a double-charged form of the peptide NGGTCDLLTLTEYK modified with *O*-fucose. The T indicates a predicted *O*-fucose modification site. (D) MS/MS fragmentation of the ion at m/z 866.4 results in formation of a major product ion at 793.5 (72.9 amu less than the parent ion, consistent with the loss of a fucose from a doubly charged peptide). The measured mass of the product ion (singly charged = 1586.0) matches the calculated mass of NGGTCDLLTLTEYK + H$^+$ (monoisotopic mass = 1584.8). Confirmation that this is the correct peptide comes from additional fragments seen in the spectra that correspond to y-ions (and occasionally b-ions) of the fragmented peptide. Several of the y-ions are indicated in the spectra.

TABLE I

MONOISOTOPIC MASSES OF O-FUCOSE AND O-GLUCOSE MODIFICATIONS

Modification	Monoisotopic mass
O-Fucose	
Fuc-	146.0579
GlcNAc-β1,3-Fuc-	349.1373
Gal-β1,4-GlcNAc-β1,3-Fuc-	511.1901
NANA-α2,3-Gal-β1,4-GlcNAc- β1,3-Fuc-	802.2855
Glc-β1,3-Fuc-	308.1107
O-Glucose	
Glc-	162.0528
Xyl-α1,3-Glc-	294.0951
Xyl-α1,3-Xyl-α1,3-Glc-	426.1274

be performed for elongated forms of O-fucose or O-glucose glycans. If desired, MS3 can be performed on the major product ions to obtain additional sequence information from the peptide. We have used this method to successfully map most of the O-fucose and O-glucose modification sites on both mouse Notch1 and D. melanogaster Notch (Nita-Lazar et al., in preparation).

Acknowledgments

The authors thank Dr. Michael P. Myers (Cold Spring Harbor Laboratory) and members of his laboratory for training and technical assistance in the mass spectral techniques. They also thank members of the Haltiwanger laboratory for critically reading this manuscript. Original work was supported by NIH Grant GM61126.

References

Adams, J. C., and Tucker, R. P. (2000). The thrombospondin type 1 repeat (TSR) superfamily: Diverse proteins with related roles in neuronal development. Develop. Dynamics 218, 280–299.
Arboleda-Velasquez, J. F., Rampal, R., Fung, E., Darland, D. C., Liu, M., Martinez, M. C., Donahue, C. P., Navarro-Gonzalez, M. F., Libby, P., D'Amore, P. A., Aikawa, M., Haltiwanger, R. S., and Kosik, K. S. (2005). CADASIL mutations impair Notch3 glycosylation by Fringe. Hum. Mol. Genet. 14, 1631–1639.
Artavanis-Tsakonas, S., Rand, M. D., and Lake, R. J. (1999). Notch Signaling: Cell Fate Control and Signal Integration in Development. Science 284, 770–776.
Bruckner, K., Perez, L., Clausen, H., and Cohen, S. (2000). Glycosyltransferase activity of Fringe modulates Notch-Delta interactions. Nature 406, 411–415.

Campbell, I. D., and Bork, P. (1993). Epidermal growth factor-like modules. *Curr. Opin. Struct. Biol.* **3**, 385–392.

Deutscher, S. L., and Hirschberg, C. B. (1986). Mechanism of galactosylation in the Golgi apparatus. A Chinese hamster ovary cell mutant deficient in translocation of UDP-galactose across Golgi vesicle membranes. *J. Biol. Chem.* **261**, 96–100.

Duelli, D. M., Hearn, S., Myers, M. P., and Lazebnik, Y. (2005). A primate virus generates transformed human cells by fusion. *J. Cell. Biol.* **171**, 493–503.

Gonzalez de Peredo, A., Klein, D., Macek, B., Hess, D., Peter-Katalinic, J., and Hofsteenge, J. (2002). C-mannosylation and O-fucosylation of thrombospondin type 1 repeats. *Mol. Cell Proteomics* **1**, 11–18.

Haines, N., and Irvine, K. D. (2003). Glycosylation regulates notch signaling. *Nat. Rev. Mol. Cell Biol.* **4**, 786–797.

Haltiwanger, R. S. (2002). Regulation of signal transduction pathways in development by glycosylation. *Curr. Opin. Struct. Biol.* **12**, 593–598.

Haltiwanger, R. S., Holt, G. D., and Hart, G. W. (1990). Enzymatic addition of *O*-GlcNAc to nuclear and cytoplasmic proteins. Identification of a uridine diphospho-*N*-acetylglucosamine: Peptide B-*N*-acetylglucosaminyltransferase. *J. Biol. Chem.* **265**, 2563–2568.

Haltiwanger, R. S., and Lowe, J. B. (2004). Role of glycosylation in development. *Annu. Rev. Biochem.* **73**, 491–537.

Haltiwanger, R. S., and Philipsberg, G. A. (1997). Mitotic arrest with nocodazole induces selective changes in the level of O-GlcNAc and accumulation of incompletely-processed N-glycans on proteins from HT29 cells. *J. Biol. Chem.* **272**, 8752–8758.

Haltiwanger, R. S., and Stanley, P. (2002). Modulation of receptor signaling by glycosylation: Fringe is an *O*-Fucose-β1,3-*N*-acetylglucosaminyltransferase. *Biochem. Biophys. Acta* **1573**, 328–335.

Hardy, M. R., and Townsend, R. R. (1994). High-pH anion-exchange chromatography of glycoprotein-derived carbohydrates. *Methods Enzymol.* **230**, 208–225.

Harris, R. J., and Spellman, M. W. (1993). O-linked fucose and other post-translational modifications unique to EGF modules. *Glycobiology* **3**, 219–224.

Hase, S., Nishimura, H., Kawabata, S., Iwanaga, S., and Ikenaka, T. (1990). The structure of (xylose)$_2$glucose-*O*-serine 53 found in the first epidermal growth factor-like domain of bovine blood clotting factor IX. *J. Biol. Chem.* **265**, 1858–1861.

Hofsteenge, J., Huwiler, K. G., Macek, B., Hess, D., Lawler, J., Mosher, D. F., and Peter-Katalinic, J. (2001). C-mannosylation and O-fucosylation of the thrombospondin type 1 module. *J. Biol. Chem.* **276**, 6485–6498.

Jacob, G. S., and Scudder, P. (1994). Glycosidases in structural analysis. *Methods Enzymol.* **230**, 280–299.

Kobata, A. (1994). Size fractionation of oligosaccharides. *Methods Enzymol.* **230**, 200–208.

Lin, A. I., Philipsberg, G. A., and Haltiwanger, R. S. (1994). Core fucosylation of high-mannose-type oligosaccharides in GlcNAc transferase I-deficient (Lec1) CHO cells. *Glycobiology* **4**, 895–901.

Luo, Y., and Haltiwanger, R. S. (2005). O-fucosylation of notch occurs in the endoplasmic reticulum. *J. Biol. Chem.* **280**, 11289–11294.

Luo, Y., Koles, K., Vorndam, W., Haltiwanger, R. S., and Panin, V. M. (2006a). Protein O-fucosyltransferase 2 adds O-fucose to thrombospondin type 1 repeats. *J. Biol. Chem.* **281**, 9393–9399.

Luo, Y., Nita-Lazar, A., and Haltiwanger, R. S. (2006b). Two distinct pathways for O-fucosylation of epidermal growth factor-like or thrombospondin type 1 repeats. *J. Biol. Chem.* **281**, 9385–9392.

Macek, B., Hofsteenge, J., and Peter-Katalinic, J. (2001). Direct determination of glycosylation sites in O-fucosylated glycopeptides using nano-electrospray quadrupole time-of-flight mass spectrometry. *Rapid Commun. Mass Spectrom.* **15,** 771–777.

Minamida, S., Aoki, K., Natsuka, S., Omichi, K., Fukase, K., Kusumoto, S., and Hase, S. (1996). Detection of UDP-D-xylose: D-xyloside 1,3xylosyltransferase activity in human hepatoma cell line HepG2. *J. Biochem.(Tokyo).* **120,** 1002–1006.

Moloney, D. J., and Haltiwanger, R. S. (1999). The O-linked fucose glycosylation pathway: Identification and characterization of a UDP-glucose: O-fucose B1,3-glucosyltransferase. *Glycobiology* **9,** 679–687.

Moloney, D. J., Lin, A. I., and Haltiwanger, R. S. (1997). The O-linked fucose glycosylation pathway: Evidence for protein specific elongation of O-linked fucose in Chinese hamster ovary cells. *J. Biol. Chem.* **272,** 19046–19050.

Moloney, D. J., Panin, V. M., Johnston, S. H., Chen, J., Shao, L., Wilson, R., Wang, Y., Stanley, P., Irvine, K. D., Haltiwanger, R. S., and Vogt, T. F. (2000a). Fringe is a glycosyltransferase that modifies notch. *Nature* **406,** 369–375.

Moloney, D. J., Shair, L., Lu, F. M., Xia, J., Locke, R., Matta, K. L., and Haltiwanger, R. S. (2000b). Mammalian Notch1 is modified with two unusual forms of O-linked glycosylation found on epidermal growth factor-like modules. *J. Biol. Chem.* **275,** 9604–9611.

Nishimura, H., Kawabata, S., Kisiel, W., Hase, S., Ikenaka, T., Takao, T., Shimonishi, Y., and Iwanaga, S. (1989). Identification of a disaccharide (Xyl-Glc) and a trisaccharide (Xyl₂-Glc) O-glycosidically linked to a serine residue in the first epidermal growth factor-like domain of human factors VII and IX and protein Z and bovine protein Z. *J. Biol. Chem.* **264,** 20320–20325.

Okajima, T., and Irvine, K. D. (2002). Regulation of notch signaling by O-linked fucose. *Cell* **111,** 893–904.

Omichi, K., Aoki, K., Minamida, S., and Hase, S. (1997). Presence of UDP-D-xylose: B-D-glucoside 1,3-D-xylosyltransferase involved in the biosynthesis of the Xyl1-3GlcB-Ser structure of glycoproteins in the human hepatoma cell line HepG2. *Eur. J. Biochem.* **245,** 143–146.

Panin, V. M., Shao, L., Lei, L., Moloney, D. J., Irvine, K. D., and Haltiwanger, R. S. (2002). Notch ligands are substrates for EGF protein O-fucosyltransferase and Fringe. *J. Biol. Chem.* **277,** 29945–29952.

Rampal, R., Arboleda-Velasquez, J. F., Nita-Lazar, A., Kosik, K. S., and Haltiwanger, R. S. (2005a). Highly conserved O-fucose sites have distinct effects on Notch1 function. *J. Biol. Chem.* **280,** 32133–32140.

Rampal, R., Li, A. S., Moloney, D. J., Georgiou, S. A., Luther, K. B., Nita-Lazar, A., and Haltiwanger, R. S. (2005b). Lunatic fringe, manic fringe, and radical fringe recognize similar specificity determinants in O-fucosylated epidermal growth factor-like repeats. *J. Biol. Chem.* **280,** 42454–42463.

Sasamura, T., Sasaki, N., Miyashita, F., Nakao, S., Ishikawa, H. O., Ito, M., Kitagawa, M., Harigaya, K., Spana, E., Bilder, D., Perrimon, N., and Matsuno, K. (2003). neurotic, a novel maternal neurogenic gene, encodes an O-fucosyltransferase that is essential for Notch-Delta interactions. *Development* **130,** 4785–4795.

Shao, L., Luo, Y., Moloney, D. J., and Haltiwanger, R. (2002). O-Glycosylation of EGF repeats: Identification and initial characterization of a UDP-glucose: Protein O-glucosyltransferase. *Glycobiology* **12,** 763–770.

Shao, L., Moloney, D. J., and Haltiwanger, R. S. (2003). Fringe modifies O-fucose on mouse notch1 at epidermal growth factor-like repeats within the ligand-binding site and the abruptex region. *J. Biol. Chem.* **278,** 7775–7782.

Shi, S., and Stanley, P. (2003). Protein O-fucosyltransferase I is an essential component of Notch signaling pathways. *Proc. Natl. Acad. Sci. USA* **100**, 5234–5239.

Stanley, P. (1981). Selection of Specific wheat germ agglutinin-resistant (Wga^R) phenotypes from Chinese hamster ovary cell populations containing numerous *lec*^R genotypes. *Mol. Cell. Biol.* **1**, 687–696.

Stanley, P. (1992). Glycosylation engineering. *Glycobiology* **2**, 99–107.

Stanley, P., and Siminovitch, L. (1977). Complementation between mutants of CHO cells resistant to a variety of plant lectins. *Somat. Cell Genet.* **3**, 391–405.

Sturla, L., Rampal, R., Haltiwanger, R. S., Fruscione, F., Etzioni, A., and Tonetti, M. (2003). Differential terminal fucosylation of N-linked glycans versus protein O-fucosylation in leukocyte adhesion deficiency type II (CDG IIc). *J. Biol. Chem.* **278**, 26727–26733.

Varki, A. (1994). Metabolic radiolabeling of glycoconjugates. *Methods Enzymol.* **230**, 16–32.

Yan, Y. T., Liu, J. J., Luo, Y., E. C., Y., Haltiwanger, R. S., Abate-shen, C., and Shen, M. M. (2002). Bi-functional activity of Cripto as a ligand and co-receptor in the nodal signaling pathway. *Mol. Cell. Biol.* **22**, 4439–4449.

[9]   Roles of *O*-Fucosyltransferase 1 and *O*-Linked Fucose in Notch Receptor Function

By TETSUYA OKAJIMA and TSUKASA MATSUDA

Abstract

O-fucosyltransferase 1 (*Ofut1*) is a soluble endoplasmic reticulum protein that directly transfers fucose onto serine or threonine residues of EGF domain-containing proteins such as Notch receptors. Genetic analysis indicates that *Ofut1* is essential for Notch receptor activation. To explore the molecular basis for the absolute requirement of *Ofut1* for Notch function, biochemical and cell biological approaches were used. Ligand-binding assay revealed that *Ofut1* is essential for Notch receptors to physically interact with their ligands. In addition, secretion assay and cell surface staining showed that secretion of Notch receptors is impaired in OFUT1-depleted cells, indicating that the structure of Notch receptors is altered. Interestingly, promotion of Notch secretion by OFUT1 does not require its enzyme activity. Together with the fact that OFUT1 physically associates with Notch and that OFUT1 prevents misfolding of Notch mutants, it is proposed that OFUT1 acts as a chaperone that promotes the folding of the EGF repeat of Notch receptors. This chapter focuses on the methods used to analyze the roles of *Ofut1* in Notch receptor structure and function.

METHODS IN ENZYMOLOGY, VOL. 417
0076-6879/06 $35.00
DOI: 10.1016/S0076-6879(06)17009-3

Overview

Notch receptors mediate an evolutionarily conserved intercellular signal that regulates a wide range of cell fate decision processes (Artavanis-Tsakonas et al., 1999). For precise control of Notch activity during development, various modulators act at different levels of the signaling pathway. One of them, which directly modifies Notch, is modification of sugars such as O-fucose glycans. O-fucose glycans were initially found as a rare type of posttranslational modification on EGF domain–containing plasma proteins including blood clotting factors VII and IX (Harris and Spellman, 1993). Recently, membrane proteins such as Notch receptors (containing approximately 36 tandemly arrayed EGF domains) and their ligands, Delta and Jagged/Serrate, have also been reported to be O-fucosylated (Moloney et al., 2000b; Panin et al., 2002).

The β O-fucose exists as either a monosaccharide, or tetrasaccharide; fucose-α1,3-GlcNAc-β1,4-Gal-α2.3/2.6-Sia (where GlcNAc is N-acetylglu cosamine, Gal is galactose, Sia is sialic acid), or di- or trisaccharide intermediate. The glycosylation process is mediated by sequential action of enzymes known as glycosyltransferases. For synthesizing the unique structure of O-fucose glycans, specific glycosyltransferases are used. One of them is a polypeptide O-fucosyltransferase1 (OFUT1), which initially acts in the glycosylation pathway and transfers fucose onto serine or threonine residues within the consensus sequence (and therefore called O-linked fucose or O-fucose) (Wang et al., 1996, 2001; Wang and Spellman, 1998). Elongation of O-fucose monosaccharide is dependent on an O-fucose-specific GlcNAc transferase, Fringe (Bruckner et al., 2000; Moloney et al., 2000a). OFUT1 and Fringe have been shown to function at a distinct compartment of the secretion pathway. OFUT1 is a soluble endoplasmic reticulum (ER) protein that carries a KDEL sequence at the carboxyl-terminus, which serves in retrograde transport from the cis-Golgi to the ER (Okajima et al., 2005). Although Fringe was originally identified as a secreted molecule, it is reported to localize in the Golgi apparatus (Munro and Freeman, 2000).

By use of genetic and biochemical techniques, the biological roles of Ofut1 and fringe for Notch receptors are being investigated. However, their roles for other substrates including Notch ligands are largely unknown. In both mouse and Drosophila, Ofut1 and its mammalian homolog Pofut1 have been demonstrated to be essential for Notch signaling (Okajima and Irvine, 2002; Sasamura et al., 2003; Shi and Stanley, 2003). As a molecular basis for this absolute requirement, it is proposed that Ofut1 is needed for proper conformation of Notch receptors for ligand binding (Okajima et al., 2005). Therefore, without Ofut1, Notch receptors are misfolded, leading to loss of expression at the cell surface and ability to bind their ligands. On the other hand, Fringe differentially modulates Delta and Serrate binding to Notch receptors; Fringe facilitates Notch-Delta binding, whereas it inhibits Notch-Serrate binding (Bruckner et al., 2000; Okajima et al., 2003). This effect of Fringe is considered to be basic mechanisms of local Notch activation at the Fringe-expression boundary (Panin and Irvine, 1998).

Despite the great advances in clarification of the roles of glycosyltransferases for Notch function, the molecular mechanisms by which an individual O-glycan exerts regulatory functions still remain largely unknown. On the basis of deletion mapping experiments using cultured Drosophila cells, it has been shown that 11th and 12th EGF domains (EGF11 and EGF12) are the minimal binding units required for physical interaction with the ligands (Rebay et al., 1991). Among them, only EGF12 carries the glycosylation site for O-fucose modification. Therefore, the simplest scenario is that an O-fucose glycan on EGF12 regulates Notch–ligand interaction. However, removal of the consensus sequence on EGF12 does not lead to complete loss of Notch activity or the Fringe effect (Lei et al., 2003). This implies that multiple O-fucose glycans cooperate in glycosylation-dependent Notch-ligand binding. Alternately, multiple O-fucose glycans might mediate the conformational change of entire EGF repeats, thus indirectly affecting Notch-ligand interaction (Haines and Irvine, 2003). On the other hand, OFUT1 has recently been reported to possess a chaperone activity that is separable from the enzyme activity of OFUT1, possibly indicating that O-fucose modification is not essential for Notch function, but rather modulates Notch activity similarly to Fringe (Okajima et al., 2005). This chapter mainly describes the biochemical methods used to analyze the roles of Ofut1 in Notch receptor structure and function.

Enzyme Assay for O-Fucosyltransferase 1 Activity

Ofut1 was isolated on the basis of its homology to mammalian Pofut1. To confirm the enzyme activity of OFUT1, it was expressed as a secreted form. OFUT1 is a soluble ER protein with a cognate signal peptide at the amino-terminus and a KDEL-type ER retrieval signal at the carboxyl-terminus. To facilitate the secretion of OFUT1 into culture medium, a V5His tag was added at the carboxyl-terminal end, which was expected to mask the ER retrieval sequence. The construct, pRmHa3-Ofut1:V5His, was transfected into S2 cells, and the culture medium was recovered and used as an enzyme source. The EGF domain from factor VII and GDP-[^{14}C]fucose were used as the acceptor substrate and the donor substrate, respectively. The enzyme products were purified by reverse-phase liquid chromatography; the incorporated radioactivity was measured using a liquid scintillation counter.

As a negative control, an inactive mutant for OFUT1 was engineered. A comparative study of the sequence from α1,2-fucosyltransferases and α1,6-fucosyltransferases and O-fucosyltransferases identified a GXHXR (R/H) motif that is implicated in donor substrate binding (Chazalet et al., 2001; Martinez-Duncker et al., 2003; Takahashi et al., 2000). The first R in this motif is invariant among all the above fucosyltransferase families and is essential for the activity of α1,6- and α1,2-fucosyltransferases in vitro and in vivo. As expected, the change of R to A completely abolished the enzyme activity.

Materials

   Ofut1 expression vector (pRmHa3-Ofut1: V-5H or pRmHa3-Ofut1[R245A])
   Serum-free media (HyQ-CCM3; Hyclone)
   Cellfectin (Invitrogen)
   2× Assay buffer A (200 mM imidazole-HCl, pH 7.0, 100 mM MnCl$_2$)
      or buffer B (100 mM HEPES, pH 7.0, 100 mM MnCl$_2$)
   0.1 mM GDP-[^{14}C]fucose (20,000 dpm/nmol)
   Ultrafree-4 Filter Units (Millipore)
   EGF domain of clotting factor VII (Obtained from Genentech)
   LC-18 SPE tube (Supelco)
   Anti-V5 antibody-conjugated agarose (Sigma)

Methods

   1. S2 cells are plated in 10-cm plates (2 × 10^6 cells/ml, 10 ml/plate).
   2. After 2–3 h, *Ofut1* expression vectors are transfected into S2 cells using Cellfectin.
   3. 24 h after transfection, the medium is replaced with serum-free medium containing 0.7 mM CuSO$_4$. After 3–4 days of incubation, the culture medium is collected.
   4a. The medium is concentrated 100-fold using Ultrafree-4 Filter Units, followed by buffer exchange with 100 mM imidazole-HCl, pH 7.0 (for Assay buffer A), or 50 mM HEPES, pH 7.0 (for Assay buffer B) and used as an enzyme source. OFUT1:V5His can be further purified using a Ni-NTA Purification System (Invitrogen).
   4b. Alternately, OFUT1:V5His is purified on anti-V5 antibody–conjugated agarose. After washing with 1× assay buffer, the beads are used as an enzyme source.
   5. Fucosyltransferase assays are performed in 1× assay buffer A or B containing 0.1 mM GDP-[^{14}C]fucose (20000 dpm/nmol), and 20 μM EGF factor VII substrate and the enzyme in a volume of 20 μl.

6. After incubation at 37° for 2 h, the reaction mixture is applied to a LC-18 SPE tube.
7. The tube is washed twice with H_2O, then labeled substrates are eluted with 1 ml of 80% acetonitrile, 0.052% TFA. Radioactivity in the eluate is measured using a liquid scintillation counter.

Downregulation of *Ofut1* by Use of RNAi Techniques

Ofut1 is endogenously expressed in S2 cells. To explore the roles of *Ofut1* in the function of Notch receptors in S2 cells, the RNAi technique was successfully applied. In *Drosophila*, simple treatment of cells with dsRNA specifically depletes the corresponding gene products (Clemens *et al.*, 2000). Reduction of the targeted gene expression can be confirmed by measuring changes in mRNA levels (e.g., using RT-PCR), or detecting changes in protein levels by immunocytochemistry or Western blot analysis. The suppression effect will vary depending on the target, cell line, and experimental conditions.

RNAi techniques are also applied to *Drosophila* tissues during development. This can be done by constructing plasmids harboring the inverted repeat sequence of the gene of interest under control of the UAS promoter. In combination with various Gal4 lines available, dsRNA expression is induced in a temporally and spatially controlled manner. Although the genetic procedures are beyond the scope of this chapter, the inducible RNAi techniques for downregulation of glycosyltransferase genes have been successfully used.

Materials

Serum-free medium (Drosophila SFM; Invitrogen)
Complete growth medium (Drosophila SFM supplemented with 5% FCS)
MEGAscript RNAi Kit, MEGAscript T7 Kit, MEGAclear (Ambion)

Methods

Preparation of dsRNA for Ofut1. A *Bgl*II/ *Eco*RI fragment from *UAS-iOfut1*, corresponding to nucleotides 154–967, was cloned into the multiple cloning site of the L4440 vector, so that the *Ofut1* sequence was flanked by the T7 promoter at both ends. The template for *in vitro* transcription was amplified by PCR using T7 primers (TAATACGACTCACTATAGGG). dsRNA was prepared using a MEGAscript RNAi Kit, or a combination of the MEGAscript T7 Kit with MEGAclear, according to the manufacturer's instructions.

RNAi Procedure

RNAi in Cell Culture Dishes

1. S2 cells are plated in 6-cm plates (.05 × 10⁶ cells/ml, 4 ml/plate) in serum-free medium without antibiotics.
2. After 2–3 h, the culture medium is replaced with 2 ml of serum-free medium.
3. 40 μg of dsRNA is added to the medium and the plates are shaken vigorously.
4. After 1 h of incubation, 2 ml of complete growth medium is added to the plates (dsRNA does not have to be removed).
5. The cells are incubated for an additional 3–5 days to allow turnover of OFUT1.

RNAi in a 96-well Plate

1. S2 cells are plated in a 96-well plate (.05 × 10⁶ cells/ml, 100 μl/well) in serum-free medium without antibiotics.
2. After 2–3 h, the culture medium is replaced with 100 μl of serum-free medium containing 1.2 μg of dsRNA.
3. After 1 h of incubation, 100 μl of complete growth medium is added to the plates (dsRNA does not have to be removed).
4. The cells are incubated for an additional 3–5 days to allow turnover of OFUT1.

Quantification of the O-Fucose Level

The decreased level of O-fucosylation on Notch in cells treated with *Ofut1* dsRNA was confirmed by *in vitro* glycosylation assay. This assay takes advantage of the fact that Fringe specifically modifies O-fucose on EGF domains, and thus the enzyme activity of Fringe toward the EGF repeat of Notch will reflect the amount of O-fucose on the EGF domains. Although metabolic labeling with radioactive fucose is successfully applied to mammalian cells, this strategy cannot be adopted to the *Drosophila* system because the components for the salvage pathway for GDP-fucose synthesis are deficient in *Drosophila* genome.

Materials

Fringe:6×His (Moloney *et al.*, 2000a)
Anti-human placental alkaline phosphatase antibody-conjugated agarose (Sigma)
Anti-FLAG M2 antibody-conjugated agarose (Sigma)
Hanks' balanced solution (HBSS; Invitrogen)

$2\times$ Fringe assay buffer (100 mM HEPES, pH 7.0, 300 mM NaCl, 100 mM MnCl$_2$).

18 μM UDP-[^{14}C]GlcNAc (266 mCi/mmol; Amersham Biosciences)

Methods

1. Culture medium containing either N:AP or N-EGF:FLAG is incubated with anti-human placental alkaline phosphatase or anti-FLAG antibody-conjugated agarose, respectively, at room temperature for more than 2 h.
2. The beads are washed with HBSS three times and then washed once with $1\times$ Fringe assay buffer.
3. Fringe assay is performed in $1\times$ Fringe assay buffer containing the enzyme (Fringe:6$\times$His), 18 μM UDP-[^{14}C]GlcNAc and 10 μl of beads in a total volume of 50 μl.
4. After 4 h of incubation at 30° with constant gentle mixing, the beads are washed four times with HBSS and the radioactivity is measured using a liquid scintillation counter.

Cell Surface Notch Staining

Genetic studies have indicated that OFUT1 is a core component of the Notch signaling pathway. To explore the role of *Ofut1* in Notch trafficking, cell surface expression of Notch receptors was examined in OFUT1-depleted S2 cells. For cell surface staining of the Notch receptor, S2 cells were stained without detergent using anti-extracellular Notch antibody. To confirm Notch expression, the cells were sequentially stained with anti-intracellular Notch antibody with detergent, reflecting the total Notch expression in the cells. In OFUT1-depleted cells, the cell surface expression of Notch was diminished. The expression was restored by the introduction of mouse *Pofut1*. These results showed that *Ofut1* is essential for cell surface expression of Notch. Interestingly, a *Pofut1*-inactivated mutant, *Pofut1*^{R245A}, can also rescue the deficit in the cell surface presentation of Notch, suggesting the enzyme activity-independent role of *Pofut1/Ofut1* in the Notch maturation process.

Materials

Tissue culture slides (8 chamber; Falcon#354118)
Concanavalin A (from *Canavalia ensiformis*; Sigma#L7647)
16% Formaldehyde (Polysciences Inc.#18814),
Rat anti-extracellular Notch antibody (Rat8; 1:2000 dilution in PBS; obtained from S. Artavanis-Tsakonas)

Alexa 555-conjugated goat anti-rat IgG (Molecular Probes; 1:50 dilution in PBS)

PBT (PBS with 0.1% Triton X-100, 1% bovine serum albumin (BSA), and 0.01% sodium azide)

Mouse anti-intracellular Notch antibody (C17.9C6; 1:1000 dilution in PBT)

Alexa488-conjugated goat anti-mouse IgG antibody (Molecular Probes; 1:200 dilution in PBT)

Vectorshield mounting medium (Vector Laboratories)

Methods

1. To better visualize the subcellular localization in S2 cells, tissue culture slides are coated for at least 2 h with 0.5 mg/ml of Concanavalin A.

2. S2 cells are plated on ConA-treated slides and allowed to spread out on the slide for 1–2 h (Rogers *et al.*, 2003).

3. The cells are fixed for 10 min with 4% formaldehyde in PBS, and the samples are washed with PBS three times for 5 min in total.

4. Samples are incubated with rat anti-extracellular Notch antibody in PBS for 40 min and then washed with PBS three times for 5 min. They are then incubated with Alexa555-conjugated goat anti-rat IgG antibody in PBS for 30 min and washed with PBS for 5 min.

5. The cells are permeabilized with PBT for 5 min.

6. Samples are incubated with mouse anti-intracellular Notch antibody in PBT for 40 min and then washed with PBT three times for 5 min in total. They are then incubated with Alexa488-conjugated goat anti-mouse IgG in PBT for 30 min and washed with PBT three times for 5 min in total.

7. The slide is mounted in 80% glycerol in PBS or Vectorshield mounting medium.

Notch Secretion Assay

The preceding experiments demonstrated that *Ofut1* is essential for cell surface presentation of Notch. Together with the fact that OFUT1 is localized in the ER, it is reasonable to suggest that OFUT1 acts in the ER to promote Notch secretion. However, it was formally possible that *Ofut1* might prevent Notch endocytosis; without *Ofut1,* Notch cannot be retained on the cell surface. To exclude this possibility, Notch secretion was monitored by the secretion assay. In this assay, a N:AP construct, in which the EGF repeat of Notch is carboxyl-terminally tagged with human placental alkaline phosphatase (AP), was transfected into S2 cells, and the secreted protein was quantified by AP activity in the culture medium.

As expected, OFUT1-depleted cells exhibited a 10-fold decrease in N:AP secretion, whereas Dl:AP and Ser:AP secretion were unaffected. Consistent with the observation that OFUT1 possesses an enzyme activity-independent role, wild-type *Pofut1*, as well as inactive *Pofut1*^{R245A}, was able to restore Notch:AP secretion in OFUT1-depleted cells.

Materials

pRmHA-N:AP, pRmHA-Dl:AP, pRmHA-Ser:AP (Bruckner *et al.*, 2000)
Copia-*Renilla* (Lum *et al.*, 2003)
4-Methylumbelliferyl phosphate (4-MUP) liquid substrate system (Sigma; M3168)
Fusion universal microplate analyzer (Packard)

Methods

1. S2 cells are plated in a 96-well plate (2×10^6 cells/ml, 100 μl/well) in serum-free medium without antibiotics.
2. After 2–3 h, the cells are transfected with a vector encoding the AP fusion construct and the copia-*Renilla* control vector (for normalization of transfection efficiency).
3. The cells are incubated for 1–2 days, and 50 μl of the culture medium is collected.
4. Endogenous AP activity is heat-inactivated by incubation at 65° for 30 min.
5. 100 μl of 4-MUP liquid substrate is added to each sample, and the mixture is incubated for more than 1 h. The incubation time can be varied from several hours to overnight, depending on the signal intensity.
6. The fluorescent products are measured in a microplate analyzer.
7. To normalize the transfection activity, sea pansy luciferase activity from the cell lysate is measured, and the value for AP activity is divided by that for luciferase.

Western Blotting for Notch

During the maturation process, Notch receptors are subject to proteolytic processing by a furin-like convertase in the trans-Golgi apparatus, which cleaves at the luminal side of the transmembrane domain of Notch and generates heterodimers composed of extracellular and transmembrane/intracellular domains. On ligand binding, Notch receptors are further processed by TACE/ADAM metalloproteinases and the $\tilde{\gamma}$-secretase complex, which removes the extracellular domains and releases intracellular domains, respectively (Mumm and Kopan, 2000). Because *Ofut1* is required for Notch trafficking from the ER to the cell surface, OFUT1-depleted cells are expected to show lack of Notch processing. To test this possibility, Western blotting was performed using mouse anti-intracellular Notch antibody, which detects the cleaved fragments of Notch. The results showed absence of the cleaved fragment of Notch in OFUT1-depleted cells. Expression of wild-type *Pofut1*, as well as inactive *Pofut1*[R245A], was able to restore the production of the cleaved fragment of Notch, again indicating that the role of OFUT1 is separable from enzyme activity.

Materials

> Urea-SDS buffer (8 *M* urea, 5% SDS, 40 m*M* Tris-HCl, pH 6.8,
> 0.1 m*M* EDTA and protease inhibitors)
> Immobilon-P membrane (Millipore)
> Mouse anti-intracellular Notch antibody (C17.9C6; 1:1000 dilution)
> Horseradish peroxidase–conjugated anti-mouse antibody (1:50,000
> dilution; Amersham Bioscience)
> SuperSignal west femto maximum sensitivity substrate (Pierce)

Methods

1. Cells expressing Notch are lysed using urea-SDS buffer for preparation of the total cell lysate.
2. The samples are boiled with DTT and applied to a 5–15% SDS-polyacrylamide gradient gel.
3. Proteins are transferred to an Immobilon-P membrane and stained with mouse anti-intracellular Notch antibody.
4. Immunoreactive protein bands are detected using a horseradish peroxidase–conjugated anti-mouse antibody and SuperSignal west femto maximum sensitivity substrate.

Notch-Ligand Binding Assay

Down-regulation of OFUT1 greatly decreases the expression of Notch on the cell surface. To investigate the potential conformational change in Notch receptors in the absence of *Ofut1*, a Notch–ligand binding assay was performed. Two different assays are described in the following. The one described first is a cell-based Notch–ligand binding assay, which was developed to demonstrate the effect of Fringe on Delta-Notch binding (Bruckner *et al.*, 2000). Recently, an *in vitro* Notch–ligand binding assay has been developed. By use of these assays, it was demonstrated that *Ofut1* is required for normal Notch–ligand binding. In contrast, overexpression of

Ofut1 suppressed Notch–Delta binding, but promoted Notch–Serrate binding (Okajima *et al.*, 2003).

On the basis of the observation that OFUT1 is localized in the ER and can promote Notch secretion independently of its enzyme activity, it is hypothesized that OFUT1 acts as a chaperone that associates with EGF domains and promotes folding of the entire EGF repeats of Notch. To explore this possibility, two types of Notch mutants, which are prone to misfolding, were engineered. One was N:AP-EGF23-32f, which carries multiple mutations that result in loss of O-fucosylation sites from the 23rd to 32nd EGF domains. The other was N:AP^{C599Y}, which mimics a mutation found in the human genetic disease CADASIL. This mutation alters a conserved cysteine residue that forms a disulfide bond, thus causing misfolding of the EGF domain. These mutants showed decreased ability to bind Notch ligands. However, expression of the OFUT1^{R245A} mutant greatly recovered the binding ability of the Notch mutants, supporting the hypothesis that OFUT1 is a Notch chaperone.

Cell-Based Notch-Ligand Binding Assay

S2 cells are versatile experimental tools for studies of Notch–ligand binding and its influence by glycosyltransferases, because these cells do not endogenously express Notch, Delta, Serrate, and Fringe. Moreover, endogenous *Ofut1* expression is easily depleted by dsRNA treatment. In this assay, EGF repeats of Notch were prepared as a soluble form of the protein, making it possible to adjust the secretion level of Notch fragments, which is greatly influenced by the level of *Ofut1*.

Materials

Serum-free medium (HyQ-CCM3: Hyclone)
Amicon Ultra centrifugal filter devices (Millipore, 10,000 molecular weight cut-off)
HBSS (Invitrogen)
Plasmid encoding AP-tagged Notch EGF repeats (pMT-WB-N:AP or pRmHa3-N:AP)
Plasmid encoding Delta or Serrate (pRmHa-Dl or pRmHa-Ser)
p-Nitrophenyl phosphate (Sigma)
AP reaction buffer (1 M diethanolamine, pH 9.8, 0.5 mM MgCl$_2$)

Methods

PREPARATION OF N:AP CONDITIONED MEDIUM   The plasmid encoding N:AP is transfected into S2 cells using Cellfectin. Then 24 h after transfection, the medium is changed to serum-free medium containing 0.7 mM CuSO$_4$

for induction of expression; 2–4 days after induction, the culture medium is collected and centrifuged to remove debris. The medium is concentrated approximately 10-fold with centrifugal filter devices. To quantify the level of AP-tagged proteins in the concentrated medium, the AP reaction is performed in the mixture containing 1 mg/ml p-nitrophenyl phosphate in AP reaction buffer for several hours at 37°. The reaction is stopped by addition of 1 M NaOH. The enzyme-reaction products are detected by spectrophotometry by measuring the absorbance at 405 nm. This value is divided by the incubation time and expressed as milliabsorbance units of product formed per minute (mOD405/min). The amount of AP fusion proteins in the conditioned medium is equalized typically to 600 mOD405/min by dilution using control medium prepared from mock-transfected cells.

PREPARATION OF S2 CELLS EXPRESSING DELTA OR SERRATE S2 cells are transiently transfected with pRmHa-Dl or pRmHa-Ser; 24 h after transfection, the culture medium is replaced with new medium containing 0.7 mM CuSO$_4$. Then 2 days after induction, the cells are washed with HyQ CCM3 medium and resuspended in HyQ CCM3 supplemented with 0.5% BSA.

BINDING REACTION 0.3 ml of conditioned medium and 0.3 ml of cells are mixed in Eppendorf tubes, and incubated for 1 h with gentle agitation at room temperature. The cells are washed four times with HBSS containing 0.5% BSA, and then lysed in 10 mM Tris-HCl, pH 8.0, 1% Triton X-100 at 4° for 1 h. After centrifugation, the supernatant is heated at 65° for 30 min to inactivate endogenous alkaline phosphatase activity. The AP reaction is performed in AP reaction buffer containing 1 mg/ml p-nitrophenyl phosphate, and the activity is measured using a spectrophotometer as described previously.

In Vitro *Notch-Ligand Binding Assay*

Materials

Plasmid encoding FLAG-tagged Notch EGF repeats (pMT(1B)/N-EGF:FLAG)
Plasmid encoding a soluble form of Notch ligand tagged with AP (pRmHa-Dl:AP or pRmHa-Ser:AP)
Ezview Red anti-FLAG M2 affinity gel (Sigma)
Tris-buffered saline (TBS)

Methods

PREPARATION OF N-EGF:FLAG BEADS pMT(1B)/N-EGF:FLAG is transfected into S2 cells using Cellfectin; 24 h after transfection, the medium is replaced with HyQ-CCM3 medium containing 0.7 mM CuSO$_4$ for induction of expression. Then 2–4 days after induction, the culture medium is collected and centrifuged to remove debris, and 5 ml of the conditioned medium is incubated with 50 μl of Ezview Red anti-FLAG M2 affinity gel at 4° overnight. For the samples from OFUT1-depleted cells, an increased amount of medium is applied because of the decreased level of secretion in the medium. The gel beads are washed three times with 5 ml of TBS containing 5 mM CaCl$_2$.

PREPARATION OF SOLUBLE LIGANDS S2 cells are transiently transfected with pRmHa-Dl:AP or pRmHa-Ser:AP. 24 h after transfection, the medium is replaced with serum-free medium containing 0.7 mM CuSO$_4$. 2–4 days after induction, the culture medium is collected and centrifuged to remove debris. To quantify the level of AP-tagged proteins in the concentrated medium, the AP reaction is performed as described previously. The amount of AP fusion protein in the conditioned medium is equalized to 3000 mOD405/min by dilution with control medium prepared from mock-transfected cells.

BINDING REACTION Five microliters of FLAG beads prepared as described above is blocked for 1 h at room temperature with 1 ml of TBS containing 1% BSA and 5 mM CaCl$_2$. The beads are incubated with 50 μl

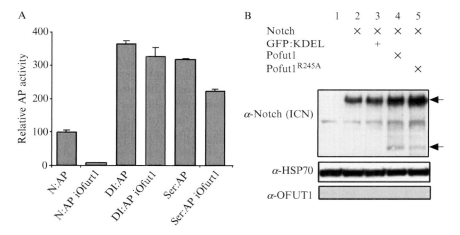

FIG. 1. Notch secretion and generation of cleaved Notch fragment are dependent on OFUT1. A, The N:AP, Dl:AP or Ser:AP construct was transfected into S2 cells and the secreted protein was quantified by alkaline phosphatase activity in the culture medium. In S2 cells in which OFUT1 was depleted by RNAi, N:AP secretion was decreased 10-fold in comparison with wild-type S2 cells, whereas Dl:AP and Ser:AP secretion were unaffected. B, Notch receptors are expressed in OFUT1-depleted cells in the presence of either a vector control (lane 2), GFP:KDEL (control, lane 3), Ofut1 (lane 4) or Ofut1^{R245A} (lane 5). A cell lysate was prepared from each transfectant and subjected to Western blotting analysis with anti-intracellular Notch antibody. The upper arrow indicates full-length Notch; the lower arrow indicates a 120-kDa fragment whose appearance requires OFUT1 activity. Lack of OFUT1 expression was confirmed by probing with anti-OFUT1 antibody. Anti-HSP70 blotting indicates equal protein loading in each lane.

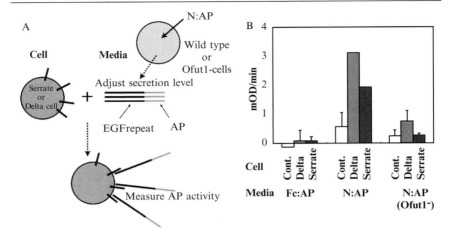

FIG. 2. Cell-based Notch ligand binding assay. A, schema showing the procedure for the cell-based Notch ligand binding assay. The plasmid encoding N:AP was transfected into S2 cells and protein expression was induced. AP activity in each conditioned medium was equalized by dilution with control medium prepared from mock-transfected cells. The conditioned medium was incubated with ligand-expressing cells. After washing, bound AP activity was measured. B, binding of Fc:AP (control) or N:AP to either control cells, Delta-expressing cells or Serrate-expressing cells. N:AP prepared from OFUT1-depleted cells exhibits a significant decrease in its ability to bind ligands.

of conditioned medium containing soluble ligands for 1 h at room temperature. After five washings with HBSS, AP activity on the beads is assayed.

Notch–OFUT1 Binding Assay

If OFUT1 acts as a chaperone for Notch, then direct binding between Notch and OFUT1 can be expected. In fact, a mammalian O-fucosyltransferase 1 (O-FucT-1) has affinity for EGF domains and has been partially purified by affinity chromatography using EGF domains (Wang and Spellman, 1998). To confirm that *Drosophila* OFUT1 also associates with Notch receptors, a Notch-OFUT1 binding assay was conducted. In this assay, S2 cells were transfected with vectors encoding Notch:AP and either OFUT1:V5His or OFUT1[R245A]:V5His. As a negative control, Boca, another KDEL-type chaperone that promotes folding of low-density lipoprotein (LDL) receptors, was used. Formation of the specific complex between Notch:AP and OFUT1:V5His or OFUT1[R245A]:V5His was observed.

Materials

pRmHa-Fc:AP (control), pRmHa-Dl:AP, pRmHa-Ser:AP, pRmHa-Notch:AP pRmHa-Boca:V5His (control), pRmHa-Ofut1:V5His, pRmHa-Ofut1^{R245A}:V5His

Monoclonal anti-human placental alkaline phosphatase-agarose (8B6; Sigma)

Mouse anti-V5 antibody (Invitrogen)

Methods

1. The AP fusion construct (pRmHa-Fc:AP, pRmHa-Dl:AP, pRmHa-Ser:AP or pRmHa-Notch:AP) is co-transfected into S2 cells with either pRmHa-Boca:V5His, pRmHa-Ofut1:V5His or pRmHa-Ofut1^{R245A}:V5His.
2. The culture medium is recovered 20 h after induction.
3. Cell debris is removed by centrifugation.
4. The supernatant is incubated at room temperature with monoclonal anti-human placental alkaline phosphatase antibody-conjugated agarose.
5. After 1 h of incubation, the beads are pelleted and briefly washed three times with HBSS.
6. Protein on the beads is eluted with SDS-loading buffer with DTT.
7. Co-purification of V5His-tagged proteins is detected by Western blotting using mouse anti-V5 antibody.

Acknowledgments

We thank Koichi Furukawa for help in the arrangement of the manuscript. T. O. is supported by the Human Frontier Science Program and the Nakajima Foundation.

References

Artavanis-Tsakonas, S., Rand, M. D., and Lake, R. J. (1999). Notch signaling: Cell fate control and signal integration in development. *Science* **284**, 770–776.

Bruckner, K., Perez, L., Clausen, H., and Cohen, S. (2000). Glycosyltransferase activity of Fringe modulates Notch-Delta interactions. *Nature* **406**, 411–415.

Chazalet, V., Uehara, K., Geremia, R. A., and Breton, C. (2001). Identification of essential amino acids in the *Azorhizobium caulinodans* fucosyltransferase NodZ. *J. Bacteriol.* **183**, 7067–7075.

Clemens, J. C., Worby, C. A., Simonson-Leff, N., Muda, M., Maehama, T., Hemmings, B. A., and Dixon, J. E. (2000). Use of double-stranded RNA interference in *Drosophila* cell lines to dissect signal transduction pathways. *Proc Natl. Acad. Sci. USA* **97**, 6499–6503.

Haines, N., and Irvine, K. D. (2003). Glycosylation regulates Notch signalling. *Nat. Rev. Mol. Cell. Biol.* **4**, 786–797.

Harris, R. J., and Spellman, M. W. (1993). O-linked fucose and other post-translational modifications unique to EGF modules. *Glycobiology* **3**, 219–224.

Lei, L., Xu, A., Panin, V., and Irvine, K. D. (2003). An O-fucose site in the ligand binding domain inhibits Notch activation. *Development* **130**, 6411–6421.

Lum, L., Yao, S., Mozer, B., Rovescalli, A., Von Kessler, D., Nirenberg, M., and Beachy, P. A. (2003). Identification of Hedgehog pathway components by RNAi in *Drosophila* cultured cells. *Science* **299**, 2039–2045.

Martinez-Duncker, I., Mollicone, R., Candelier, J. J., Breton, C., and Oriol, R. (2003). A new superfamily of protein-O-fucosyltransferases, alpha2-fucosyltransferases, and alpha6-fucosyl transferases: Phylogeny and identification of conserved peptide motifs. *Glycobiology* **13,** 1C–5C.

Moloney, D. J., Panin, V. M., Johnston, S. H., Chen, J., Shao, L., Wilson, R., Wang, Y., Stanley, P., Irvine, K. D., Haltiwanger, R. S., and Vogt, T. F. (2000a). Fringe is a glycosyltransferase that modifies Notch. *Nature* **406,** 369–375.

Moloney, D. J., Shair, L. H., Lu, F. M., Xia, J., Locke, R., Matta, K. L., and Haltiwanger, R. S. (2000b). Mammalian Notch1 is modified with two unusual forms of O-linked glycosylation found on epidermal growth factor-like modules. *J. Biol. Chem.* **275,** 9604–9611.

Mumm, J. S., and Kopan, R. (2000). Notch signaling: From the outside in. *Dev. Biol.* **228,** 151–165.

Munro, S., and Freeman, M. (2000). The notch signalling regulator fringe acts in the Golgi apparatus and requires the glycosyltransferase signature motif DXD. *Curr. Biol.* **10,** 813–820.

Okajima, T., and Irvine, K. D. (2002). Regulation of notch signaling by O-linked fucose. *Cell* **111,** 893–904.

Okajima, T., Xu, A., and Irvine, K. D. (2003). Modulation of notch-ligand binding by protein O-fucosyltransferase 1 and fringe. *J. Biol. Chem.* **278,** 42340–42345.

Okajima, T., Xu, A., Lei, L., and Irvine, K. D. (2005). Chaperone activity of protein O-fucosyltransferase 1 promotes Notch receptor folding. *Science* **307,** 1599–1603.

Panin, V. M., and Irvine, K. D. (1998). Modulators of Notch signaling. *Semin. Cell Dev. Biol.* **9,** 609–617.

Panin, V. M., Shao, L., Lei, L., Moloney, D. J., Irvine, K. D., and Haltiwanger, R. S. (2002). Notch ligands are substrates for protein O-fucosyltransferase-1 and Fringe. *J. Biol. Chem.* **277,** 29945–29952.

Rebay, I., Fleming, R. J., Fehon, R. G., Cherbas, L., Cherbas, P., and Artavanis-Tsakonas, S. (1991). Specific EGF repeats of Notch mediate interactions with Delta and Serrate: Implications for Notch as a multifunctional receptor. *Cell* **67,** 687–699.

Rogers, S. L., Wiedemann, U., Stuurman, N., and Vale, R. D. (2003). Molecular requirements for actin-based lamella formation in *Drosophila* S2 cells. *J. Cell. Biol.* **162,** 1079–1088.

Sasamura, T., Sasaki, N., Miyashita, F., Nakao, S., Ishikawa, H. O., Ito, M., Kitagawa, M., Harigaya, K., Spana, E., Bilder, D., Perrimon, N., and Matsuno, K. (2003). neurotic, a novel maternal neurogenic gene, encodes an O-fucosyltransferase that is essential for Notch-Delta interactions. *Development* **130,** 4785–4795.

Shi, S., and Stanley, P. (2003). Protein O-fucosyltransferase 1 is an essential component of Notch signaling pathways. *Proc. Natl. Acad. Sci. USA* **100,** 5234–5239.

Takahashi, T., Ikeda, Y., Tateishi, A., Yamaguchi, Y., Ishikawa, M., and Taniguchi, N. (2000). A sequence motif involved in the donor substrate binding by alpha1,6-fucosyltransferase: The role of the conserved arginine residues. *Glycobiology* **10,** 503–510.

Wang, Y., and Spellman, M. W. (1998). Purification and characterization of a GDP-fucose: Polypeptide fucosyltransferase from Chinese hamster ovary cells. *J. Biol. Chem.* **273,** 8112–8118.

Wang, Y., Lee, G. F., Kelley, R. F., and Spellman, M. W. (1996). Identification of a GDP-L-fucose:polypeptide fucosyltransferase and enzymatic addition of O-linked fucose to EGF domains. *Glycobiology* **6,** 837–842.

Wang, Y., Shao, L., Shi, S., Harris, R. J., Spellman, M. W., Stanley, P., and Haltiwanger, R. S. (2001). Modification of epidermal growth factor-like repeats with O-fucose. Molecular cloning and expression of a novel GDP-fucose protein O-fucosyltransferase. *J. Biol. Chem.* **276,** 40338–40345.

[10] Roles of O-Fucose Glycans in Notch Signaling Revealed by Mutant Mice

By LINCHAO LU and PAMELA STANLEY

Abstract

Notch receptor signaling is important for many developmental processes in the metazoa. Insights into how Notch receptor signaling is regulated have been obtained from the characterization of mutants of model organisms in which Notch signaling is perturbed. Here we describe the effects of mutations that alter the glycosylation of Notch receptors and Notch ligands in the mouse. The extracellular domain of Notch receptors and Notch ligands carries N-glycans and O-glycans, including O-fucose and O-glucose glycans. Mutations in several genes that inhibit the synthesis of O-fucose glycans, and one that also affects the maturation of N-glycans, cause Notch signaling defects and disrupt development.

Overview

Notch receptors belong to a family of single transmembrane glycoproteins containing 29–36 EGF repeats in their extracellular domain (ECD). Notch receptor signaling is critical for cell fate determination, cell growth control, and development in metazoans. In mammals, there are four Notch receptors (Notch1–Notch4) and five Notch ligands (Jagged1, Jagged2, Delta1, Delta3, and Delta4). Each Notch receptor is synthesized in the endoplasmic reticulum (ER) as a single polypeptide and later cleaved in the trans-Golgi by a furin-like convertase and expressed on the cell surface as a heterodimer (Blaumueller *et al.*, 1997; Logeat *et al.*, 1998). On the binding of Notch ligands expressed in neighboring cells, Notch receptors are successively cleaved—first by a cell surface metalloprotease (ADAM10/ADAM17), leaving approximately 12 amino acids on the extracellular side of the Notch transmembrane portion and subsequently by the presenilin complex that has a γ-secretase activity that cleaves within the membrane. The latter cleavage releases Notch intracellular domain (ICD), which binds to the CSL transcriptional repressor (RBP-Jκ in mammals) and recruits other transcriptional co-activators to turn the CSL/NICD complex into a transcriptional activator that induces the expression of target genes, among which are the Hairy-enhancer-of-split or *Hes* transcriptional factors. The overall pathway is presented in a simplified form in Fig. 1, with a focus on the location of the predicted N- and O-glycans of Notch1. The Notch receptor ECD becomes glycosylated as it transits the

METHODS IN ENZYMOLOGY, VOL. 417
0076-6879/06 $35.00
DOI: 10.1016/S0076-6879(06)17010-X

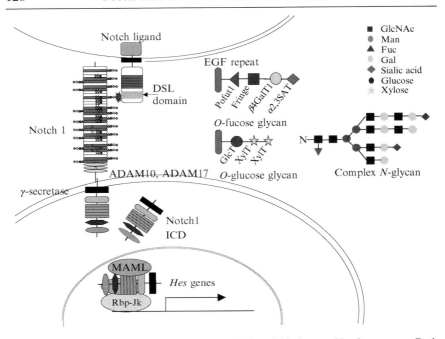

FIG. 1. Notch signaling occurs when a Notch ligand binds to a Notch receptor. Both ligands and receptors have multiple EGF repeats in their extracellular domain. The EGF repeats of both Notch receptors and ligands contain consensus sites for the addition of *O*-fucose and *O*-glucose glycans (see text) and AsnXSer/Thr consensus sequons for the addition of *N*-glycans. The EGF repeats that carry these glycans are identified for the Notch1 receptor because the glycosylation of this receptor is best characterized. EGF repeats with an *N*-glycan consensus are marked N; *O*-fucose and *O*-glucose glycans are shown with symbols. Notch/ligand binding is indicated by a red symbol at the ligand-binding domain of Notch1 (EGF repeats 11 and 12). When ligand binds, the dimeric Notch1 receptor is cleaved at a position 12 aa residues outside the membrane by an ADAM protease. The released ECD is endocytosed by the ligand-expressing cell, and there is an intramembrane cleavage of Notch by a complex with γ-secretase activity. The released NICD forms a complex with the transcriptional repressor CSL/Rbp-Jκ and activators (master-mind like; MAML) and induces the expression of target genes such as *Hes* genes. (See color insert.)

ER and Golgi compartments. Experiments in *Drosophila*, mice, and cultured cells have revealed functional roles for individual sugars on the ECD of Notch receptors (Haines and Irvine, 2003; Haltiwanger and Lowe, 2004).

Glycosylation of Notch Receptors and Notch Ligands

The extracellular domain of Notch receptors is glycosylated with *N*-glycans (Johansen *et al.*, 1989) and *O*-glycans, including *O*-fucose and *O*-glucose glycans (Moloney *et al.*, 2000b). *O*-glucosylation occurs at Ser or

TABLE I
MOUSE MUTANTS IN THE O-FUCOSE GLYCAN PATHWAY

Gene	Chromosome position	Mutation	References
Pofut1	Chr 2	Floxed exon 2 is removed. No Pofut1 transcripts.	Shi and Stanley, 2003
Lfng	Chr 5	Replacement of exon1 with neo selection cassette.	Evrard *et al.*, 1998
		Replacement of partial exon1 with LacZ fused with neo selection cassette.	Zhang and Gridley, 1998
Rfng	Chr 11	Replacement of entire protein exon 1–8 with PGKneo selection cassette, then removed PGKneo cassette.	Moran *et al.*, 1999
		Deletion of exon 2–7 and part of exon1 replaced by PGKneo cassette.	Zhang *et al.*, 2002
B4galt1	Chr 4	Deletion of exon1, replaced with PGKneo selection cassette	Lu *et al.*, 1997
		Replacement of exon1 with PGKneo selection cassette	Asano *et al.*, 1997

Thr in the consensus sequence C_1XXPS/TC_2 between the first and second Cys residue in Notch EGF repeats. A trisaccharide may be present at O-glucose sites. Although the structure of the trisaccharide is unknown, it is predicted to contain xylose in the structure Xylβ1,3Xylβ1,3Glc (Fig. 1). O-fucosylation occurs at Ser or Thr in the consensus sequence C_2X_{4-5}-S/ TC_3 between the second and third Cys of Notch EGF repeats. Fucose is transferred by the enzyme protein O-fucosyltransferase-1 (Pofut1), and the O-fucose may be extended with N-acetylglucosamine (GlcNAc) transferred by a Fringe β1,3GlcNAcT, and subsequently by galactose and sialic acid in mammals (Moloney *et al.*, 2000b). Notch ligands also contain EGF repeats with the consensus sequence for O-fucose and O-glucose and the O-fucose residues are modified by Fringe (Panin *et al.*, 2002; Shao *et al.*, 2002). Targeted mutations of glycosyltransferases responsible for the synthesis of O-fucose glycans have been made in the mouse (Table I).

Pofut1 Is an Essential Component of the Canonical Notch Signaling Pathway

O-Fucose on Notch receptors was first shown to play a role in Notch signaling in Lec13 Chinese hamster ovary (CHO) cells that make very low amounts of GDP-fucose (Moloney *et al.*, 2000a; Chen *et al.*, 2001). Jagged1-induced Notch signaling in a co-culture reporter assay is markedly reduced in Lec13 cells. These cells transfer little fucose to glycoproteins but can be corrected by exogenous fucose and thus have normal function of fucosyltransferases including Pofut1. The fact that Lec13 cells exhibit little Notch signaling suggests that fucose is necessary for optimal Notch signaling in mammalian cells. In the fly, knockdown of OFUT1 causes Notch to accumulate intracellularly in the endoplasmic reticulum, and an inactive form of OFUT1 facilitates the secretion of Notch ECD fragments leading to the conclusion that *Drosophila* OFUT1 functions as a Notch chaperone (Okajima *et al.*, 2005). By contrast, Notch receptors are equivalently expressed on the surface of mammalian cells lacking Pofut1 (Stahl, M. Uemura, K. Shi, S. Ge, C., and Stanley, P., manuscript in preparation).

Targeted mutation of the mouse *Pofut1* gene leads to embryonic lethality at ~E9.5 with severe defects in somitogenesis, vasculogenesis, cardiogenesis, and neurogenesis, a phenotype typical of mutants lacking downstream effectors that are required for signaling through all four Notch receptors (Shi and Stanley, 2003). The *Pofut1* gene has also been specifically ablated in mouse oocytes (Shi *et al.*, 2005). However, pre-implantation embryos lacking maternal and zygotic *Pofut1* gene transcripts develop through blastogenesis, implant, and progress through the formation of the three germ layers (Shi *et al.*, 2005). This is surprising, because lower organisms use Notch signaling in the determination of cell fates during pregastrulation development, suggesting co-option of Notch signaling during evolution for use in early cell fate decisions and embryogenesis (Shi and Stanley, 2006).

Mice unable to synthesize GDP-fucose would also be expected to have a phenotype that reflects inhibition of Notch signaling. However, embryonic lethality in homozygous null mutants is observed at variable stages *in utero* with the earliest being E12.5 (Smith *et al.*, 2002). Homozygous mutant pups are also born, but survival depends on the provision of fucose in the diet. The variability in the phenotype of mice with a null mutation in the FX gene is strain dependent (Becker *et al.*, 2003). Thus, there is one or more modifier locus that corrects the GDP-fucose deficiency. In strains with highly penetrant embryonic lethality, it seems that Notch signaling is at least partially rescued by fucose obtained from maternal or fraternal sources (Becker *et al.*, 2003; Smith *et al.*, 2002).

Fringe Is a Modulator of Notch Signaling

Fringe is a β1,3N-acetylglucosaminyltransferase (β1,3GlcNAcT) that transfers GlcNAc to *O*-fucose on EGF repeats of Notch receptors (Haines and Irvine, 2003). In mammals, there are three Fringe homologs,

Lunatic, Manic, and Radical Fringe (Lfng, Mfng, and Rfng). They function in the Golgi, but all of them are also secreted.

Lfng is expressed in many cell types during embryonic stages, but its dynamic expression in mouse somites indicated a functional role during segmentation (Johnston *et al.*, 1997). Lunatic fringe is the only fringe gene expressed during somitogenesis (Johnston *et al.*, 1997). Targeted mutations that inactivate the mouse *Lfng* gene cause somites to have irregular sizes and shapes and their anterior-posterior patterning to be disrupted. A severely disorganized axial skeleton is the result (Evrard *et al.*, 1998; Zhang and Gridley, 1998). Mutants die perinatally or in early adulthood. The expression patterns of somitogenic genes are altered, such as Myogenin, *Mox1*, *Pax1*, *Pax3*, *Pax9*, and *Uncx4.1*. Furthermore, the expression of Notch signaling pathway genes, such as *Dll1*, *Dll3*, *Notch1,* and *Notch4*, is reduced. Some *Lfng* null mutants may survive to adulthood, and in at least one mutant strain *Lfng* null females are infertile (Hahn *et al.*, 2005). These females exhibit aberrant folliculogenesis with altered expression of Notch pathway genes, and their eggs are arrested in meiosis II and thus are not developmentally competent (Hahn *et al.*, 2005). The two strains of $Lfng^{-/-}$ mutant mice differ in the degree of their somitic defects and in the penetrance of their defect in female infertility (Hahn *et al.*, 2006; Xu *et al.*, 2006).

In mice lacking both the Notch ligand Jagged2 and Lfng, the generation of supernumerary hair cells in the inner hair cell row that occurs when Jagged2 is not present is suppressed by the concomitant loss of Lfng, whereas supernumerary hair cells in the outer hair cell rows are unaffected (Zhang *et al.*, 2000). *Lfng* modulation of Notch signaling is also thought to be important in tooth development. Although $Lfng^{-/-}$ mice do not show abnormal tooth development, Lfng is expressed in the epithelium surrounding the enamel knot signaling center that controls tooth size and shape (Mustonen *et al.*, 2002). A human patient with spondylocostal dysostoses (SCDs) with vertebral malsegmentation has been found to carry an autosomal recessive mutation (F188L) in the LFNG gene (Sparrow *et al.*, 2006). This mutation of a conserved amino acid in LFNG causes mouse Lfng F187L to be mislocalized in transfected cells, to have no enzyme activity, and to be unable to modulate Notch signaling in a co-culture assay (Sparrow *et al.*, 2006).

Interestingly, the targeted mutation of the mouse *Rfng* gene did not result in an obvious abnormal phenotype (Moran *et al.*, 1999; Zhang *et al.*, 2002). Double null mice null for both *Lfng* and *Rfng* resemble *Lfng* mutant mice. The double mutants exhibited no obvious synergistic or additive effects to the $Lfng^{-/-}$ somitogenic phenotype, showing there is no functional redundancy between mouse Lfng and Rfng during somitogenesis (Zhang *et al.*, 2002).

Lfng is expressed in a dynamic, repetitive, and complex wave pattern within the mouse presomitic mesoderm (PSM; [Forsberg *et al.*, 1998]). A wave takes 2 h. The mechanism of Lfng cyclic gene expression in the PSM has been investigated using LacZ reporter transgenes fused with *Lfng* promoter regions (Cole *et al.*, 2002; Morales *et al.*, 2002). A conserved *Lfng* promoter fragment is required for *Lfng* cyclic gene expression in the PSM. The oscillatory expression of the *Lfng* gene is controlled by a negative feedback that seems to be regulated by the *Hes7* gene (Bessho *et al.*, 2003; Chen *et al.*, 2005). Cyclic Lfng expression is critical for somitogenesis to proceed normally. Thus, an *Lfng* transgene under the control of a portion of the *Dll1* promoter termed msd exhibits constitutive expression in the PSM (Cordes *et al.*, 2004; Serth *et al.*, 2003). Mice carrying this transgene in a background with or without endogenous Lfng have a similar phenotype to *Lfng* null mice exhibiting severe defects in skeleton formation. Although the noncyclic exogenous expression of Lfng did not abolish cyclic expression of endogenous Lfng in the PSM, the fact that constitutive expression in the anterior PSM causes defective somitogenesis shows that transcriptional oscillation of Lfng is essential for somitogenesis (Cordes *et al.*, 2004; Serth *et al.*, 2003). Overexpression of Lfng by the SPC-Lfng transgene in distal epithelial cells of the developing mouse lung did not affect spatial or temporal expression of the *Hes1* Notch target gene or other differentiation markers (van Tuyl *et al.*, 2005). These mice have no detectable lung defects, suggesting that Lfng does not play a significant role in determining cell fate in fetal airway epithelium (van Tuyl *et al.*, 2005). Misexpression of Lfng in the thymus under the lck promoter inhibits T-cell development that is dependent on Notch signaling and results in the generation of B cells in the thymus (Koch *et al.*, 2001). The increased Lfng expression in T cells enhances their binding to stromal epithelial cells expressing Delta ligands and thus inhibits their ability to mature as T cells (Tan *et al.*, 2005; Visan *et al.*, 2006).

The mechanisms by which Fng genes modulate canonical Notch signaling have been investigated in model organisms including *Drosophila, Xenopus,* and Zebrafish (Haines and Irvine, 2003). In mammals, mechanisms of Fng actions have been investigated in co-culture Notch signaling assays using reporter constructs that respond to NICD. Overexpression of mammalian Fng proteins has been shown to differentially modulate ligand-induced Notch signaling. Lfng and Mfng inhibit Jagged1-induced Notch1 signaling and potentiate Delta1-induced Notch1 signaling in co-culture systems (Hicks *et al.*, 2000). However, Rfng was found to enhance Notch1 signaling induced by either Delta1 or Jagged1 (Yang *et al.*, 2005). Interestingly, it seems Fng proteins may have different effects on different Notch receptors. For example, Lfng potentiates both

Jagged1- and Delta1-induced signaling by Notch2 (Hicks *et al.*, 2000). Binding assays with soluble ligands and Notch fragemnts or to cells expressing transfected Notch receptors generally correlate with predictions from signaling assays; that is, Fng effects that inhibit Notch signaling correlate with decreased ligand binding, whereas Fng effects that potentiate Notch signaling correlate with increased ligand binding (Hicks *et al.*, 2000; Shimizu *et al.*, 2001; Yang *et al.*, 2005). The three Fng β1,3GlcNAcTs have different *in vitro* catalytic efficiencies and may also have different specificities for particular EGF repeats of Notch1 (Rampal *et al.*, 2005a,b; Shao *et al.*, 2003). Thus *in vivo* effects of Fng actions in the modulation of Notch signaling by all four Notch receptors induced by the five Notch ligands may be extremely complex.

β4GalT-1 Is a Novel Regulator of Notch Signaling

The elongation of the GlcNAcβ3Fuc disaccharide on Notch EGF repeats by β4GalT-1 was found to be required for Lfng and Mfng inhibition of Jagged1-induced Notch signaling in a co-culture assay (Chen *et al.*, 2001). Neither the Lec20 CHO mutant that lacks β4GalT-1 and β4GalT-6 nor the Lec8 CHO mutant that cannot transport UDP-Gal into the Golgi exhibits Lfng or Mfng modulation of Notch signaling. When corrected with a cDNA encoding β4GalT-1, Lec20 cells are rescued for Fng effects. To investigate *in vivo* mouse embryos lacking β4GalT-1 were examined (Chen *et al.*, 2006). Although mutant embryos do not have obvious skeletal defects as would be predicted if Fng modulation of Notch signaling were inhibited, many have severely reduced expression of certain Notch target genes such as *Hes5* and *Mesp2* and altered expression of the Notch ligand genes *Dll1* and *Dll3*. Furthermore, the number of lumbar vertebrae in 12 of 13 perinatal mutant embryos differed significantly from control litter mates (Chen *et al.*, 2006). This finding is consistent with known effects of Notch signaling on *Hox* gene functions during mouse skeletal development (Cordes *et al.*, 2004). The subtlety of the defect in mice lacking β4GalT-1 may reflect the fact that there are six β4GalT genes, several of which are expressed during embryogenesis.

Mutations in Other Glycosyltransferase Genes

There are now many strains of mice that have a null mutation in a glycosyltransferases gene that affects the synthesis of *N*-glycans or mucin *O*-glycans or *O*-Mannose glycans (Lowe and Marth, 2003). Although Notch receptors and their ligands carry *N*-glycans, none of the mutant mice defective in *N*-glycan synthesis with the exception of the mice lacking

β4GalT-1 described previously have been found to exhibit an overt phenotype consistent with defective Notch signaling. However, they may well have a subtle defect in Notch signaling that affects functions that would require specific analyses or biological challenges to uncover. Mucin *O*-glycans or *O*-mannose glycans may also be present on Notch receptors and their ligands, but no mutants in the glycosyltransferases that synthesize these *O*-glycans have given a Notch signaling phenotype. Similarly, mice with defective proteoglycan synthesis have not been described to have Notch signaling defects.

Acknowledgments

This work was supported by NCI grant RO1 95022 to PS.

References

Asano, M., Furukawa, K., Kido, M., Matsumoto, S., Umesaki, Y., Kochibe, N., and Iwakura, Y. (1997). Growth retardation and early death of beta-1,4-galactosyltransferase knockout mice with augmented proliferation and abnormal differentiation of epithelial cells. *EMBO J.* **16**, 1850–1857.

Becker, D. J., Myers, J. T., Ruff, M. M., Smith, P. L., Gillespie, B. W., Ginsburg, D. W., and Lowe, J. B. (2003). Strain-specific modification of lethality in fucose-deficient mice. *Mamm. Genome.* **14**, 130–139.

Bessho, Y., Hirata, H., Masamizu, Y., and Kageyama, R. (2003). Periodic repression by the bHLH factor Hes7 is an essential mechanism for the somite segmentation clock. *Genes Dev.* **17**, 1451–1456.

Blaumueller, C. M., Qi, H., Zagouras, P., and Artavanis-Tsakonas, S. (1997). Intracellular cleavage of Notch leads to a heterodimeric receptor on the plasma membrane. *Cell* **90**, 281–291.

Chen, J., Kang, L., and Zhang, N. (2005). Negative feedback loop formed by Lunatic fringe and Hes7 controls their oscillatory expression during somitogenesis. *Genesis* **43**, 196–204.

Chen, J., Lu, L., Shi, S., and Stanley, P. (2006). Expression of Notch signaling pathway genes in mouse embryos lacking beta4galactosyltransferase-1. *Gene Expr. Patterns* **6**, 376–382.

Chen, J., Moloney, D. J., and Stanley, P. (2001). Fringe modulation of Jagged1-induced Notch signaling requires the action of beta4galactosyltransferase-1. *Proc. Natl. Acad. Sci. USA* **98**, 13716–13721.

Cole, S. E., Levorse, J. M., Tilghman, S. M., and Vogt, T. F. (2002). Clock regulatory elements control cyclic expression of Lunatic fringe during somitogenesis. *Dev. Cell.* **3**, 75–84.

Cordes, R., Schuster-Gossler, K., Serth, K., and Gossler, A. (2004). Specification of vertebral identity is coupled to Notch signalling and the segmentation clock. *Development* **131**, 1221–1233.

Evrard, Y. A., Lun, Y., Aulehla, A., Gan, L., and Johnson, R. L. (1998). lunatic fringe is an essential mediator of somite segmentation and patterning. *Nature* **394**, 377–381.

Forsberg, H., Crozet, F., and Brown, N. A. (1998). Waves of mouse Lunatic fringe expression, in four-hour cycles at two-hour intervals, precede somite boundary formation. *Curr. Biol.* **8**, 1027–1030.

Hahn, K. L., Johnson, J., Beres, B. J., Howard, S., and Wilson-Rawls, J. (2005). Lunatic fringe null female mice are infertile due to defects in meiotic maturation. *Development* **132**, 817–828.

Hahn, K. L., Johnson, J., Beres, B. J., and Wilson-Rawls, J. (2006). A loss of lunatic fringe is associated with female infertility. *Development* **133**, 579–580.

Haines, N., and Irvine, K. D. (2003). Glycosylation regulates Notch signalling. *Nat. Rev. Mol. Cell. Biol.* **4**, 786–797.

Haltiwanger, R. S., and Lowe, J. B. (2004). Role of glycosylation in development. *Annu. Rev. Biochem.* **73**, 491–537.

Hicks, C., Johnston, S. H., diSibio, G., Collazo, A., Vogt, T. F., and Weinmaster, G. (2000). Fringe differentially modulates Jagged1 and Delta1 signalling through Notch1 and Notch2. *Nat. Cell. Biol.* **2**, 515–520.

Johansen, K. M., Fehon, R. G., and Artavanis-Tsakonas, S. (1989). The notch gene product is a glycoprotein expressed on the cell surface of both epidermal and neuronal precursor cells during *Drosophila* development. *J. Cell. Biol.* **109**, 2427–2440.

Johnston, S. H., Rauskolb, C., Wilson, R., Prabhakaran, B., Irvine, K. D., and Vogt, T. F. (1997). A family of mammalian Fringe genes implicated in boundary determination and the Notch pathway. *Development* **124**, 2245–2254.

Koch, U., Lacombe, T. A., Holland, D., Bowman, J. L., Cohen, B. L., Egan, S. E., and Guidos, C. J. (2001). Subversion of the T/B lineage decision in the thymus by lunatic fringe-mediated inhibition of Notch-1. *Immunity* **15**, 225–236.

Logeat, F., Bessia, C., Brou, C., LeBail, O., Jarriault, S., Seidah, N. G., and Israel, A. (1998). The Notch1 receptor is cleaved constitutively by a furin-like convertase. *Proc. Natl. Acad. Sci. USA* **95**, 8108–8112.

Lowe, J. B., and Marth, J. D. (2003). A genetic approach to Mammalian glycan function. *Annu. Rev. Biochem.* **72**, 643–691.

Lu, Q., Hasty, P., and Shur, B. D. (1997). Targeted mutation in beta1,4-galactosyltransferase leads to pituitary insufficiency and neonatal lethality. *Dev. Biol.* **181**, 257–267.

Moloney, D. J., Panin, V. M., Johnston, S. H., Chen, J., Shao, L., Wilson, R., Wang, Y., Stanley, P., Irvine, K. D., Haltiwanger, R. S., and Vogt, T. F. (2000a). Fringe is a glycosyltransferase that modifies Notch. *Nature* **406**, 369–375.

Moloney, D. J., Shair, L. H., Lu, F. M., Xia, J., Locke, R., Matta, K. L., and Haltiwanger, R. S. (2000b). Mammalian Notch1 is modified with two unusual forms of O-linked glycosylation found on epidermal growth factor-like modules. *J. Biol. Chem.* **275**, 9604–9611.

Morales, A. V., Yasuda, Y., and Ish-Horowicz, D. (2002). Periodic lunatic fringe expression is controlled during segmentation by a cyclic transcriptional enhancer responsive to notch signaling. *Dev. Cell* **3**, 63–74.

Moran, J. L., Levorse, J. M., and Vogt, T. F. (1999). Limbs move beyond the radical fringe. *Nature* **399**, 742–743.

Mustonen, T., Tummers, M., Mikami, T., Itoh, N., Zhang, N., Gridley, T., and Thesleff, I. (2002). Lunatic fringe, FGF, and BMP regulate the Notch pathway during epithelial morphogenesis of teeth. *Dev. Biol.* **248**, 281–293.

Okajima, T., Xu, A., Lei, L., and Irvine, K. D. (2005). Chaperone activity of protein O-Fucosyltransferase 1 promotes notch receptor folding. *Science* **307**, 1599–1603.

Panin, V. M., Shao, L., Lei, L., Moloney, D. J., Irvine, K. D., and Haltiwanger, R. S. (2002). Notch ligands are substrates for protein O-fucosyltransferase-1 and Fringe. *J. Biol. Chem.* **277**, 29945–29952.

Rampal, R. A. S. Y. L., Moloney, D. J., Georgiou, S. A., Luther, K. B., Nita-Lazar, A., and Haltiwanger, R. S. (2005a). Lunatic fringe, manic fringe, and radical fringe recognize similar specificity determinants in O-fucosylated epidermal growth factor-like repeats. *J. Biol. Chem.* **280**, 42454–42463.

Rampal, R., Arboleda-Velasquez, J. F., Nita-Lazar, A., Kosik, K. S., and Haltiwanger, R. S. (2005b). Highly conserved O-fucose sites have distinct effects on Notch1 function. *J. Biol. Chem.* **280**, 32133–32140.

Serth, K., Schuster-Gossler, K., Cordes, R., and Gossler, A. (2003). Transcriptional oscillation of lunatic fringe is essential for somitogenesis. *Genes Dev.* **17,** 912–925.

Shao, L., Luo, Y., Moloney, D. J., and Haltiwanger, R. (2002). O-glycosylation of EGF repeats: Identification and initial characterization of a UDP-glucose: Protein O-glucosyltransferase. *Glycobiology* **12,** 763–770.

Shao, L., Moloney, D. J., and Haltiwanger, R. (2003). Fringe modifies O-fucose on mouse Notch1 at epidermal growth factor-like repeats within the ligand-binding site and the Abruptex region. *J. Biol. Chem.* **278,** 7775–7782.

Shi, S., and Stanley, P. (2003). Protein O-fucosyltransferase 1 is an essential component of Notch signaling pathways. *Proc. Natl. Acad. Sci. USA* **100,** 5234–5239.

Shi, S., and Stanley, P. (2006). Evolutionary Origins of Notch Signaling in Early Development. *Cell Cycle* **5,** 274–278.

Shi, S., Stahl, M., Lu, L., and Stanley, P. (2005). Canonical Notch signaling is dispensable for early cell fate specifications in mammals. *Mol. Cell. Biol.* **25,** 9503–9508.

Shimizu, K., Chiba, S., Saito, T., Kumano, K., Takahashi, T., and Hirai, H. (2001). Manic fringe and lunatic fringe modify different sites of the Notch2 extracellular region, resulting in different signaling modulation. *J. Biol. Chem.* **276,** 25753–25758.

Smith, P. L., Myers, J. T., Rogers, C. E., Zhou, L., Petryniak, B., Becker, D. J., Homeister, J. W., and Lowe, J. B. (2002). Conditional control of selectin ligand expression and global fucosylation events in mice with a targeted mutation at the FX locus. *J. Cell. Biol.* **158,** 801–815.

Sparrow, D. B., Chapman, G., Wouters, M. A., Whittock, N. V., Ellard, S., Fatkin, D., Turnpenny, P. D., Kusumi, K., Sillence, D., and Dunwoodie, S. L. (2006). Mutation of the Lunatic Fringe gene in humans causes spondylocostal dysostosis with a severe vertebral phenotype. *Am. J. Hum. Genet.* **78,** 28–37.

Tan, J. B., Visan, I., Yuan, J. S., and Guidos, C. J. (2005). Requirement for Notch1 signals at sequential early stages of intrathymic T cell development. *Nat. Immunol.* **6,** 671–679.

van Tuyl, M., Groenman, F., Kuliszewski, M., Ridsdale, R., Wang, J., Tibboel, D., and Post, M. (2005). Overexpression of lunatic fringe does not affect epithelial cell differentiation in the developing mouse lung. *Am. J. Physiol. Lung Cell. Mol. Physiol.* **288,** L672–L682.

Visan, I., Yuan, J. S., Tan, J. B., Cretegny, K., and Guidos, C. J. (2006). Regulation of intrathymic T-cell development by Lunatic Fringe–Notch1 interactions. *Immunol. Rev.* **209,** 76–94.

Xu, J., Norton, C. R., and Gridley, T. (2006). Not all lunatic fringe null female mice are infertile. *Development* **133,** 579; author reply 579–580.

Yang, L. T., Nichols, J. T., Yao, C., Manilay, J. O., Robey, E. A., and Weinmaster, G. (2005). Fringe glycosyltransferases differentially modulate Notch1 proteolysis induced by Delta1 and Jagged1. *Mol. Biol. Cell* **16,** 927–942.

Zhang, N., and Gridley, T. (1998). Defects in somite formation in lunatic fringe-deficient mice. *Nature* **394,** 374–377.

Zhang, N., Norton, C. R., and Gridley, T. (2002). Segmentation defects of Notch pathway mutants and absence of a synergistic phenotype in lunatic fringe/radical fringe double mutant mice. *Genesis* **33,** 21–28.

Zhang, N., Martin, G. V., Kelley, M. W., and Gridley, T. (2000). A mutation in the Lunatic fringe gene suppresses the effects of a Jagged2 mutation on inner hair cell development in the cochlea. *Curr. Biol.* **10,** 659–662.

[11]   Defect in Glycosylation that Causes
Muscular Dystrophy

By TAMAO ENDO and HIROSHI MANYA

Abstract

Muscular dystrophies are a diverse group of inherited disorders characterized by progressive muscle weakness and wasting. The dystrophin–glycoprotein complex is composed of α-, β-dystroglycan (DG), dystrophin and some other molecules. α- and β-DG stabilize the sarcolemma by acting as an axis through which the extracellular matrix is tightly linked to the cytoskeleton. The relative molecular weights of α-DG differ in different tissues as a result of differential glycosylation. New findings indicate that disrupted glycosylation of α-DG results in a loss of ligand binding, giving rise to both progressive muscle degeneration and abnormal neuronal migration in the brain. This article discusses methods, including purification of α-DG and glycosyltransferase assays involved in α-DG glycosylation.

Overview

Muscular dystrophies are genetic diseases that cause progressive muscle weakness and wasting (Emery, 2002). The causative genes of several muscular dystrophies have been identified in the past 15 years. Recent data suggest that the aberrant protein glycosylation of a specific glycoprotein, α-DG, is the primary cause of some forms of congenital muscular dystrophy (Endo and Toda, 2003; Michele and Campbell, 2003).

Dystroglycan is encoded by a single gene and is cleaved into two proteins, α-DG and β-DG, by posttranslational processing. α-DG is an extracellular peripheral membrane glycoprotein anchored to the cell membrane by binding to a transmembrane glycoprotein, β-DG. The α-DG–β-DG complex is expressed in a broad array of tissues and is thought to stabilize the plasma membrane by acting as an axis through which the extracellular matrix is tightly linked to cytoskeleton. α-DG is heavily glycosylated, and its sugars have a role in binding to laminin, neurexin, and agrin (Michele and Campbell, 2003; Montanaro and Carbonetto, 2003). The relative molecular weights of α-DG differ in different tissues as a result of differential glycosylation. We previously demonstrated that a sialyl O-mannosyl glycan, Siaα2-3Galβ1-4GlcNAcβ1-2Man, is a laminin-binding ligand of α-DG (Chiba *et al.*, 1997). Mammalian O-mannosylation

METHODS IN ENZYMOLOGY, VOL. 417   0076-6879/06 $35.00
          DOI: 10.1016/S0076-6879(06)17011-1

is a rare type of protein modification that is observed in a limited number of glycoproteins of brain, nerve, and skeletal muscle (Endo, 2004).

We have identified and characterized glycosyltransferases, protein O-mannose β1,2-N-acetylglucosaminyltransferase (POMGnT1) (Yoshida et al., 2001), and protein O-mannosyltransferase 1 (POMT1) and POMT2 (Manya et al., 2004), involved in the biosynthesis of O-mannosyl glycans (Fig. 1). We subsequently found that loss of function of the *POMGnT1* gene is responsible for muscle-eye-brain disease (MEB) (Yoshida et al., 2001). It has also been reported that the *POMT1* gene and the *POMT2* gene are responsible for Walker–Warburg syndrome (WWS) (Beltran-Valero de Bernabe et al., 2002; van Reeuwijk et al., 2005). MEB and WWS are autosomal recessive disorders characterized by congenital muscular dystrophies with abnormal neuronal migration. Like MEB and WWS, some muscular dystrophies have been suggested to be caused by abnormal glycosylation of α-DG (e.g., Fukuyama-type congenital muscular dystrophy [FCMD], congenital muscular dystrophy type 1C [MDC1C], limb-girdle muscular dystrophy2I [LGMD2I], congenital muscular dystrophy type 1D [MDC1D], and the myodystrophy [LARGE^{myd}] mouse) (Table I). Highly glycosylated α-DG was found to be selectively deficient in the skeletal muscle of these patients, and the gene products were suggested to be glycosyltransferases (Endo and Toda, 2003; Michele and Campbell, 2003). Therefore, the ability to assay enzyme activities of O-mannosylation would facilitate progress in the identification of other O-mannosylated proteins, the explanation of their functional roles, and the understanding of muscular dystrophies. This chapter describes the biochemical protocols used to analyze α-DG glycosylation.

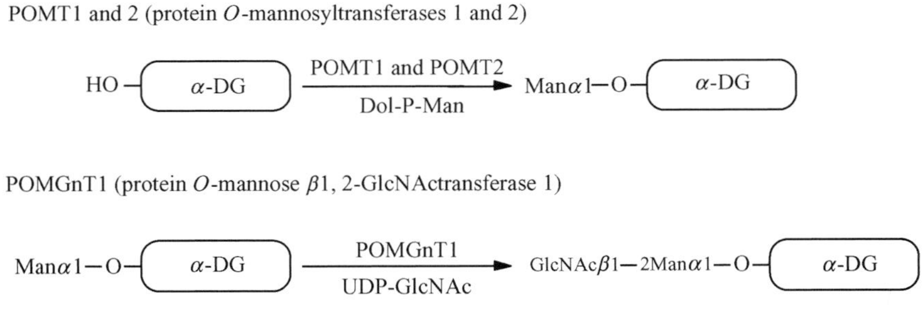

POMT1 and 2 (protein O-mannosyltransferases 1 and 2)

POMGnT1 (protein O-mannose β1, 2-GlcNActransferase 1)

FIG. 1. Biosynthetic pathway of O-mannosylglycan of α-DG. Two homologs, POMT1 and POMT2, are responsible for protein O-mannosylation, and POMGnT1 is responsible for GlcNAcβ1-2Man linkage of O-mannosylglycan.

TABLE I
MUSCULAR DYSTROPHIES POSSIBLY CAUSED BY ABNORMAL GLYCOSYLATION

Condition	Gene	Protein function	Gene locus
Muscle-eye-brain disease (MEB)	*POMGnT1*	GlcNActransferase	1p33
Fukuyama congenital muscular dystrophy (FCMD)	*Fukutin*	Putative glycosyltransferase	9q31
Walker-Warburg syndrome (WWS)	*POMT1* *POMT2*	*O*-Mannosyltransferase	9q34.1 14q24.3
MDC1C Limb-girdle muscular dystrophy 2I (LGMD2I)	*FKRP* (Fukutin-related protein)	Putative glycosyltransferase	19q13.3
MDC1D	*LARGE*	Putative glycosyltransferase	22q12.3
Myodystrophy (LARGE^{myd}) mouse	*large*		8 (mouse)

Purification of α-DG

Reagents

> Protease inhibitor cocktail (Complete, EDTA-free, Roche Diagnostics, Basel, Switzerland)
> Silver Stain "DAIICHI" reagent kit (Daiichi Pure Chemicals, Tokyo, Japan)

Solutions

> Buffer A (50 mM Tris-HCl, pH 7.4, 0.75 mM benzamidine, and 0.1 mM PMSF)
> Buffer B (buffer A containing 300 mM GlcNAc)
> Buffer C (50 mM Tris-HCl, pH 7.4, 140 mM NaCl, 1 mM CaCl$_2$, 0.75 mM benzamidine, 0.1 mM PMSF, and 0.1% Triton X-100)
> Buffer D (50 mM Tris-HCl, pH 7.4, 140 mM NaCl, 2 mM EGTA, 0.75 mM benzamidine, 0.1 mM PMSF, and 0.1% Triton X-100)

Materials

> Wheat germ agglutinin (WGA)-Sepharose (GE Healthcare Bio-Sciences Corp., Piscataway, NJ)

Laminin-Sepharose (Mouse Engelbreth-Holm-Swarm [EHS] laminin [Biomedical Technologies Inc., Stoughton, MA]) was coupled to CNBr-activated Sepharose 4B (GE Healthcare)]
Centricon 30 and Microcon-100 (Millipore Corp., Billerica, MA)

Method for Purification of α-DG

1. Crude rabbit skeletal muscle membranes were suspended at a protein concentration of 5 mg/ml in 50 mM Tris-HCl, pH 7.4, containing 0.5 M NaCl and a protease inhibitor cocktail.
2. The suspension was titrated to pH 12 by slowly adding 1 N NaOH, extracted for 1 h, and then centrifuged at 140,000g for 30 min at 25°.
3. The supernatant was titrated to pH 7.4 and centrifuged at 140,000g for 30 min at 4°.
4. The supernatant was circulated over the WGA-Sepharose overnight at 4°. After extensive washing with buffer A, the proteins bound to the WGA-Sepharose were eluted with the buffer B.
5. The eluate was added to 10 mM CaCl$_2$ and 10% Triton X-100 (final concentrations of 1 mM and 0.1%, respectively), then circulated over laminin-Sepharose overnight at 4° in the presence of 1 mM CaCl$_2$. After extensive washing with the buffer C, the proteins bound to the laminin–Sepharose were eluted with the buffer D. The eluate was collected and concentrated using Centricon 30 down to 200 μl.
6. This concentrated sample was further separated by 5–30% sucrose gradient centrifugation. This sample was concentrated and desalted by using Microcon-100.

The purity of the sample was determined by SDS-PAGE with 7.5% gel followed by silver staining using the Silver Stain "DAIICHI" reagent kit. To increase the sensitivity of glycoprotein detection, the periodic acid-silver stain method (Dubray and Bezard, 1982) was used with a modification. The gel was treated with 1% periodic acid for 10 min at room temperature before fixation instead of with 0.2% periodic acid treatment for 1 h at 4° after fixation in the original method. By this staining method, rabbit skeletal muscle α-DG was detected as a single prominent broad band around 150 kDa. Bovine peripheral nerve α-DG was also purified by the similar procedures by the sequential WGA-Sepharose and laminin-Sepharose columns without sucrose gradient centrifugation. Thus, obtained bovine peripheral nerve α-DG migrated at approximately 116 kDa.

Method for Glycoprotein Enrichment and Western Blot Analysis
(Laminin Overlay Assay)

Functional glycosylation of α-DG (purified α-DG or enrichment of α-DG from tissues) was examined by laminin binding and by dystroglycan antibodies.

Reagents

Anti-α-DG antibodies from mouse (VIA4-1 and IIH6, Upstate Biotechnology, Lake Placid, NY)
Anti-laminin antibody from rabbit (Sigma-Aldrich Corp., St. Louis, MO)
Horseradish peroxidase (HRP)–conjugated anti-mouse or rabbit Ig secondary antibodies (anti-mouse Ig-HRP, anti-rabbit Ig-HRP, GE Healthcare)
Enhanced chemiluminescent reagent kit (ECL Plus, GE Healthcare)

Solutions

Laminin-binding buffer (LBB: 10 mM triethanolamine, 140 mM NaCl, 1 mM MgCl$_2$, 1 mM CaCl$_2$, pH 7.6)

Materials

Polystyrene ELISA microplates (Corning Inc., Corning, NY)

1. For enrichment of α-DG from skeletal muscle, skeletal muscle was disrupted with a polytron followed by Dounce homogenization and incubation in 50 mM Tris-HCl, pH 7.4, 500 mM NaCl, 1% Triton X-100 and protease inhibitors.
2. The solubilized fraction was incubated with WGA-Sepharose beads for 16 h. Pellets formed from the beads, and these were washed three times in TBS containing 0.1% Triton X-100 and protease inhibitors.
3. The beads were then either directly boiled for 2 min in SDS-PAGE loading buffer (Western blotting and laminin overlay assay) or eluted with TBS containing 0.1% Triton X-100, protease inhibitors and 300 mM GlcNAc (solid-phase binding assay).
4. Proteins were separated by 3–15% SDS-PAGE and transferred to polyvinylidene fluoride (PVDF) membranes and probed with dystroglycan antibodies (VIA4-1 and IIH6), and then incubated with anti-mouse Ig-HRP.

Method for Laminin Overlay Assay

Laminin overlay assay was performed on PVDF membranes using EHS laminin. PVDF membranes were blocked in LBB containing 5% nonfat dry milk followed by incubation with anti-laminin antibody followed by anti-rabbit Ig-HRP. Blots were developed by enhanced chemiluminescence.

Solid-Phase Binding Assay

WGA eluate was diluted 1:50 in TBS and coated on polystyrene ELISA microplates for 16 h at 4°. Plates were washed in LBB and blocked for 2 h in 3% BSA in LBB. EHS laminin was diluted in LBB and applied for 2 h. Wells were washed with 3% BSA in LBB, incubated for 30 min with 1:10,000 anti-laminin antibody followed by anti-rabbit HRP. Plates were developed with o-phenylenediamine dihydrochloride and H_2O_2, reactions were stopped with 2 N H_2SO_4, and values were obtained on a microplate reader at 492 nm.

Assay for POMT and POMGnT1 Activities

Preparation of Enzyme Sources from HEK293T Cells

Reagents

pcDNA3.1-POMGnT1, pcDNA3.1-POMT1 and *pcDNA3.1-POMT2* expression plasmids: Human cDNAs encoding *POMGnT1* (Yoshida *et al.*, 2001), *POMT1* (Jurado *et al.*, 1999) (the most common splicing variant of human *POMT1,* which lacks bases 700–765, corresponding to amino acids 234–255) and *POMT2* (Willer *et al.*, 2002) are inserted into mammalian expression vectors, pcDNA3.1/Zeo or pcDNA3.1/ Hygro (Invitrogen Corp., Carlsbad, CA).
Lipofectamine transfection reagent and Plus reagent (Invitrogen).
Antibodies: Rabbit antisera specific to the human POMT1, POMT2, and POMGnT1 are produced by using synthetic peptides corresponding to residues 348–362 (YPMIYENGRGSSH) of POMT1, 390–403 (HNTN SDPLDPSFPV) of POMT2 and 649–660 (KEEGAPGAPEQT) of POMGnT1, respectively.
Anti-rabbit IgG conjugated with HRP (GE Healthcare).
Enhanced chemiluminescent reagent kit (ECL, GE Healthcare)
Amplify fluorographic reagent (GE Healthcare)
Kodak BioMax MS X-ray film (GE Healthcare)

Solutions

Phosphate-buffered saline (PBS): 137 mM NaCl, 2.7 mM KCl, 10 mM Na$_2$HPO$_4$, 1.8 mM KH$_2$PO$_4$, pH 7.4, and store at 4°.

Homogenization buffer: 10 mM Tris-HCl, pH 7.4, 1 mM EDTA, 250 mM sucrose (SET buffer) with protease inhibitor cocktail (Roche Diagnostics). SET buffer is stored at 4°. Addition of protease inhibitor cocktail is before use.

Materials

Silicone blade cell scraper (SUMILON, SUMITOMO BAKELITE Co., Tokyo, Japan)

Method for Preparation of Membrane Fraction from HEK293T Cells

HEK293T cells are maintained in DMEM supplemented with 10% FBS, 2 mM L-glutamine, and 100 units/ml penicillin–50 μg/ml streptomycin at 37° with 5% CO$_2$. The expression plasmids of human *pcDNA3.1-POMT1* and *pcDNA3.1-POMT2* are transfected into HEK293T cells using Lipofectamine PLUS reagent (Fig. 2A).

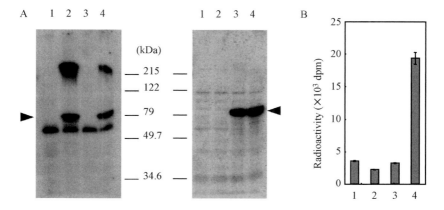

FIG. 2. Western blot analysis of POMT1 and POMT2 (A) and POMT activity of POMT1 and POMT2 (B) expressed in HEK293T cells. Lanes 1, cells transfected with vector alone; lanes 2, cells transfected with human *POMT1*; lanes 3, cells transfected with human *POMT2*; lanes 4, cells co-transfected with *POMT1* and *POMT2*. The proteins (20 μg of membrane fraction) were subjected to SDS-PAGE (10% gel), and the separated proteins were transferred to a PVDF membrane. The PVDF membrane was stained with anti-POMT1 (left panel of A) or anti-POMT2 (right panel of A). Arrowheads indicate the positions of the corresponding molecules. Molecular weight standards are shown in the middle. Reprinted with permission from (Manya *et al.*, 2004). Copyright 2004 National Academy of Sciences, USA.

1. For transient expression of POMT1 and POMT2, subconfluent HEK293T cells are plated on 100-mm culture dish with antibiotic-free 10% FBS-DMEM 1 day before transfection. Cells are transfected at 60–70% confluence. Avoid antibiotics during transfection.

2. Dilute 4 μg of DNA with 750 μl of serum-free DMEM, add the 20 μl of Plus reagent and let stand at room temperature for 15 min (reagent A). In another tube, dilute 30 μl of Lipofectamine reagent with 750 μl of serum-free DMEM (reagent B).

3. Mix reagent A with reagent B and let stand at room temperature for 15 min (reagent C).

4. During step 3, replace medium on the cells with 5 ml of serum-free DMEM.

5. Add reagent C to the cells (from step 4) and incubate at 37° with 5% CO_2 for 3 h.

6. Add 5 ml of 20% FBS-DMEM to the cells (from step 5) and culture for 2–3 days.

7. The culture supernatants are aspirated and the cells are rinsed gently with cold-PBS. Then, 5 ml of cold-PBS is added, and the cells are harvested with a cell scraper and then are collected by centrifugation at 1000g for 10 min at 4°.

8. The cell pellet is broken with a tip type sonicator in 500 μl of homogenization buffer (Typical sonication conditions to reach semi-translucent cell suspensions are: 10 cycles of 0.6-sec pulse with 0.4-sec interval, and these procedures are repeated again.). After centrifugation at 900g for 10 min, the supernatant is dispensed in halves and subjected to ultracentrifugation at 100,000g for 1 h. The precipitates thus obtained are used as microsomal membrane fraction.

9. One-half is used to determine protein concentration and is subjected to Western blotting, and the remaining is used to assess the enzymatic activity.

10. Western blot is performed for detection of products. The microsomal fraction (20 μg) is separated by SDS-PAGE (10% gel), and proteins are transferred to a PVDF membrane. The membrane, after blocking in PBS containing 5% skim milk and 0.5% Tween 20, is incubated with each antibody and then the membrane is treated with anti-rabbit IgG conjugated with HRP. Proteins bound to antibody are visualized with ECL.

Preparation of Membrane Fraction for Enzyme Sources from Brain

1. Brain is harvested from newborn rat (F344/N, Nihon SLC, Shizuoka, Japan) and rinsed with cold-PBS. Immediately, 9 ml of homogenization

buffer for every gram of brain is added and homogenized on ice using a potter's homogenizer (800 rpm, 8 strokes).

2. Nuclei, cellular debris, and connective tissues are removed by centrifugation at 900g for 10 min.

3. For preparation of microsomal membranes, the postnuclear supernatant is subjected to ultracentrifugation at 100,000g for 1 h. The pellet fraction is aliquoted and stored at $-80°$ until used.

Preparation of GST-α-DG

Reagents and Solutions

LB broth (Invitrogen) supplemented with 50 μg/ml ampicillin.
LB agar plate (1.5 w/v% agar) supplemented with 50 μg/ml ampicillin (Nacalai Tesque, Kyoto, Japan).
Isopropyl-D-thiogalactopyranoside (IPTG) (Invitrogen): prepare 1 M stock solution in water, sterilize by filtration, and store at $-20°$.
50 mM (NH$_4$)HCO$_3$, pH 7.0.

Materials

pGEX-GST-α-DG: The region corresponding to amino acids 313–483 sites of α-DG (Ibraghimov-Beskrovnaya *et al.*, 1992) was amplified from mouse brain total RNA by RT-PCR using the primer set 5′-GGGAATTCCACGCCACACCTACAC-3′ (sense) and 5′-GGG TCTAGAACTGGTGGTAGTACGGATTCG-3′ (antisense), and subcloned it into pGEX-4T-3 vector to express the peptide as a glutathione-S-transferase (GST)-fusion protein (GE Healthcare).
Glutathione-Sepharose column (GSTrap, 1 ml) (GE Healthcare).

Method for Preparation of GST-α-DG

1. BL21(DE3) *E. coli* cells are transformed with *pGEX-GST-α-DG*. Cultures are prepared by growing single colony overnight in LB broth at 37°. The overnight culture is then used to inoculate a fresh 50-ml culture, which is grown at 37° to $A620 = 0.5$. At this point, 1 mM IPTG is added to the culture to induce GST-α-DG expression. The induced cells are grown for an additional 4 h and harvested by centrifugation at 6000g for 15 min at 4°.

2. The cell pellet is suspended in 10 ml of PBS, pH 7.4, and broken with a tip type sonicator (Semi-translucent cell suspensions are obtained by 3-sec sonication with 3-sec intervals for 5–10 min.). The cell supernatant is recovered by ultracentrifugation at 100,000g for 1 h.

3. Recombinant GST-α-DG proteins are purified from the supernatant with a GSTrap column as follows. Pre-equilibrate the GSTrap column with 10 ml of PBS. Load the supernatant onto the column and wash with PBS at a flow rate 0.2 ml/min. The absorbed recombinant GST-α-DG proteins are eluted with 10 ml of 10 mM reduced glutathione in PBS at a flow rate 1 ml/min.
4. The purified GST-α-DG is dialyzed with 50 mM (NH$_4$)HCO$_3$, pH 7.0.
5. Protein concentration is determined by BCA assay (PIERCE, Rockford, IL), and the purity of GST-α-DG is checked by SDS-PAGE.
6. The GST-α-DG aliquots are dispensed by 10 μg in microcentrifugal tubes, dried up with a centrifugal evaporator and kept at −80°.

POMT Assay

The POMT activity is based on the amount of [^{3}H]-Man transferred from Dol-P-Man to GST-α-DG (Manya *et al.*, 2004). The reaction product is purified with a glutathione-Sepharose column and radioactivity of mannosyl GST-α-DG is measured by a liquid scintillation counter. Although whole cells instead of membrane fractions may be used as an enzyme source, we recommend using membrane fractions, because mammalian tissues and cells have a low specific activity (Akasaka-Manya *et al.*, 2004).

Reagents

n-Octyl-β-D-thioglucoside and CHAPS (DOJINDO LABORA-TORIES, Kumamoto, Japan): prepare 10% (w/v) stock solution in water and store at −20°.
Mannosylphosphoryldolichol$_{95}$: [Mannose-6-^{3}H] Dol-P-Man (1.48-2.22 TBq/mmol, American Radiolabeled Chemical, Inc., St. Louis, MO). 1.85 MBq of solution in chloroform and methanol is transferred into screw-cap centrifugal tube and evaporated with centrifugal evaporator (Do not dry completely, ∼10 μl solvent should remain.). Add 1 ml of 20 mM Tris-HCl (pH 8.0), 0.5% CHAPS and dissolve by sonication with bath type sonicator in ice-cold water (10 cycles of 15-sec pulse with 30-sec interval). Measure radioactivity and then adjust to 40,000 cpm/μl with 20 mM Tris-HCl (pH 8.0), 0.5% CHAPS. Aliquot and store at −80°.
Jack bean-α-mannosidase (Seikagaku Corp., Tokyo, Japan): 0.8 U enzyme is dissolved in 50 μl of 0.1 M ammonium acetate buffer (pH 4.5). The enzyme solution is dried up with centrifugal evaporator and stored at −20°. The dried enzyme is dissolved with 50 μl of 1 mM ZnCl$_2$ before use (Li and Li, 1972).

Solutions

> POMT reaction buffer: 10 mM Tris-HCl, pH 8.0, 2 mM 2-mercaptoethanol, 10 mM EDTA, 0.5% *n*-octyl-β-D-thioglucoside. Store at $-20°$.
> PBS containing 1% Triton X-100 (1% Triton-PBS): Store at 4°.
> 0.5% Triton-Tris buffer: 20 mM Tris-HCl (pH 7.4) containing 0.5% Triton X-100. Store at 4°.

Materials

Glutathione-Sepharose 4B (GE Healthcare): Prepare 25% slurry working suspension as follows. Suspension (1 ml, equivalent to 0.75-ml beads) is put in centrifugal tube. Water (9 ml) is added to the suspension and vortexed. After centrifugation at 1000g for 1 min, the supernatant is removed by aspiration. The beads are rinsed with 10 ml of PBS and collected by centrifugation. 1% Triton-PBS (2.25 ml) is added and stored at 4°.

Method for POMT Assay

1. The POMT reaction buffer is added to the microsomal membrane fraction at a protein concentration of 4 mg/ml. The fraction is suspended by moderate pipetting and solubilized for 30 min on ice with mild stirring occasionally.
2. 20 μl of solubilized fraction and 2 μl of Dol-P-Man solution are added to the dried GST-α-DG, vortexed, and spun down gently. Immediately, the reaction mixture is incubated at 25° for 1 h. The reaction is stopped by adding 200 μl of 1% Triton-PBS (POMT activity is inactivated in the presence of Triton X-100.).
3. The reaction mixture is centrifuged at 10,000g for 10 min. The supernatant is transferred into a screw-cap tube with a packing seal. 400 μl of 1% Triton-PBS and 40 μl of 25% slurry glutathione-Sepharose beads are mixed with the supernatant, and rotated with rotary mixer at 4° for 1 h.
4. After centrifugation at 1000g for 1 min, the supernatant is removed by aspiration, and the beads are washed three times with 0.5% Triton-Tris buffer. 2% SDS is added to the beads and boiled at 100° for 3 min. The suspension is cooled down to room temperature and mixed with liquid scintillation cocktail. The radioactivity adsorbed to the beads is measured using a liquid scintillation counter.
5. The incorporation of radioactive mannose into GST-α-DG can be detected by SDS-PAGE and subsequent autoradiography as follows. Instead of 2% SDS in step 4, 20 μl of 2× loading buffer is added to

the beads followed by boiling at 100° for 3 min. After centrifugation at 1000g for 1 min, the supernatant is subjected to SDS-PAGE. Gel is stained with CBB to visualize GST-α-DG, soaked in Amplify fluorographic reagent for 30 min to enhance detection efficiency of tritium, dried with vacuum gel dryer, and exposed to x-ray film.

6. The linkage of the mannosyl residue to peptide is determined as follows. Instead of 2% SDS in step 4, 50 μl of jack bean-α-mannosidase (0.8 U) is added to the beads and incubated with at 37°. α-Mannosidase (0.8 U) is added freshly every 24 h and is incubated for up to 60 h. Inactivated α-mannosidase, prepared by heating the enzyme (100° for 5 min), is used as a control. After incubation, the radioactivities of the supernatant and the beads are measured using a liquid scintillation counter.

Expression and Enzymatic Activity of POMT1 Mutants

An expression vector encoding each mutant of *POMT1* was prepared by site-directed mutagenesis (Akasaka-Manya *et al.*, 2004). For each mutation that found patients with WWS, the *POMT1* gene was modified with a QuickChange Site-Directed Mutagenesis Kit (Stratagene Corp., La Jolla, CA) according to the manufacturer's instructions. The expression plasmids of POMT1 mutants and wild-type POMT2 were co-transfected into HEK293T cells, and the transfected cells were harvested and homogenized after being cultured for 2 days. POMT activity was assayed as described.

POMGnT1 Assay

The POMGnT1 activity is based on the amount of [^{3}H]-GlcNAc transferred from UDP-GlcNAc to benzyl-α-mannose (Benzyl-Man) (Zhang *et al.*, 2003) or mannosylpeptide (Ac-Ala-Ala-Pro-Thr(Man)-Pro-Val-Ala-Ala-Pro-NH$_2$) (Takahashi *et al.*, 2001). The reaction product is purified with a reverse-phased HPLC, and radioactivity is measured. The mannosylpeptide is not commercially available, but it is possible to use Benzyl-Man, which is commercially available, as a substitute. Although whole cells instead of membrane fractions may be used as an enzyme source, we recommend using membrane fractions, because mammalian tissues and cells have a low specific activity.

Membrane fraction from transiently transfected HEK293T cells (Fig. 3A) and brain were prepared similar to POMT1 and POMT2 as described previously.

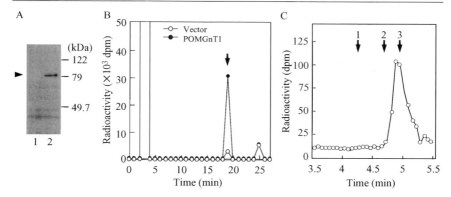

FIG. 3. Western blot analysis of POMGnT1 (A) expressed in HEK293T cells and POMGnT activity (B) expressed in HEK293T cells. (A), Lane 1, cells transfected with vector alone; lane 2, cells transfected with human *POMGnT1*. The proteins (20 μg of membrane fraction) were subjected to SDS-PAGE (10% gel), and the separated proteins were transferred to a PVDF membrane. The PVDF membrane was stained with anti-POMGnT1 antibody. Arrowhead indicates the positions of the corresponding molecule. Molecular weight standards are shown on the right. (B) UDP-[^{3}H]GlcNAc and mannosylpeptide were reacted with membrane fraction in POMGnT1 reaction buffer and then subjected to reversed-phase HPLC. Arrow indicates the elution position of the mannosylpeptide. Vector (open circle), cells transfected with vector alone; POMGnT1 (closed circle), cells transfected with human *POMGnT1*. (C), Analysis of the β-eliminated product by HPAEC-PAD. The radioactive component in B was coinjected with standard disaccharides into a CarboPac PA-1 column. Arrows 1, 2, and 3 indicate the elution positions of authentic standard disaccharide alditols: 1, GlcNAcβ1-6Man$_{OH}$; 2, GlcNAcβ1-4Man$_{OH}$, and GlcNAcβ1-3Man$_{OH}$; 3, GlcNAcβ1-2Man$_{OH}$.

Reagents

UDP-GlcNAc (Sigma-Aldrich): Prepare 1 mM stock solution in water and store at $-20°$.

UDP-GlcNAc [glucosamine-6-^{3}H(N)] (UDP-[^{3}H]-GlcNAc, 0.74–1.66 TBq/mmol, PerkinElmer, Inc., Wellesley, MA): Store at $-20°$.

Benzyl-α-D-mannopyranoside (Sigma-Aldrich): Prepare 100 mM stock solution in 20% ethanol and store at $-20°$.

Mannosylpeptide (Ac-Ala-Ala-Pro-Thr(Man)-Pro-Val-Ala-Ala-Pro-NH$_2$) (Takahashi *et al.*, 2001): Prepare 2 mM stock solution in water and store at $-20°$.

Solutions

POMGnT reaction buffer: 140 mM MES (adjust to pH 7.0 with NaOH), 2% Triton X-100, 5 mM AMP, 200 mM GlcNAc, 10% glycerol, 10 mM MnCl$_2$. Store at $-20°$ without MnCl$_2$. MnCl$_2$ is added just before use.

Streptococcal β-N-acetylhexosaminidase (HEXaseI, Prozyme, San
Leandro, CA): 50 mU of enzyme is dissolved with 50 µl of 0.3 M
citrate phosphate buffer (pH 5.5) and stored at −20°.
0.05 N NaOH, 1 M NaBH₄ and 4 N acetic acid solution in water.
0.1% Trifluoroacetic acid (TFA) in water (Solvent A): Add 1 ml of
TFA to 1000 ml of HPLC grade water and degas with aspirator
before use.
0.1% TFA in acetonitrile (Solvent B): Add 1 ml of TFA to 1000 ml of
HPLC grade acetonitrile and degas by sonication before use.

Materials

Reverse phase column: Wakopak 5C18-200 column (4.6 × 250 mm,
Wako Pure Chemical Industries, Osaka, Japan).
AG-50W-X8 (H⁺ form, Bio-Rad Laboratories, Hercules, CA)

POMGnT1 Assay

1. 10 µl of 1 mM UDP-GlcNAc, 10 µl of UDP-[³H]GlcNAc (100,000
   dpm/nmol) and 10 µl of 2 mM mannosylpeptide (or 100 mM Benzyl-
   Man) are mixed in microcentrifugal tube and dried up with a
   centrifugal evaporator.
2. The POMGnT reaction buffer is added to the microsomal membrane
   fraction at a protein concentration of 2 mg/ml. The fraction is
   suspended with a bath type sonicator on ice and solubilized by
   moderate pipetting until transparent. After centrifugation at 10,000g
   for 10 min, 20 µl of the supernatant is added to dried substrate
   (prepared in step 1), vortexed gently and incubated at 37° for 2 h.
3. The reaction is stopped by boiling at 100° for 3 min. Water (180 µl)
   is added to the reaction mixture and filtered with a centrifugal filter
   device.
4. The filtrate is analyzed by reversed phase HPLC on the condition as
   follows. The gradient solvents are aqueous 0.1% TFA (solvent A)
   and acetonitrile containing 0.1% TFA (solvent B). The mobile
   phase consists of (1) 100% A for 5 min, (2) a linear gradient to 75% A,
   25% B for 20 min, (3) a linear gradient to 100% B for 1 min, and (4)
   100% B for 5 min. The peptide separation is monitored by measuring
   the absorbance at 214 nm, and the radioactivity of each fraction (1 ml)
   is measured by liquid scintillation counting (Fig. 3B).

The reaction product is characterized by two different methods as
follows.

(1) The product is dried up by an evaporator and then incubated with streptococcal β-N-acetylhexosaminidase (50 mU) at 37° for 48 h. After incubation, the enzyme is inactivated by boiling at 100° for 3 min. The enzyme-digested sample is re-chromatographed as described previously.

(2) β-Elimination is performed as described below. The product is dissolved in 500 μl of 0.05 N NaOH and 1 M NaBH$_4$ and incubated for 18 h at 45°. After adjusting the pH to 5 by adding 4 N acetic acid, the solution is applied to a column containing 1 ml of AG-50W-X8 (H$^+$ form), and the column is then washed with 10 ml of water. The effluent and the washing are combined and evaporated. After the remaining borate is removed by repeated evaporation with methanol, the residue is analyzed by high-pH anion-exchange chromatography with pulsed amperometric detection (HPAEC-PAD) (Fig. 3C) (Hardy and Townsend, 1994; Takahashi et al., 2001).

Expression and Enzymatic Activity of POMGnT1 Mutants

An expression vector encoding each mutant of *POMGnT1* was prepared by site-directed mutagenesis. Template cDNA for site-directed mutagenesis encoding full-length *POMGnT1* tagged with the His-tag and Xpress epitope was cloned into pcDNA 3.1 as described previously. Site-directed mutagenesis for mutants found in patients with MEB were performed using a QuickChange Site-Directed Mutagenesis Kit (Stratagene) according to the manufacturer's instructions. The expression plasmids of POMGnT1 mutants were transfected into HEK293T cells, and the transfected cells were harvested and homogenized after being cultured for 2 days. POMGnT1 activity was assayed as described.

References

Akasaka-Manya, K., Manya, H., and Endo, T. (2004). Mutations of the *POMT1* gene found in patients with Walker-Warburg syndrome lead to a defect of protein O-mannosylation. *Biochem. Biophys. Res. Commun.* **325,** 75–79.

Beltran-Valero de Bernabe, D., Currier, S., Steinbrecher, A., Celli, J., van Beusekom, E., van der Zwaag, B., Kayserili, H., Merlini, L., Chitayat, D., Dobyns, W. B., Cormand, B., Lehesjoki, A. E., Cruces, J., Voit, T., Walsh, C. A., van Bokhoven, H., and Brunner, H. G. (2002). Mutations in the O-mannosyltransferase gene *POMT1* give rise to the severe neuronal migration disorder Walker-Warburg syndrome. *Am. J. Hum. Genet.* **71,** 1033–1043.

Chiba, A., Matsumura, K., Yamada, H., Inazu, T., Shimizu, T., Kusunoki, S., Kanazawa, I., Kobata, A., and Endo, T. (1997). Structures of sialylated O-linked oligosaccharides of bovine peripheral nerve dystroglycan. The role of a novel O-mannosyl-type oligosaccharide in the binding of dystroglycan with laminin. *J. Biol. Chem.* **272,** 2156–2162.

Dubray, G., and Bezard, G. (1982). A highly sensitive periodic acid-silver stain for 1,2-diol groups of glycoproteins and polysaccharides in polyacrylamide gels. *Anal. Biochem.* **119,** 325–329.

Emery, A. E. (2002). The muscular dystrophies. *Lancet.* **359,** 687–695.

Endo, T. (2004). Structure, function and pathology of O-mannosyl glycans. *Glycoconj. J.* **21,** 3–7.

Endo, T., and Toda, T. (2003). Glycosylation in congenital muscular dystrophies. *Biol. Pharm. Bull.* **26,** 1641–1647.

Hardy, M. R., and Townsend, R. R. (1994). High-pH anion-exchange chromatography of glycoprotein-derived carbohydrates. *Methods Enzymol.* **230,** 208–225.

Ibraghimov-Beskrovnaya, O., Ervasti, J. M., Leveille, C. J., Slaughter, C. A., Sernett, S. W., and Campbell, K. P. (1992). Primary structure of dystrophin-associated glycoproteins linking dystrophin to the extracellular matrix. *Nature* **355,** 696–702.

Jurado, L. A., Coloma, A., and Cruces, J. (1999). Identification of a human homolog of the *Drosophila rotated abdomen* gene (*POMT1*) encoding a putative protein O-mannosyl-transferase, and assignment to human chromosome 9q34.1. *Genomics* **58,** 171–180.

Li, Y.-T., and Li, S.-C. (1972). Mannosidase, N-acetylhexosaminidase and galactosidase from jack bean meal. *Methods Enzymol.* **28,** 702–713.

Manya, H., Chiba, A., Yoshida, A., Wang, X., Chiba, Y., Jigami, Y., Margolis, R. U., and Endo, T. (2004). Demonstration of mammalian protein O-mannosyltransferase activity: Coexpression of POMT1 and POMT2 required for enzymatic activity. *Proc. Natl. Acad. Sci. USA* **101,** 500–505.

Michele, D. E., and Campbell, K. P. (2003). Dystrophin-glycoprotein complex: Post-translational processing and dystroglycan function. *J. Biol. Chem.* **278,** 15457–15460.

Montanaro, F., and Carbonetto, S. (2003). Targeting dystroglycan in the brain. *Neuron* **37,** 193–196.

Takahashi, S., Sasaki, T., Manya, H., Chiba, Y., Yoshida, A., Mizuno, M., Ishida, H., Ito, F., Inazu, T., Kotani, N., Takasaki, S., Takeuchi, M., and Endo, T. (2001). A new 1,2-N-acetylglucosaminyltransferase that may play a role in the biosynthesis of mammalian O-mannosyl glycans. *Glycobiology* **11,** 37–45.

van Reeuwijk, J., Janssen, M., van den Elzen, C., Beltran-Valero de Bernabe, D., Sabatelli, P., Merlini, L., Boon, M., Scheffer, H., Brockington, M., Muntoni, F., Huynen, M. A., Verrips, A., Walsh, C. A., Barth, P. G., Brunner, H. G., and van Bokhoven, H. (2005). POMT2 mutations cause alpha-dystroglycan hypoglycosylation and Walker-Warburg syndrome. *J. Med. Genet.* **42,** 907–912.

Willer, T., Amselgruber, W., Deutzmann, R., and Strahl, S. (2002). Characterization of POMT2, a novel member of the *PMT* protein O-mannosyltransferase family specifically localized to the acrosome of mammalian spermatids. *Glycobiology* **12,** 771–783.

Yoshida, A., Kobayashi, K., Manya, H., Taniguchi, K., Kano, H., Mizuno, M., Inazu, T., Mitsuhashi, H., Takahashi, S., Takeuchi, M., Herrmann, R., Straub, V., Talim, B., Voit, T., Topaloglu, H., Toda, T., and Endo, T. (2001). Muscular dystrophy and neuronal migration disorder caused by mutations in a glycosyltransferase, POMGnT1. *Dev. Cell* **1,** 717–724.

Zhang, W., Vajsar, J., Cao, P., Breningstall, G., Diesen, C., Dobyns, W., Herrmann, R., Lehesjoki, A. E., Steinbrecher, A., Talim, B., Toda, T., Topaloglu, H., Voit, T., and Schachter, H. (2003). Enzymatic diagnostic test for Muscle-Eye-Brain type congenital muscular dystrophy using commercially available reagents. *Clin. Biochem.* **36,** 339–344.

Section IV

Glycolipid Function

[12] Identification and Analysis of Novel Glycolipids in Vertebrate Brains by HPLC/Mass Spectrometry

By YASUKO NAGATSUKA, HIROMASA TOJO, and YOSHIO HIRABAYASHI

Abstract

Glycosphingolipids are a major component of microdomains or lipid rafts in biological membranes. A new member of raft glycolipids, phosphatidylglucoside (PtdGlc), as well as 6-O-Ac-PtdGlc, a form of PtdGlc O-acetylated at position 6 of its glucopyranose ring, is present in central nervous system tissues. Because the glycolipids represent a minor constituent of lipid rafts and because their mass numbers are the same as that of phosphatidylinositol (PI), the glycolipids are difficult to detect and purify. Here we describe methods to purify and identify glycolipids from rodent brain and methods to discriminate PtdGlc from PI in chick spinal cord using HPLC/electrospray ionization ion-trap mass spectrometry.

Overview

Glycosphingolipids (GSLs) are ubiquitous in mammalian tissues and are particularly abundant in nervous system tissues. They form aggregates or clusters with cholesterol, glycosylphosphatidylinositol (GPI)-anchored proteins, and acylated proteins (Brown and London, 2000; Simonk and Ikonen, 1997) in biological membranes. GSL-enriched microdomains or rafts in the CNS play important roles in nonreceptor NSrc signaling (Mukherjee *et al.*, 2003), chemotropic axonal guidance of nerve growth cones (Guirland *et al.*, 2004; Herincs *et al.*, 2005), neuron–glia interactions (Schaeren-Wiemers *et al.*, 2004; Vinson *et al.*, 2003), and amyloid precursor protein processing (Kakio *et al.*, 2002; Yu *et al.*, 2005).

Using an anti-blood group "i" monoclonal antibody, Nagatsuka *et al.* (2001) identified a novel glucosylated phosphoglycerolipid in human cord red blood cells. The promyelocytic leukemia cell line HL60 also possesses a similar phospholipid, which we call phosphatidylglucoside (PtdGlc). PtdGlc, which is present in a detergent-insoluble membrane (DIM) fraction, is possibly involved in cellular differentiation.

Its chemical structure, however, has not been fully studied (Nagatsuka *et al.*, 2003). A specific probe to detect PtdGlc is essential if we are to fully understand the functional roles of PtdGlc. Because the raft-like microdomain acts as an effective immunogen (Katagiri, 2001, 2002), many monoclonal

METHODS IN ENZYMOLOGY, VOL. 417
0076-6879/06 $35.00
DOI: 10.1016/S0076-6879(06)17012-3

A   Phosphatidyl-β-D-glucopyranoside (PtdGlc)

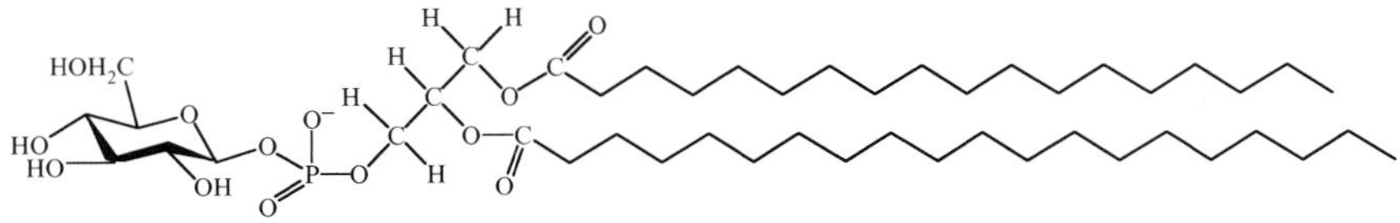

B   Phosphatidyl-β-D-(6-O-acetyl) glucopyranoside
      (6-O-Ac-PtdGlc)

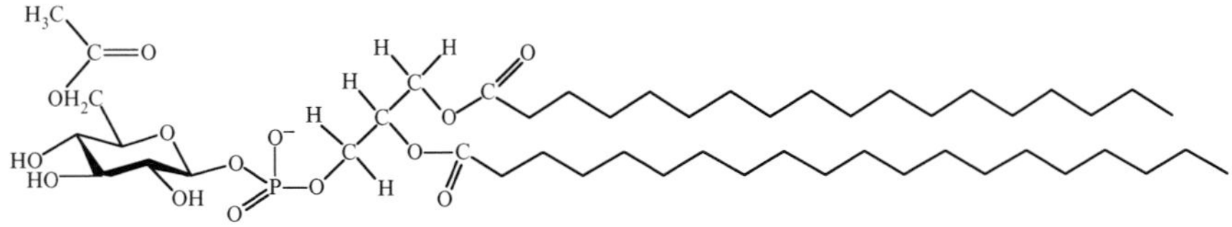

FIG. 1. Chemical structure of phosphatidyl-β-D-glucopyranoside (PtdGlc) and phosphatidyl-β-D-(6-O-acetyl) glucopyranoside (6-O-Ac-PtdGlc) isolated from embryonic day 21 fetal rat brain.

antibodies (mAbs) to lipid microdomains have been generated by immunizing mice with raft-like microdomains isolated from differentiated HL60 cells. One of these mAbs, DIM21, preferentially reacts with PtdGlc (Yamazaki *et al.*, 2006). DIM21 immunostaining of mouse brain sections shows that the DIM21-immunoreactive antigen is present in the murine central nervous system and that its expression is regulated during development. DIM21-immunoreactive lipid antigens have also been isolated from fetal rat brains, and analyses of their complete structures reveal that these antigens are PtdGlc and 6-O-Ac-PtdGlc, a form acetylated at position 6 of the PtdGlc glucopyranose ring (Fig. 1) (Nagatsuka *et al.*, 2006). Both glycolipids are composed of a single molecular species of acyl chains—C18:0 at the *sn*-1 position and C20:0 at the *sn*-2 position of the glycerol backbone (Fig. 1). This saturated fatty acyl chain composition strongly suggests that PtdGlc may reside in raft-like lipid microdomains. Here we describe biochemical methods to identify and purify glycolipids from central nervous system tissues.

Isolation of PtdGlc from Developing Brains

Isolating pure PtdGlc from adult rat brains is difficult, because adult rat brain contains very low levels of PtdGlc and because similar acidic glycolipids such as sulfatide are also present. A good source of PtdGlc is fetal rat brain, because sulfatide is not detectable before myelination, which occurs postnatally. At embryonic day 14 (E14), the brain contains lipids that react with DIM21. The total isolatable amount of PtdGlc in E14 rat brain, however, is quite low, because the dry brain tissue weight of even 100 E14 brains is only 500 mg. Therefore, a more preferable source of PtdGlc is the fetal brain at E21, the stage just before birth. The dry weight of 100 E21 brains is approximately 2.1–2.4 g, from which we can isolate approximately 200–500 nmole/g of PtdGlc. During the isolation procedure, it is important to monitor PtdGlc with orcinol reagent, which detects GSLs, and with mAb DIM21. PtdGlc stains positively with orcinol reagent and mAb DIM21.

Materials for Immunochemical Detection

- POLYGRAM SIL G plate (No. 805 013 [20 × 20 cm] or No. 805 012 [5 × 20 cm]; Macherey-Nagel GmbH & Co. KG, Germany)
- Can Get Signal Immunoreaction Enhancer solution (NKB-101; Toyobo Co., Ltd., Japan)
- HRP-conjugated anti-mouse immunoglobulins (No. 55556; ICN-Cappel, USA)
- ImmumO 3, 3'-diaminobenzidine tetrahydrochloride (DAB) Kit (No. 980571; MP Biomedicals, Inc., USA) or 4-chloro-1-naphthol (No. 08527–64; Nacalai Tesque, Inc., Japan)

Preparation of DIM21 Monoclonal Antibody

The DIM fraction isolated from rGL-7–stimulated HL60 cells is used as an antigen for preparing mAb DIM21, as described previously (Yamazaki et al., 2006). DIM21-producing cells are cultured in RPMI 1640 medium supplemented with 10% fetal calf serum, 100 μg/ml of streptomycin, and 100 U/ml of penicillin at 37° in a 5% CO_2 humidified atmosphere. The cells (2×10^6/ml) are then transferred to dishes containing SFM-101 serum-free medium (Nissui Pharmaceutical Co., Ltd, Japan) and cultured for 2 weeks. After harvesting the culture supernatant, IgM is precipitated by treating the supernatant with ammonium sulfate (50% saturation) for 1–2 weeks at 4°. The IgM is then dissolved in Tris-buffered saline (TBS), dialyzed against the same buffer, then purified by using an Immunoassist MGPP gel column (Kanto Chemical Co., Inc., Japan), according to the supplier's recommendations.

TLC Immunostaining

Glycolipid samples are applied to a Polygram Sil G plastic TLC plate, which is developed with a solvent system containing chloroform (C), methanol (M), and water (W) (C:M:W; 65:35:8, v/v/v). To reduce nonspecific staining, the plate is then soaked in a solution of 1% ovalbumin in TBS. Next, the plate is soaked in 20 μg/ml of DIM21 prepared in Can Get Signal solution 1 and incubated overnight at 4°. After washing with TBS (2 × 5 min and 1 × 15 min) with gentle shaking, the plate is reacted with HRP-conjugated anti-mouse immunoglobulins (1:1000; ICN-Cappel) prepared in Can Get Signal solution 2 for 3 hrs at room temperature. After washing with TBS, the antigen is visualized by adding chromogen. Although 3,3′-diaminobenzidine tetrahydrochloride (DAB; ICN Biochemicals, Inc., USA) enhanced with of Co^{2+} (0.15 μmole/ml) has proven to be a very sensitive means of visualizing PtdGlc, this chromogen cross-reacts with PI. Thus, for the specific visualization of PtdGlc, we recommend using 4-chloro-1-naphtol as a chromogen, although its sensitivity for PtdGlc is weaker than that of DAB.

Extraction of Lipids

We typically use brains from E21 rats for PtdGlc isolation. Ten pregnant female Wistar rats (SLC, Shizuoka, Japan) are deeply anesthetized, fetuses are removed, and fetal rat brains are extirpated immediately. The brains are put into liquid nitrogen and then lyophilized. Total lipids are extracted from 1 g of lyophilized brain tissue with 50 ml of C:M (2:1, v/v) twice, then with C:M:W (5:8:3, v/v/v) twice in a Polytron homogenizer (Polytron Aggregate, Switzerland). After evaporation, the lipid extract is suspended in 300 μl of TBS containing 0.5% Triton X-100 and treated with 250 mU of PI-specific phospholipase C (PIPLC; Sigma) (Ikezawa *et al.*, 1976) at 37° for 1 h. Additional PIPLC (250 mU) is added to the reaction mixture, which is incubated overnight to remove PI. The PIPLC-treated sample is then desalted by Folch's partition (C:M:W; 8:4:3, v/v/v). PtdGlc partitions into the organic layer.

Purification of PtdGlc

PtdGlc is a class of glycerophospholipids that degrades under alkaline conditions. Thus, treating the lipid extract with either sodium hydroxide or potassium hydroxide should be avoided. Passing the extract through a phenylboronate affinity column is essential for removing abundant membrane lipids, such as cholesterol, phosphatidylcholine (PC), and phosphatidylethanolamine (PE). PtdGlc is unique in that it is not completely soluble in organic solvents. For example, PtdGlc does not readily dissolve in chloroform–methanol solvents. If insoluble materials are present during the isolation procedure, adding a few drops of water often improves solubility.

Materials for Purification

- Phenylboronate agarose (PBA 60, Amicon; Millipore, USA)
- Q-Sepharose Fast Flow (No. 17-0510-01; Amersham Biosciences, UK)

• Senshu-pack Aquasil column (4.6 ϕ × 250 mm; Senshu Scientific Co., Ltd., Japan)

Phenyboronate Agarose Chromatography

After Folch's partition, the organic layer is thoroughly dried, and the dried material is suspended in C:M (8:2, v/v) and loaded onto a PBA column (1 × 6 cm) to remove non-glycosylated lipids. Before use, the PBA is washed with methanol and pre-equilibrated with C:M (8:2, v/v). Chloroform-methanol–insoluble lipids are also transferred to the column. Unbound lipids (PB-1) are washed from the column by adding 5 column volumes of C:M (8:2, v/v), then the PBA-bound fraction (PB-2) is eluted with 5 column volumes of C:M:W (5:5:1, v/v/v). PtdGlc is eluted in the PB-2 fraction. The PB-2 fraction is dried under a continuous stream of N_2 gas.

Q-Sepharose Chromatography—Method 1 (Open column)

The dried PB-2 fraction is dissolved in 1–2 ml of C:M (2:1, v/v) then loaded onto an open Q-Sepharose column (1 cm × 10 cm), previously activated with 1 M sodium acetate in M:W (1:1, v/v), washed with 5 column volumes of methanol, and pre-equilibrated with 5 column volumes of C:M:W (30:60:8, v/v/v). Unbound lipids are removed by washing with C:M:W (30:60:8, v/v/v). PtdGlc is eluted with 50 ml of C:M:aqueous 0.1 M sodium acetate (30:60:8, v/v/v). The lipids remaining in the column are then washed out with C:M:aqueous 2 M sodium acetate (30:60:8, v/v/v). PtdGlc and PI are co-eluted with C:M:aqueous 0.1 M sodium acetate.

Q-Sepharose Chromatography—Method II (HPLC)

The dried PB-2 fraction is dissolved in 1 ml of C:M (2:1, v/v) then loaded onto a packed Q-Sepharose column (7.6 ϕ × 250 mm), previously equilibrated with C:M:W (30:60:8, v/v/v). After washing the column with C:M (2:1, v/v) for 10 min, the lipids retained within the column are eluted for 90 min with a gradient of solvents, starting with C:M:W (30:60:8, v/v/v) and ending with C:M:aqueous 0.25 M sodium acetate (30:60:8, v/v/v). An HPLC system is used set at a flow rate of 1 ml/min.

For both methods, the elution profile of the lipids is monitored by TLC and TLC-immunostaining with DIM21. The lipids (mainly gangliosides) remaining in the column are then washed out with C:M:aqueous 2 M sodium acetate (30:60:8, v/v/v). The packed column must be washed with M:W (1:1, v/v) for more than 1 h before re-use.

Aquasil Chromatography Using HPLC

PtdGlc-containing fractions are pooled, partitioned with the Folch's solvent system, then the organic solvent layer is dried with a N_2 gas evaporator.

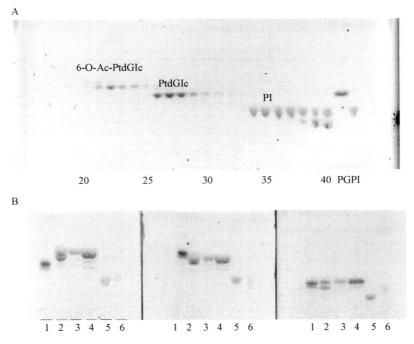

FIG. 2. TLC analysis. (A) The glycolipids eluted using Aquasil column chromatography. The solvent system of I in (B) was used. (B) Lane 1, LacCer; lane 2, sulfatide; lane 3, 6-O-Ac-PtdGlc; lane 4, PtdGlc; lane 5, PI; and lane 6, PC. TLC was developed using the solvent system of I (C:M:2.5N NH$_4$OH; 60:35:8, v/v/v), II (C:M:W; 60/35/8, v/v/v), and III (C:Acetone:M:HOAc:W; 5:2:1:1:0.5, v/v/v/v). The lipids were visualized with orcinol reagent. Abbreviations: LacCer, lactosylceramide; 6-O-Ac-PtdGlc, phosphatidyl-β-D-(6-O-acetyl) glucopyranoside; PtdGlc, phosphatidyl-β-D-glucopyranoside; PI, phosphatidylinositol; PC, phosphatidylcholine. (See color insert.)

The dried material is then dissolved in 1 ml of C:M (2:1, v/v) and loaded onto a Senshu-pack Aquasil column in an HPLC system. After washing with C:M (9:1, v/v), elution is done with a gradient of solvents, starting with C:M (9:1, v/v) and ending with C:M (7:3, v/v) for 90 min (Fig. 2A).

Identification of PtdGlc by Normal-phase HPLC/Electrospray Ionization Ion-trap Mass Spectrometry

Mass spectrometry (MS) is one of the most powerful and sensitive methods to identify lipid classes and their molecular species on a structural basis. PtdGlc is a structural isomer of PI. The mass-to-charge ratio (m/z)

values of their polar head groups, phosphoglucose and phosphoinositol, are the same, but their chemical natures differ in that glucose in PtdGlc is an aldose and inositol in PI is a cyclic polyalcohol. Normal-phase HPLC can recognize slight differences in the polarity of functional groups. To differentiate between PtdGlc and PI contained within crude lipid extracts, both HPLC separation and tandem mass spectrometry are required. Ion-trap mass spectrometry is suitable for such a task, because it has better connectivity for HPLC, has greater full-scan sensitivity, and has data-dependent MS^n capability. Finnigan ion- or linear-trap mass spectrometers fulfill the requirements for discriminating between PtdGlc and PI through automatic gain control, optimal adjustment of collision energy on the basis of m/z values of analytes, and software-controlled data-dependent MS^n acquisition. Next, we describe how to use a combined normal phase HPLC–electrospray ionization (ESI) LCQdeca-XP mass spectrometer (Thermo Electron Corp., USA) apparatus to analyze the PtdGlc molecular species contained within crude lipid extracts.

One of the authors (H. T.) has developed a method capable of analyzing a wider range of lipid classes from neutral lipids to phospholipids on a single chromatographic run with three solvent gradients and post-column mixing of solvents that aids the ESI of lipids eluted earlier with nonpolar solvents. This method has been used to analyze phospholipids (Ito et al., 2002) and ceramides (Takagi et al., 2003). The method we describe in the following is essentially based on the phospholipid analysis mode (Ito et al., 2002) of the comprehensive method.

HPLC

In the phospholipid analysis mode, a binary gradient was operated with intelligent pulse-less pumps (OmniSeparo-TJ, Hyogo, Japan) without post-column mixing of ESI-facilitating solvents.

Elution Solvents

- Solvent A: hexane:2-propanol (4:6, v/v)
- Solvent B: hexane:2-propanol:water (4:6:1.2, v/v/v) containing 20 mM ammonium formate

A precolumn and a separation silica column (1×20 mm and 1×150 mm, respectively; OmniSeparo-TJ, Japan) fitted to a switching valve (Valco Instruments Co., USA) are sequentially connected to each other and pre-equilibrated with solvent A. An aliquot of extracted and partially purified lipids (1–2 μl) is applied onto the trap. When a large amount of less polar contaminants is present in samples, it may be necessary to disconnect the

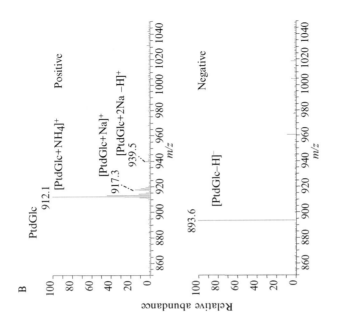

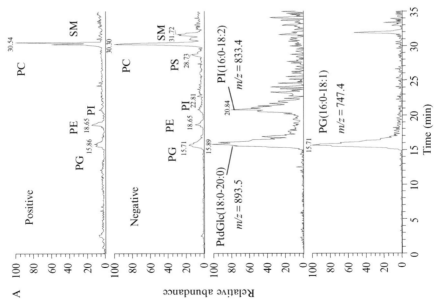

trap from the separation column, wash with the same eluent (solvent A), and then re-connect the trap by switching the valve. Phospholipids bound to the trap are eluted through the separation column with the following graded solvent sequence: 0–85% solvent B in 25 min; hold at 85% for 5 min; 100% solvent B in 2 min; hold at 100% for 5 min at a flow rate of 50 μl/min.

Mass Spectrometer

To increase ionization efficiency, the recently developed fluoropolymer-coated ESI tip, FortisTip (Tojo, 2004) (20 μm ID, 150 μm OD; OmniSeparo-TJ, Japan), is fitted onto an XYZ stage (AMR, Japan), with its exit end placed 2–3 mm in front of a heated capillary inlet. The effluent is dispersed by means of a T splitter, yielding a flow rate of 0.5–1 μl/min across the tip. The effluent is monitored with a Finnigan LCQdeca-XP spectrometer set to either data-dependent negative or positive tandem MS (MS^2) mode, as well as to alternate positive and negative ion full scan (alternate polarity switching) mode on a single run with the automatic gain control on. An ESI voltage of 1.6 kV is used.

Analysis of PtdGlc

Figure 3 shows base-peak and ion chromatograms of a mixture of synthetic PtdGlc and various standard phospholipids ($\sim$1 pmol each) in the alternate polarity-switching mode. Stearoyl-arachidoyl-PtdGlc (m/z = 893) is well separated from soybean PI, with palmitoyl-linoleoyl being the most abundant species (m/z = 833). In general, during analysis of anionic phospholipids, ESI generates abundant $[M–H]^-$ ions but very few $[M+H]^+$ ions; PtdGlc conforms to this rule (Fig. 3B). The presence of ammonium ions in elution solvents greatly promotes the formation of ammonium adduct ions from PtdGlc, although minor $[M + Na]^+$ and $[M + 2Na – H]^+$ ions are discernible. PtdGlc co-elutes with phosphatidylglycerol (PG) (m/z = 747) in normal-phase HPLC under the conditions used (Fig. 3), but the m/z values and MS^2 spectra of these two anionic phospholipids are easily discriminated.

FIG. 3. Analysis of mixtures containing standard phospholipids and synthetic PtdGlc. Standard phospholipids include palmitoyl-oleoyl-PG (phosphatidylglycerol), palmitoyl-oleoyl-PE, soybean PI, palmitoyl-oleoyl-PS (phosphatidylserine), palmitoyl-oleoyl-PC, and bovine brain sphingomyelin (SM). (A) Base-peak chromatograms in the positive and negative ion modes and ion chromatograms in the negative ion mode at the indicated *m/z* values. (B) Full-scan MS spectra of PtdGlc in the positive and negative modes. A MS^2 spectrum of *m/z* = 960.8 ions in the negative ion mode contains an intense *m/z* = 893 peak corresponding to PtdGlc.

Low-energy tandem mass spectrometry with ion-trap mass spectrometry clearly shows a difference in fragmentation between PtdGlc and PI. Fig. 4 shows a comparison of the on-line MS^2 spectra of 18:0-20:0-PtdGlc ($m/z = 893$) and 18:0-20:4-PI ($m/z = 885$). A MS^2 spectrum of PI lacks a peak derived from the neutral loss of inositol–H_2O ($m/z = 162$) but contains strong peaks formed by the neutral loss of 20:4 and 18:0 acids and subsequent loss of inositol–H_2O. In contrast, the PtdGlc spectrum contains a peak formed by the neutral loss of glucose–H_2O and subsequent loss of 20:0 and 18:0 acids, but the peak intensity of ions derived from fatty acid loss from the molecular ions is very small. These results suggest that the P–O–C bond of phosphoglucose in PtdGlc is more labile than that of phosphoinositol in PI. In PtdGlc, ions arising from the loss of 20:0 are more abundant than those from the loss of 18:0, and similarly those arising from the loss of 20:4 are more abundant in PI. This suggests that both 20:0 and 20:4 tend to

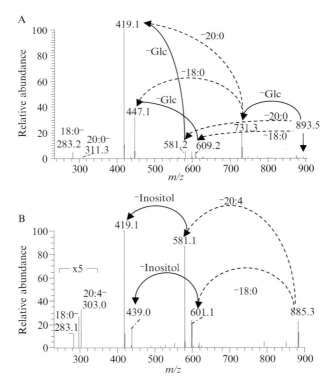

Fig. 4. Comparison of tandem mass spectra of 18:0-20:0-PtdGlc (A) and 18:0-20:4-PI (B). –Glc, loss of glucose–H_2O; –Inositol, loss of inositol–H_2O.

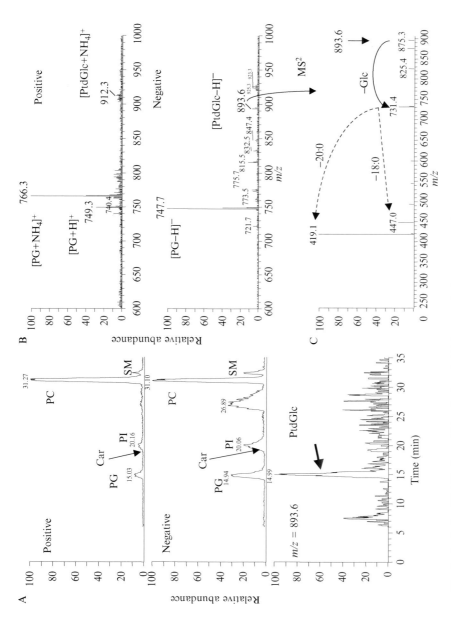

FIG. 5. Analysis of PtdGlc from chick spinal cord. (A) Base-peak chromatograms in the positive and negative ion modes, and ion chromatograms in the negative ion mode at m/z = 893.6. Car, cardiolipin. (B) Full-scan MS^2 spectra across a base peak labeled as PG in the positive and negative modes. (C) Tandem mass spectra of m/z = 893 ion in B.

be esterified at the sn-2 position in PtdGlc and PI, respectively (Larsen et al., 2001). In both spectra, carboxylate anions are less abundant, which is characteristic of collision-induced dissociation in ion-trap MS.

To test the performance of this method, we used lipids extracted from chick spinal cord. The lipids were digested by PIPLC and partially purified by phenylboronate affinity and Q-Sepharose column chromatography. The fraction that exhibited positive reactions with TLC-immunostaining using mAb DIM21 was subjected to HPLC/MS. We found that the DIM21-immunoreactive fraction contained a variety of phospholipids including the major classes of PG, cardiolipin, PI, PC, and sphingomyelin (SM). We still observed an appreciable amount of PI, despite extensive digestion with PIPLC. When assessed in negative and positive modes, note that this fraction did contain small but significant $m/z = 893$ and $m/z = 912$ peaks, respectively, in the base peak corresponding to PG ($m/z = 747$), which are suggestive of the presence of 18:0–20:0-PtdGlc. The intensity of these peaks increased in a dose-dependent manner. These peaks were not observed on a blank run. Indeed on-line tandem mass spectrometry of the $m/z = 893$ ions in the negative ion mode confirmed a structure of 18:0-20:0-PtdGlc (Fig. 5C).

References

Brown, D. A., and London, E. (2000). Structure and function of sphingolipid- and cholesterol-rich membrane rafts. J. Biol. Chem. **275**, 11721–17224.

Guirland, C., Suzuki, S., Kojima, M., Lu, B., and Zheng, J. Q. (2004). Lipid rafts mediate chemotropic guidance of nerve growth cones. Neuron **42**, 51–62.

Herincs, Z., Corset, V., Cahuzac, N., Furne, C., Castellani, V., Hueber, A. O., and Mehlen, P. (2005). DCC association with lipid rafts is required for netrin-1-mediated axon guidance. J. Cell. Sci. **118**, 1687–1692.

Ikezawa, H., Yamanegi, M., Taguchi, R., Miyashita, T., and Ohyabu, T. (1976). Studies on phosphatidylinositol phosphodiesterase (phospholipase C type) of Bacillus cereus purification, properties and phosphatase-releasing activity. Biochim. Biophys. Acta **450**, 154–164.

Ito, M., Tchoua, U., Okamoto, M., and Tojo, H. (2002). Purification and properties of a phospholipase A$_2$/lipase preferring phosphatidic and lysobisphosphatidic Acids, and monoacylgycerol from rat testis. J. Biol. Chem. **277**, 43674–43681.

Kakio, A., Nishimoto, S., Yanagisawa, K., Kozutsumi, Y., and Matsuzaki, K. (2002). Interactions of amyloid beta-protein with various gangliosides in raft-like membranes: Importance of GM1 ganglioside-bound form as an endogenous seed for Alzheimer amyloid. Biochemistry **11**, 7385–7390.

Katagiri, Y. U., Ohmi, K., Katagiri, C., Sekino, T., Nakajima, H., Ebata, T., Kiyokawa, N., and Fujimoto, J. (2001). Prominent immunogenicity of monosialosyl galactosylgloboside, carrying a stage-specific embryonic antigen-4 (SSEA-4) epitope in the ACHN human renal tubular cell line-a simple method for producing monoclonal antibodies against detergent-insoluble microdomains/raft. Glycoconj. J. **18**, 347–353.

Katagiri, Y-U., Ohmi, K., Tang, W., Takenouchi, H., Taguchi, T., Kiyokawa, N., and Fujimoto, J.. (2002). Raft.1, a monoclonal antibody raised against the raft microdomain, recognizes G-protein beta1 and 2, which assemble near nucleus after shiga toxin binding to human renal cell line. *Lab. Invest.* **82**, 1735–1745.

Larsen, A., Uran, S., Jacobsen, P. B., and Skotland, T. (2001). Collision-induced dissociation of glycerol phospholipids using electrospray ion-trap mass spectrometry. *Rapid Commun. Mass Spectrom.* **15**, 2393–2398.

Mukherjee, A., Arnaud, L., and Cooper, J. A. (2003). Lipid-dependent recruitment of neuronal Src to lipid rafts in the brain. *J. Biol. Chem.* **278**, 40806–40814.

Nagatsuka, Y., Kasama, T., Uzawa, J., Ohashi, Y., Ono, Y., and Hirabayashi, Y. (2001). A new phosphoglycerolipid, "phosphatidylglucose", found in human cord red cells. by multi-reactive monoclonal anti-i cold agglutinin, mAb GL-1/GL-2. *FEBS Lett.* **497**, 141–147.

Nagatsuka, Y., Hara-Yokoyama, M., Kasama, T., Takekoshi, M., Maeda, F., Ihara, S., Fujiwara, S., Ohshima, E., Ishii, K., Kobayashi, T., Shimizu, K., and Hirabayashi, Y. (2003). Carbohydrate-dependent signaling from the phosphatidylglucoside based microdomain induces differentiation of HL60 cells. *Proc. Natl. Acad. Sci. USA* **100**, 7454–7459.

Nagatsuka, Y., Horibata, Y., Yamazaki, Y., Kinoshita, M., Hashikawa, T., Koshino, H., Nakamura, T., and Hirabayashi, Y. (2006). Phosphatidylglucoside with unique fatty acyl chains is present in the murine central nervous system. *Biochemistry* **45**, 8742–8750.

Schaeren-Wiemers, N., Bonnet, A., Erb, M., Erne, B., Bartsch, U., Kern, F., Mantei, N, Sherman, D., and Suter, U. (2004). The raft-associated protein MAL is required for maintenance of proper axon—glia interactions in the central nervous system. *J. Cell. Biol.* **166**, 731–742.

Simons, K., and Ikonen, E. (1997). Functional rafts in cell membranes. *Nature.* **387**, 569–572.

Takagi, S., Tojo, H., Tomita, S., Sano, S., Itami, S., Hara, M., Inoue, S., Horie, K., Kondoh, G., Hosokawa, K., Gonzalez, F. J., and Takeda, J. (2003). Alteration of the 4-sphingenine scaffolds of ceramides in keratinocyte-specific Arnt deficient mice affects skin barrier function. *J. Clin. Invest.* **112**, 1372–1382.

Tojo, H. (2004). Properties of an electrospray emitter coated with material of low surface energy. *J. Chromatogr. A* **1056**, 223–228.

Vinson, M., Rausch, O., Maycox, P-R., Prinjha, R-K., Chapman, D., Morrow, R., Harper, A.J, Dingwall, C., Walsh, F. S., Burbidge, S. A., and Riddell, D. R. (2003). Lipid rafts mediate the interaction between myelin-associated glycoprotein (MAG) on myelin and MAG-receptors on neurons. *Mol. Cell. Neurosci.* **22**, 344–352.

Yamazaki, Y., Nagatsuka, Y., Oshima, E., Suzuki, Y., Hirabayashi, Y., and Hashikawa, T. (2006). Comprehensive analysis of monoclonal antibodies against detergent-insoluble membrane/lipid rafts from HL60 cells. *J. Immunol. Methods* **311**, 106–116.

Yu, W., Zou, K., Gong, J-S., Ko, M., Yanagisawa, K., and Michikawa, M. (2005). Oligomerization of amyloid beta-protein occurs during the isolation of lipid rafts. *J Neurosci. Res.* **80**, 114–119.

Further Reading

Higashi, H., Hirabayashi, Y., Fukui, Y., Naiki, M., Matsumoto, M., Ueda, S., and Kato, S. (1985). Characterization of N-glycolylneuraminic acid-containing gangliosides as tumor-associated Hanganutziu-Deicher antigen in human colon cancer. *Cancer Res.* **45**, 3796–3802.

[13]   Modulation of Growth Factor Signaling by
        Gangliosides: Positive Or Negative?

By KAREN KAUCIC, YIHUI LIU, and STEPHAN LADISCH

Abstract

Increasing evidence has implicated gangliosides, sialic acid–containing cell surface glycosphingolipids, in the biological and clinical behavior of many types of human tumors. Gangliosides are overexpressed and actively shed by tumor cells, can bind to normal cells in the tumor microenvironment, and have a number of biological properties that could conceivably alter tumor–host interactions to influence the survival of the malignant cells that carry these molecules. One major area of investigation is the modulation of cell signaling by gangliosides. Published studies have demonstrated modulation of growth factor signaling through the epidermal growth factor (EGF), fibroblast growth factor (FGF), platelet-derived growth factor (PDGF), Trk family, and insulin receptors. Studies conducted over the past 10 y have demonstrated either inhibition or enhancement of signaling by gangliosides, depending on cell type, ganglioside species, and experimental conditions. Of particular concern are conflicting studies that demonstrate opposite effects of gangliosides on the same growth factor receptor. This chapter discusses a methodological approach to addressing this apparent conflict.

Overview

Ganglioside molecules consist of a sialic acid–containing carbohydrate portion and a hydrophobic lipid backbone (ceramide) embedded in the outer leaflet of the cell membrane (Ledeen and Yu, 1982). Individual carbohydrate species can be classified according to ganglioside biosynthetic pathways. Their biosynthesis, in a sequential order of glycosylations, occurs by two main pathways, designated "a" (GM2, GM1a, GD1a) and "b" (GD3, GD2, GD1b, GT1b, GQ1b), from a common precursor (GM3) derived from lactosylceramide (van Echten and Sandhoff, 1993) (Fig. 1). Each ganglioside is structurally more complex than its precursor molecule, and the stepwise addition of monosaccharide or sialic acid residues by specific membrane-bound glycosyltransferases in the Golgi apparatus is catalyzed by the same glycosyltransferases in both pathways. Gangliosides represent a diverse group of molecules, and marked structural differences in carbohydrate and

METHODS IN ENZYMOLOGY, VOL. 417
0076-6879/06 $35.00
DOI: 10.1016/S0076-6879(06)17013-5

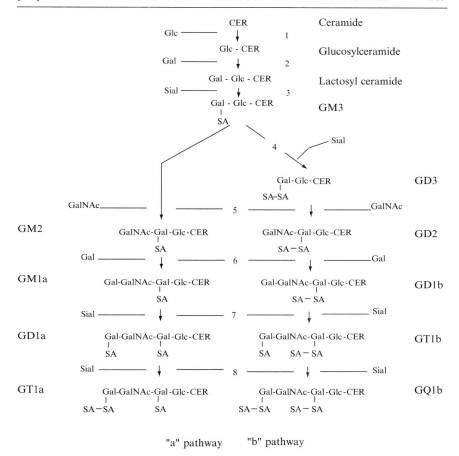

FIG. 1. Schematic representation of the major pathways of ganglioside biosynthesis. GM3, derived from lactosylceramide, is the common precursor for both "a" and "b" pathway gangliosides. Each ganglioside consists of a ceramide backbone (CER) and a carbohydrate chain (glc, glucose; gal, galactose; GalNAc, N-acetylgalactosamine) containing one or more sialic acid (SA, sialic acid) residues. Parallel steps in both pathways are catalyzed by the same glycosyltransferases: (1) glucosylceramide synthase; (2) lactosylceramide synthase; (3) GM3 synthase; (4) GD3 synthase; (5) GM2/GD2 synthase; (6) GD1b/GM1a synthase; (7) GT1b/GD1a synthase; (8) GQ1b/GT1a synthase.

ceramide structures distinguish the ganglioside complement of many tumors from that of normal tissue (Hakomori, 1996). This structural heterogeneity has implications for immunological activity, as evidenced, for example, by studies demonstrating differential immunosuppressive effects of ceramide subspecies of tumor gangliosides (Ladisch *et al.*, 1994). Furthermore, gangliosides are overexpressed by tumor cells and shed into the tumor microenvironment. Tumor gangliosides can both influence the behavior of the tumor cells that produce them and can alter biological functions of target cells to which shed gangliosides bind.

Gangliosides, which exist in glycosphingolipid-enriched domains (Hakomori *et al.*, 1998), have several important biological properties, including potent immunosuppressive activity (Bergelson *et al.*, 1989; Grayson and Ladisch, 1992; Ladisch *et al.*, 1983; Lu and Sharom, 1996), proangiogenic activity (Alessandri *et al.*, 1987; Manfredi *et al.*, 1999; Zeng *et al.*, 2000), and enhancement of growth factor–mediated fibroblast and vascular endothelial cell proliferation (Lang *et al.*, 2001; Li *et al.*, 2001; Rusnati *et al.*, 2002). These findings have in turn led to interest in the role of gangliosides in the modulation of cell signaling, resulting in studies spanning a wide range of growth factors, cell types, ganglioside species, and experimental conditions. A number of growth factor receptors, including receptors for epidermal growth factor (EGF), fibroblast growth factor (FGF), platelet-derived growth factor (PDGF), nerve growth factor (NGF), and insulin are known to be influenced by gangliosides. This diversity has, not surprisingly, led to equally wide-ranging results, including reports delineating opposing effects of gangliosides on the same receptor-ligand pair. Inhibitory effects include inhibition of EGFR phosphorylation by GM3 (Alves *et al.*, 2002; Bremer *et al.*, 1986; Miljan *et al.*, 2002; Mirkin *et al.*, 2002; Rebbaa *et al.*, 1996; Suarez Pestana *et al.*, 1997) and inhibition of PDGFR phosphorylation by GM1, GM2, GD1a, and GT1b (Farooqui *et al.*, 1999; Hynds *et al.*, 1995). In contrast, enhancement of growth factor receptor phosphorylation has been demonstrated for GD1a and GM1 in a number of systems including EGFR (Li *et al.*, 2000, 2001; Liu *et al.*, 2004), and FGF (Rusnati *et al.*, 2002). The body of evidence suggests that gangliosides exert their effects on cell surface receptors by not just one but multiple mechanisms of action (Miljan and Bremer, 2002); however, divergent opinions exist as to whether ganglioside effects are in general stimulatory or inhibitory. We have previously shown that when normal human fibroblasts are preincubated with gangliosides, followed by removal of unbound gangliosides, enhanced proliferation is observed in response to EGF, PDGF, and bFGF, whereas when gangliosides are present in the culture medium during growth factor stimulation, an overall inhibitory effect on cell proliferation is observed (Li *et al.*, 2000). Similar results have been obtained in Chinese hamster ovary (CHO) cells fibroblast growth factor-2 (FGF2) (Rusnati *et al.*, 2002). In general, a fundamental distinction may exist between the activity of soluble gangliosides present in the cellular milieu as opposed to gangliosides bound to cell membranes.

The inconsistencies in current methodological approaches to the study of ganglioside modulation of signal transduction have resulted in the comparison of studies that assess the effects of soluble gangliosides with those that assess the effects of membrane-bound gangliosides. In addition, this divergence in experimental approach has raised the question of how to best reflect the physiological conditions of the cellular microenvironment in *in vitro* experimental systems. For example, in a compilation of 11 studies investigating the role of exogenously added gangliosides in modulation of signaling through the EGF receptor, 4 different gangliosides were studied in 8 different cell lines, with growth factor concentrations ranging from 1–100 ng/ml and ganglioside concentrations ranging from 5–1000 μM (Table I). In addition, the serum content of the cell culture medium and use of cell washing to remove unbound ganglioside varied among the 11 studies, with 4 studies (Li *et al.*, 2000, 2001, Liu *et al.*, 2004; Suarez Pestana *et al.*, 1997)using low serum concentrations ($\leq 2\%$). In eight of the studies (Alves *et al.*, 2002; Bremer *et al.*, 1986; Meuillet *et al.*, 2000; Miljan *et al.*, 2002; Mirkin *et al.*, 2002; Rebbaa *et al.*, 1996; Sottocornola *et al.*, 2003; Suarez Pestana *et al.*, 1997) gangliosides inhibited signaling through the EGF receptor, in three studies (Bremer *et al.*, 1986; Mirkin *et al.*, 2002; Sottocornola *et al.*, 2003) no ganglioside effect could be demonstrated, and in four studies (Li *et al.*, 2000, 2001; Liu *et al.*, 2004; Miljan *et al.*, 2002) signaling was enhanced. Although the cell lines and the gangliosides used in these studies varied, a review of the methods used reveals striking differences in the experimental conditions used. Specifically, among the six studies in which inhibition of EGF receptor phosphorylation was observed, gangliosides were added at concentrations in excess of 100 μM, and gangliosides were present in the culture medium throughout the experiment without removal of unbound molecules. In addition, in only one of the six studies (Suarez Pestana *et al.*, 1997), phosphorylation studies were carried out in the media containing less than 5% serum. Finally, the six studies demonstrating inhibition of EGF phosphorylation used EGF concentrations of at least 10 ng/ml. In contrast, in three of four (Li *et al.*, 2000, 2001; Liu *et al.*, 2004) studies in which addition of exogenous gangliosides resulted in an enhancement of EGF phosphorylation, ganglioside concentrations ranged from 10–50 μM, unbound gangliosides were removed by washing before stimulation with EGF, studies were conducted under low serum ($\leq 2\%$) concentrations, and receptor stimulation was induced using low EGF concentrations (2 ng/ml).

TABLE I

STUDIES EXAMINING THE EFFECT OF EXOGENOUS ADDITION OF GANGLIOSIDES ON SIGNALING THROUGH THE EGFR

Ganglioside	Concentration	Cell line/cell	Modulation
GM1	100 μM	A431	Positive (Miljan et al., 2002)
	125~1000 μM	KB; NBL-W	No effect (Bremer et al., 1986; Mirkin et al., 2002)
GM3	5~1000 μM	A431; NBL-W; H125; A1S; HC11	Negative (Alves et al., 2002; Bremer et al., 1986; Meuillet et al., 2000; Miljan et al., 2002; Mirkin et al., 2002; Rebbaa et al., 1996; Sottocornola et al., 2003; Suarez Pestana et al., 1997)
	125 μM	MG1361	No affect (Sottocornola et al., 2003)
GD1a	5~1000 μM	NBL-W	Negative (Mirkin et al., 2002)
	5~100	A431; NHDF	Positive (Li et al., 2000, 2001, 2004; Miljan et al., 2002)
GT1b	5~1000 μM	NBL-W	Negative (Mirkin et al., 2002)
	100 μM	A431	Positive (Miljan et al., 2002)

    To more fully understand the role of gangliosides in modulating cell signaling, one methodological approach is the development of *in vitro* systems that mimic the *in vivo* situation to the maximum extent possible. This general approach underlies efforts to study the effects of gangliosides on signaling in cells that have been engineered to produce the ganglioside of interest by transfection (Nishio *et al.*, 2005; Zurita *et al.*, 2001). Practically, however, these systems are more difficult to develop, and systems using the exogenous addition of gangliosides will continue to be the primary venue for the study of ganglioside modulation of cell signaling. The physiological relevance of the results obtained in this latter system is directly related to the extent to which the milieu in which gangliosides exert their effects *in vivo* is replicated *in vitro*.

    The characteristics of gangliosides, their interaction with cell membranes, and the conditions existing in the cellular microenvironment are important to consider in designing a physiologically relevant *in vitro* system. First, the cellular microenvironment consists of tumor cells and target cells in close proximity with little or no free interstitial proteins. Second, gangliosides are present in low concentration on the membranes of normal cells. The membrane content of tumor cells is generally much higher, and active shedding of gangliosides results in their cell-to-cell transfer followed by binding to the cell membrane. Third, gangliosides of differing carbohydrate structures are known to have differential effects on a number of cellular functions including growth factor signaling (Miljan *et al.*, 2002; Mirkin *et al.*, 2002). Two other characteristics of gangliosides are important when considering *in vitro* systems using their exogenous addition. First, it is known that high ganglioside concentrations (in excess of 100 μM) behave as detergents, resulting in cell membrane disruption. Second, because gangliosides are known to bind to serum proteins in serum-containing systems, the effective concentrations of free unbound ganglioside is not accurately known. Finally, in experimental systems in which gangliosides and receptor ligands are present concomitantly, dissection of the mechanism of ganglioside interaction with the receptor-ligand pair is not possible.

The interaction of gangliosides with EGFR provides an excellent example not only of the association of ganglioside structure with activity but also of the multiple points in the process of signal transduction that can be modulated by gangliosides. The EGFR is actually a family of four receptors each consisting of an extracellular domain, a transmembrane domain, and a tyrosine kinase cytoplasmic domain. It is believed that the receptors undergo both ligand-dependent and ligand-independent predimer formation. Unbound receptors reside in lipid rafts until ligand stimulation. On binding of EGFR predimers to ligand, dimerization and exit from the raft domain is followed by autophosphorylation and subsequently downstream signaling (Fig. 2). As membrane-associated molecules, gangliosides can act to alter signal transduction by modulating one or more of the steps that occur before, during, or after ligand binding. Identification of the precise steps that are influenced by gangliosides requires strict adherence to rigorous methods, which are designed to closely mimic the *in vivo* physiological state and permit assessment of ganglioside interaction at discreet steps in the signal transduction process.

Taking the characteristics of both gangliosides and the cellular microenvironment together, an *in vitro* system that is designed to test the effects of membrane-associated gangliosides in an *in vitro* system that mimics the *in vivo* environment could be envisioned as using low concentrations of gangliosides in a low serum or serum-free milieu, the removal of excess unbound ganglioside before ligand exposure, and the use of low ligand (growth factor) concentrations. We have developed and successfully used a method for measuring EGFR phosphorylation and dimerization using low

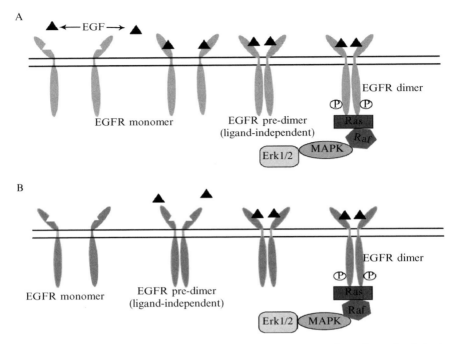

FIG. 2. The EGF–EGFR interaction at the cell membrane. Predimer formation is both ligand (EGF, ▲)-dependent and ligand-independent.

concentrations of serum, growth factor, and ganglioside, which more closely mimics the *in vitro* microenvironment.

Methods

Cell Culture

Normal human dermal fibroblasts (NHDF) purchased from Clonetics (San Diego, CA) are cultured in fibroblast complete growth medium, FGM-2 (Clonetics) that contains 2% fetal bovine serum and 0.5 ml each of insulin, hFGF, and GA1000 per 500 ml. The culture medium is changed every 3 days. All experiments will be performed using subconfluent cultures of passages 3–10 (if the storage cells are used for signaling study, the cell are cultured for at least one passage, and then seeded for the signaling study). Cell viability is assessed by trypan blue dye exclusion. For serum-free culture, fibroblast-basal medium (FBM) is used.

Preparation of Cell Lysate

$1–2 \times 10^5$ NHDF are seeded per 100-mm dish or per well in 6-well plates in FGM-2. On reaching subconfluence, the cells are washed with FBM twice and then incubated with gangliosides in FBM for 6 or 18 (low concentration available) h. Then the culture medium are removed, the cells are washed twice with FBM to remove unbound ganglioside, and exposed to 2 ng/ml EGF in FBM for 5 min at 37° (Iyer *et al.*, 1999). Then the cells are immediately washed twice with ice-cold phosphate-buffered saline (PBS) and lysed for 20 min in lysis buffer, 1 ml/100-mm dish, or 300 μl/6-well culture plate. The lysis buffer contains 20 mM Tris, pH 7.5, 150 mM NaCl, 1 mM EDTA, 1 mM EGTA, 1% Triton X-100, 2.5 mM sodium pyrophosphate, 1 mM β-glycerol phosphate, 1 mM Na$_3$VO$_4$, 1 μg/ml leupeptin, and 1 mM phenylmethylsulfo-nyl flouride (PMSF). The lysates are transferred to microcentrifuge tubes, sonicated briefly on ice, and centrifuged at 10,000g for 10 min at 4°. The supernatant is used for the kinase assays. Proteins are quantified by the Lowry method, using bovine albumin as a standard (Lowry *et al.*, 1951).

Cell Preparation for the EGFR Inhibitor Assay

To test for the specificity of the effect of ganglioside preincubation on receptor autophosphorylation, the EGFR inhibitor, AG1478, is used. NHDF will be exposed to ganglioside GD1a in FBM for 18 h. During the last 3 h, 1 μM AG1478 or DMSO is added (Jo *et al.*, 2000). Then the cells are exposed to 2 ng/ml EGF for 5 min, harvested, and the cell lysate are prepared as previously.

EGFR Autophosphorylation Assay

To preclear nonspecific binding, about 200 μl of cell lysate ($>$100 μg total protein) is mixed with 50 μl of washed protein G-Sepharose agarose bead slurry (50 μl packed beads), stirred for 2 h at 4° and microcentrifuged at 10,000g for 5 sec. The supernatant is transferred to a new Microfuge tube, mixed with 4 μg of sheep polyclonal IgG anti-EGFR antibody, and incubated overnight with gentle stirring at 4°. The immune complexes are recovered by adding 50 μl of the Protein G-Sepharose agarose bead slurry, gently rocking the mixture for 2 h at 4°, and microcentrifuging at 14,000g for 5 sec. When the supernatant is removed, the beads are washed three times with ice-cold lysis buffer, resuspended in 50 μl 2$\times$ SDS sample buffer, boiled for 5 min, and microcentrifuged; 20 μl of each supernatant ($\sim$40 μg) protein is loaded onto a 7.5% SDS-polyacrylamide gel. EGFR autopho-sphorylation is detected by Western blot using an anti-phosphotyrosine antibody p-Tyr (PY99) (Li *et al.*, 2001), and total EGFR will be detected by an anti-EGFR antibody.

Western Blotting

The samples are transferred to a polyvinylidene fluoride (PVDF) membrane (the PVDF membrane must be activated by methanol for 2 min, and then transferred to transfer buffer balanced for >5min before sample transfer), incubated in 25 ml of blocking buffer for 1 h at room temperature, and then incubated with primary antibody (rabbit polyclonal IgG; 1:1000 dilution) with gentle agitation overnight at 4°. After three washes, the membrane will be incubated in the medium containing horseradish peroxidase–conjugated anti-rabbit antibody (1:2000). Proteins will be detected by chemiluminescence and compared with standard proteins of different molecular weights.

EGFR Tyrosine Kinase Activity

EGFR tyrosine kinase activity is measured as the phosphorylation of an EGFR substrate peptide, using a tyrosine kinase assay kit (Calbiochem). The EGFR are immunoprecipitated from the cell lysate ($\sim$200 μg protein). Half of the immunoprecipitated protein (10 μl) is transferred to a Microfuge tube that contains 20 μl protein tyrosine kinase (PTK) reaction mix, containing 30 μM ATP, 50 μM substrate peptide, 1 μCi [^{32}P]ATP, and 4 μl 10$\times$ PTK reaction buffer, consisting of 200 mM magnesium chloride, 10 mM manganese chloride, 2 mM EGTA, 80 mM beta-glycerophosphate, and 80 mM imidazole hydrochloride, pH7.3. The mixture is incubated at 30° for 10 min with agitation. The reaction will be stopped by placing the tubes on ice, followed by centrifugation at 10,000g at 4° for 5 sec. The supernatant will be transferred to a new tube containing 10 μl stop solution mix (8 M guanidine hydrochloride) and briefly centrifuged; 8 μl of avidin solution is added to the supernatants, and the samples will be incubated for 5 min at room temperature; 50 μl of wash solution and 20 μl of the reaction samples are transferred into the reservoirs of centrifugal ultrafiltration units, centrifuged for 5 min at 14,000g, and washed three times with 100 μl of wash solution. The washed filters will be transferred to scintillation vials, scintillation cocktail is added, and radioactivity is quantified. Net cpm of [^{32}P] incorporated into the substrate peptide is calculated by subtracting the nonspecific binding of [^{32}P]ATP from the total cpm.

Phospho-p44/42 MAP Kinase Assays

Phosphorylation of p44/42 MAP kinase is determined by Western blot, using the phospho-p44/42 MAP kinase antibody. A MAP kinase antibody is used to detect the total MAP kinase in each sample. Equal volumes of lysate and sample buffer are mixed, boiled for 5 min, microcentrifuged for

2 min, and then subjected to SDS-PAGE electrophoresis (12% gel) and Western blotting.

Plasma Membrane Separation and In Vitro Assay of EGFR Autophosphorylation

NHDF are incubated with GD1a in FBM for 18 h. The plasma membranes are separated as previously described (Smart et al., 1995). After aspirating the medium, the cells are washed twice with the ice-cold buffer A (0.25 M sucrose, 1 mM EDTA, 20 mM Tricine, pH 7.8), scraped into 3 ml buffer A, centrifuged at 1000g for 5 min, resuspended in 1 ml buffer A, and homogenized in a 2-ml Wheaton Tissue grinder with 20 strokes. The cell homogenates are centrifuged at 1000g for 10 min. The pellets are resuspended in 1 ml buffer A, homogenized and centrifuged at 1000g for 10 min. The two supernatants are combined, layered on 30% Percoll in buffer A, ultracentrifuged at 84,000g for 30 min, and the visible bands are collected. These plasma membrane fractions are washed three times with HNG buffer (20 mM HEPES [N-2-hydroxyethylpiperazine-N-2-ethanesulfonic acid], 150 mM NaCl, 10% glycerol, pH 7.5), and the protein content is quantified. The resuspended pellets are assayed for EGFR activity (following).

In Vitro EGFR Activity Assays

Fifteen microliters protein of plasma membrane preparation (about 10 μg protein) is mixed, on ice, with 15 μl autophosphorylation buffer (HNG buffer containing 40 μM ATP, 30 mM MnCl$_2$, and 400 mM Na$_3$VO$_4$). EGF is added to a final concentration of 2 ng/ml. Then the mixture is incubated at 37° for 10 min with gentle agitation. The reaction is stopped by adding 30 μl of 2× lysis buffer, and the sample is stored on ice for 30 min. The samples are immunoprecipitated with the EGFR antibody (sheep polyclonal IgG); and the p-EGFR, total EGFR, and p-tyr kinase activities are tested as before.

Plasma Membrane Binding of GD1a

To assess ^{14}C–GD1a binding, cells are cultured in FBM with 3.5 × 10^5 cpm ^{14}C–GD1a/well in a 24-well plate for 18 h. Then the medium is removed, and the cells are washed twice with PBS, trypsinized, and harvested. The cell membranes are separated, and membrane-bound radioactivity is quantified by scintillation counting.

^{125}I-EGF Binding

Binding of ^{125}I-EGF to whole cells is assessed by modified Scatchard for 18-h analysis (Bremer et al., 1984). Three parallel sets of 10^4 NHDF are grown for 24 h in a 96-well plate in FGM-2, washed twice with FBM, and incubated with 0, 5, or 10 μM GD1a in FBM for 18 h. One set of cells is to be washed and incubated at 4° for 2 h with 0.5–20 ng/ml ^{125}I-labeled human recombinant EGF in FBM, without cold EGF, to determine the total binding. The second set, used to test for nonspecific binding, is treated as earlier, but with the addition of 300 ng/ml unlabeled EGF. After washing, 50 μl of lysis buffer is added to each well, the samples will be lysed on ice for 30 min, and the cpm of ^{125}I-EGF in 20-μl aliquots are counted. Specific binding is calculated by subtracting the nonspecific binding of ^{125}I-EGF from the total binding. The third set of cells will be trypsinized and the cell number determined. The binding curves and Scatchard analysis of the data is performed using GraphPad Prism 3.03 software.

Assessment of EGFR Dimerization and Ligand-Independent Dimerization

NHDF cells are cultured in 100×20-mm culture dishes and preincubated with GD1a in FBM for 18 h, washed, treated with (for dimerization) or without (for ligand-independent dimerization) 2 ng EGF/ml in FBM for 5 min, and washed twice with ice-cold phosphate-buffered saline (PBS); 3 ml of PBS containing 1 mg/ml bis(sulfosuccinimidylpropionate)suberate (BS^3) is added, and the cells are incubated on ice for 30 min, washed twice with ice-cold PBS, and lysed in 1 ml lysis buffer for 20 min on ice (because of BS3 sensitivity to moisture, it must be prepared before adding to the cell system). Total EGFR is immunoprecipitated with the EGFR antibody; 500 μg of lysate protein is loaded onto a 7.5% SDS-PAGE, and dimerization of the EGFR is detected by Western blotting using an anti-EGFR antibody.

Measurement of Effects of Cellular Ganglioside Enrichment on Cell Proliferation

Fibroblasts are seeded at 5×10^3 cells/well in 96-well plates (area = 0.32 cm^2; Corning Glass) FGM containing 2% FBS. In selected assays, fibroblasts are seeded at 3×10^4 cells/well in the larger 24-well plates (area = 1.9 cm^2; Corning Glass). The medium is replaced with fresh medium $\pm$ gangliosides on the following day. After an 18-h incubation, the culture medium is removed, and the cells are starved overnight in serum-free FBM. This medium is to be replaced with serum-free medium $\pm$ EGF, and the cells are further cultured for 24 h. In some experiments, the starvation period is omitted, and after incubation of the cells with gangliosides for 18 h, the culture medium is replaced directly with serum-free medium $\pm$ EGF, and the cells are further cultured for 24 h. During the final 3 h, the cells were pulsed with 50 μl of [^{3}H]thymidine (5 μCi/ml) and harvested by a cell harvester. [^{3}H]Thymidine uptake is determined by β-scintillation counting. In selected experiments, [^{3}H]thymidine incorporation into trichloroacetic acid-insoluble material is also determined as described previously. Under these same conditions, in parallel experiments, cell counts of the 24-well plates were performed after trypsinization of the cells. The data are expressed as the mean $\pm$ SD of duplicate or triplicate cultures.

Results and Conclusion

As we have described previously, gangliosides may act on cell membrane to modulate growth factor–induced proliferation by influencing important membrane-associated events, including receptor dimerization. We hypothesize that the current controversy as to the exact nature of the effect of gangliosides on cell signaling may be clarified by the adoption of a standard methodological approach that as closely as possible mimics the *in vivo* physiological environment. Using the methods outlined previously, we have demonstrated that preincubation of HDF with ganglioside (GD1a, GM3, or GD3) followed by washout of unbound ganglioside before ligand (EGF) stimulation results in enhancement of fibroblast proliferation (Fig. 3). In contrast, cell proliferation was inhibited when gangliosides were not removed from the culture media before ligand addition. Furthermore, we have demonstrated enhancement of signaling through the EGF receptor in the presence of GD1a, including receptor autophosphorylation, as well as activation of downstream molecules Ras and MAPK (Fig. 4). Finally, lending some insight into the possible mechanism of action of gangliosides in enhancement of signaling through EGFR, we have demonstrated that GD1a enhances EGFR dimerization (and hence the effective number of high-affinity receptors) both in the presence and in the absence of ligand (Fig. 5).

We have demonstrated an enhancement of signaling through the EGFR by the ganglioside GD1a, which stands in apparent contrast to previously published studies that demonstrate an overall inhibitory effect of gangliosides on signaling through EGFR (Alves *et al.*, 2002; Bremer *et al.*, 1986; Meuillet *et al.*, 2000; Miljan *et al.*, 2002; Mirkin *et al.*, 2002; Rebbaa *et al.*, 1996; Sottocornola *et al.*, 2003; Suarez Pestana *et al.*, 1997) and other growth factor receptors (Farooqui *et al.*, 1999; Hynds *et al.*, 1995). Although it is certainly true that different gangliosides may have differing effects on signaling through the same receptor, we believe that the observed differences may be attributable in part to critical differences in methodological approaches. High concentrations of gangliosides and the continuous

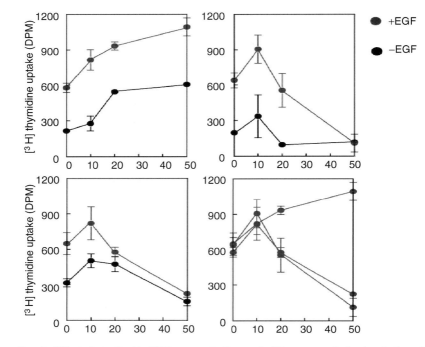

FIG. 3. Effect of ganglioside GD1a concentration, and of its presence during incubation with EGF, on cell proliferation. The effect of ganglioside GD1a on cell proliferation was evaluated under the following three condition: (1) preincubation of cells with gangliosides GD1a (*Panel A*); (ii) continuous incubation of cells with ganglioside GD1a (*Panel B*); and (iii) ganglioside preincubation, starvation without ganglioside presence, and co-incubation of the cells with ganglioside + EGF (*Panel C*). NHDF cells were seeded at 5×10^3 cells/well in a 96-well plate in FGM-2. The medium was replaced with fresh medium ± ganglioside GD1a on the following day. After an 18 hours, the medium was removed, and the cells were cultured in serum-free medium (FBM) overnight to starved the cells and were then replaced with serum-free medium ± EGF (2ng/ml). The cells were further cultured for 24 hours, and cell proliferation was measured by [^{3}H]thymidine uptake (A). In (B), ganglioside GD1a was present in the culture medium throughout the experiment (i.e., during the preincubation, the starvation period, and the incubation of the cells with EGF). In (C) the cells were preincubated with ganglioside GD1a for 18 h and then starved overnight in the absence of ganglioside. Ganglioside GD1a was added back to the culture medium during the incubation with EGF. (D) Composite of (A–C), showing the effect of ganglioside GD1a on EGF-induced cell proliferation. Each value is the mean ± SD of duplicated culture. Reproduced from Li *et al.* (2000), with permission.

presence of ganglioside (including concomitant presence of ganglioside and receptor ligand), the presence of serum proteins, and the use of high ligand (growth factor) concentrations may create a number of biological

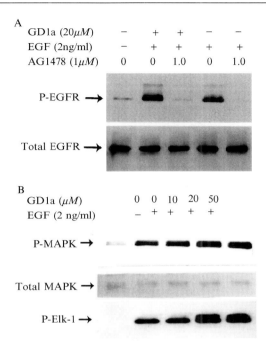

FIG. 4. Ganglioside GD1a preincubation enhances EGF-induced EGFR and MAPK phosphorylation. In (A) NHDF were cultured in FGM-2 to subconfluence, washed twice, and incubated with GD1a in FBM for 18 h. The inhibitor AG1478 (1 μM) was added to the culture system for the final 3 h of the incubation. The cells were then exposed to EGF (2 ng/ml) in FBM for 5 min, washed with ice-cold PBS, and lysed; 200 μl cell lysate containing 100 μg total protein was immunoprecipitated using an anti-EGFR sheep polyclonal IgG. The immune complex was subjected to SDS-PAGE and Western blot. EGFR phosphorylation was detected by an anti-phosphotyrosine antibody. Total EGFR was detected using an anti-EGFR sheep polyclonal IgG. (B) After the cell density reached 70% of confluence, the cells were incubated with ganglioside GD1a in FGM with 2% FBS for 18 h after removal of the culture medium by aspiration, the cells were washed twice and starved in serum free FBM overnight. The cells were then cultured in serum-free medium $\pm$ EGF (2 ng/ml) in for 5 min, immediately washed with ice-cold PBS, and lysed; 20 μg of the cell lysate was subjected to SDS-PAGE and Western blot, using either the phospho-P44/42 MAP kinase or the P44/42 MAP kinase antibody. In addition, the MAP kinase activity was determined by measuring Elk-1 phosphorylation. In these experiments, the cell lysate containing 200 μg of total protein for each sample was immunoprecipitated using a specific anti-phospho–p44/42 MAP kinase antibody. The kinase assay was performed using Elk-1 as the substrate. One third of the total product was subjected to SDS-PAGE and Western blot to visualize the phosphorylated Elk-1 bands, using anti-phospho-Elk-1 antibody. Reproduced from Liu *et al.* (2004) (A) and Li *et al.* (2001) (B), with permission.

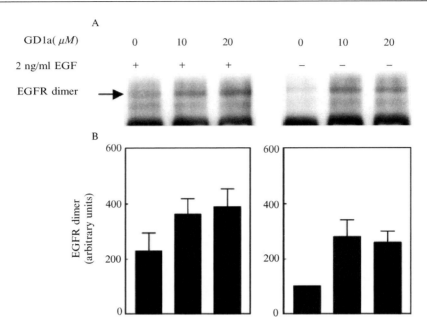

FIG. 5. Effects of GD1a pretreatment on EGFR dimerization. Subconfluent NHDF cells were washed, preincubated with GD1a, and treated with EGF as in Fig. 2A. Then, the cells were washed twice with ice-cold phosphate-buffered saline (PBS) and incubated with 3 ml of PBS containing 1 mg/ml BS^3 for 30 min on ice, washed again, and lysed in 1 ml lysis buffer for 20 min. Immunoprecipitation and detection of EGFR dimers were performed as described in "Materials and Methods." The relative optical densities of the EGFR dimer bands, representing the composite (mean $\pm$ SD) results of five separate experiments, are shown below the Western blot. (A) Ganglioside preincubation followed by EGF exposure; (B) ganglioside incubation only (no EGF exposure). Reproduced from Liu *et al.* (2004), with permission.

phenomena that one would not expect to find in the *in vivo* state. Specifically, high ganglioside concentrations that are present even during growth factor stimulation may not only intercalate into the membrane but may also bind to ligand, preventing binding to its receptor. Serum proteins present in the experimental milieu may nonspecifically bind gangliosides reducing their availability for membrane intercalation (and will, at the very least, prevent accurate knowledge of the exact ganglioside concentration). Finally, use of high ligand (growth factor) concentrations will maximally stimulate signaling, preventing detection of the subtle changes in signaling that are more likely present in the *in vivo* state.

Acknowledgments

This work was supported in part by the National Institutes of Health Grant R01 CA61010 (S. L.) and R01 CA106532 (K. K.) and by The Children's Cancer Foundation, Baltimore, MD (K. K. and S. L.)

References

Alessandri, G., Filippeschi, S., Sinibaldi, P., Mornet, F., Passera, P., Spreafico, F., Cappa, P. M., and Gullino, P. M. (1987). Influence of gangliosides on primary and metastatic neoplastic growth in human and murine cells. *Cancer Res.* **47**, 4243–4247.

Alves, F., Borchers, U., Keim, H., Fortte, R., Olschimke, J., Vogel, W. F., Halfter, H., and Tietze, L. F. (2002). Inhibition of EGF-mediated receptor activity and cell proliferation by HK1-ceramide, a stable analog of the ganglioside GM3-lactone. *Glycobiology* **12**, 517–522.

Bergelson, L. D., Dyatlovitskaya, E. V., Klyuchareva, T. E., Kryukova, E. V., Lemenovskaya, A. F., Matveeva, V. A., and Sinitsyna, E. V. (1989). The role of glycosphingolipids in natural immunity. Gangliosides modulate the cytotoxicity of natural killer cells. *Eur. J. Immunol.* **19**, 1979–1983.

Bremer, E. G., Hakomori, S., Bowen-Pope, D. F., Raines, E., and Ross, R. (1984). Ganglioside-mediated modulation of cell growth, growth factor binding, and receptor phosphorylation. *J. Biol. Chem.* **259**, 6818–6825.

Bremer, E. G., Schlessinger, J., and Hakomori, S. (1986). Ganglioside-mediated modulation of cell growth. Specific effects of GM3 on tyrosine phosphorylation of the epidermal growth factor receptor. *J. Biol. Chem.* **261**, 2434–2440.

Farooqui, T., Kelley, T., Coggeshall, K. M., Rampersaud, A. A., and Yates, A. J. (1999). GM1 inhibits early signaling events mediated by PDGF receptor in cultured human glioma cells. *Anticancer Res.* **19**, 5007–5013.

Grayson, G., and Ladisch, S. (1992). Immunosuppression by human gangliosides. II. Carbohydrate structure and inhibition of human NK activity. *Cell Immunol.* **139**, 18–29.

Hakomori, S. (1996). Tumor malignancy defined by aberrant glycosylation and sphingo(glyco) lipid metabolism. *Cancer Res.* **56**, 5309–5318.

Hakomori, S., Handa, K., Iwabuchi, K., Yamamura, S., and Prinetti, A. (1998). New insights in glycosphingolipid function: "glycosignaling domain," a cell surface assembly of glyco-sphingolipids with signal transducer molecules involved in cell adhesion coupled with signaling. *Glycobiology* **8**, xi–xix.

Hynds, D. L., Summers, M., Van Brocklyn, J., O'Dorisio, M. S., and Yates, A. J. (1995). Gangliosides inhibit platelet-derived growth factor-stimulated growth, receptor phosphorylation, and dimerization in neuroblastoma SH-SY5Y cells. *J. Neurochem.* **65**, 2251–2258.

Iyer, V. R., Eisen, M. B., Ross, D. T., Schuler, G., Moore, T., Lee, J. C., Trent, J. M., Staudt, L. M., Hudson, J., Jr., Boguski, M. S., Lashkari, D., Shalon, D., Botstein, D., and Brown, P. O. (1999). The transcriptional program in the response of human fibroblasts to serum. *Science* **283**, 83–87.

Jo, M., Stolz, D. B., Esplen, J. E., Dorko, K., Michalopoulos, G. K., and Strom, S. C. (2000). Cross-talk between epidermal growth factor receptor and c-Met signal pathways in transformed cells. *J. Biol. Chem.* **275**, 8806–8811.

Ladisch, S., Gillard, B., Wong, C., and Ulsh, L. (1983). Shedding and immunoregulatory activity of YAC-1 lymphoma cell gangliosides. *Cancer Res.* **43**, 3808–3813.

Ladisch, S., Li, R., and Olson, E. (1994). Ceramide structure predicts tumor ganglioside immunosuppressive activity. *Proc. Natl. Acad. Sci. USA* **91**(5), 1974–1978.

Lang, Z., Guerrera, M., Li, R., and Ladisch, S. (2001). Ganglioside GD1a enhances VEGF-induced endothelial cell proliferation and migration. *Biochem. Biophys. Res. Commun.* **282**, 1031–1037.

Ledeen, R. W., and Yu, R. K. (1982). Gangliosides: Structure, isolation, and analysis. *Methods Enzymol.* **83**, 139–191.

Li, R., Liu, Y., and Ladisch, S. (2001). Enhancement of epidermal growth factor signaling and activation of SRC kinase by gangliosides. *J. Biol. Chem.* **276**, 42782–42792.

Li, R., Manela, J., Kong, Y., and Ladisch, S. (2000). Cellular gangliosides promote growth factor-induced proliferation of fibroblasts. *J. Biol. Chem.* **275**, 34213–34223.

Liu, Y., Li, R., and Ladisch, S. (2004). Exogenous ganglioside GD1a enhances epidermal growth factor receptor binding and dimerization. *J. Biol. Chem.* **279**, 36481–36489.

Lowry, O. H., Rosebrough, N. J., Farr, A. L., and Randall, R. J. (1951). Protein measurement with the Folin phenol reagent. *J. Biol. Chem.* **193**, 265–275.

Lu, P., and Sharom, F. J. (1996). Immunosuppression by YAC-1 lymphoma: Role of shed gangliosides. *Cell. Immunol.* **173**, 22–32.

Manfredi, M. G., Lim, S., Claffey, K. P., and Seyfried, T. N. (1999). Gangliosides influence angiogenesis in an experimental mouse brain tumor. *Cancer Res.* **59**, 5392–5397.

Meuillet, E. J., Mania-Farnell, B., George, D., Inokuchi, J. I., and Bremer, E. G. (2000). Modulation of EGF receptor activity by changes in the GM3 content in a human epidermoid carcinoma cell line, A431. *Exp. Cell Res.* **256**, 74–82.

Miljan, E. A., and Bremer, E. G. (2002). Regulation of growth factor receptors by gangliosides. *Sci, STKE* **2002**, RE15.

Miljan, E. A., Meuillet, E. J., Mania-Farnell, B., George, D., Yamamoto, H., Simon, H. G., and Bremer, E. G. (2002). Interaction of the extracellular domain of the epidermal growth factor receptor with gangliosides. *J. Biol. Chem.* **277**, 10108–11013.

Mirkin, B. L., Clark, S. H., and Zhang, C. (2002). Inhibition of human neuroblastoma cell proliferation and EGF receptor phosphorylation by gangliosides GM1, GM3, GD1A and GT1B. *Cell Prolif.* **35**, 105–115.

Nishio, M., Tajima, O., Furukawa, K., Urano, T., and Furukawa, K. (2005). Over-expression of GM1 enhances cell proliferation with epidermal growth factor without affecting the receptor localization in the microdomain in PC12 cells. *Int. J. Oncol.* **26**, 191–199.

Rebbaa, A., Hurh, J., Yamamoto, H., Kersey, D. S., and Bremer, E. G. (1996). Ganglioside GM3 inhibition of EGF receptor mediated signal transduction. *Glycobiology* **6**, 399–406.

Rusnati, M., Urbinati, C., Tanghetti, E., Dell'Era, P., Lortat-Jacob, H., and Presta, M. (2002). Cell membrane GM1 ganglioside is a functional coreceptor for fibroblast growth factor 2. *Proc. Natl. Acad. Sci. USA* **99**, 4367–4372.

Smart, E. J., Ying, Y. S., Mineo, C., and Anderson, R. G. (1995). A detergent-free method for purifying caveolae membrane from tissue culture cells. *Proc. Natl. Acad. Sci. USA* **92**, 10104–10108.

Sottocornola, E., Berra, B., and Colombo, I. (2003). GM3 content modulates the EGF-activated p185c-neu levels, but not those of the constitutively activated oncoprotein p185neu. *Biochim. Biophys. Acta.* **1635**, 55–66.

Suarez Pestana, E., Greiser, U., Sanchez, B., Fernandez, L. E., Lage, A., Perez, R., and Bohmer, F. D. (1997). Growth inhibition of human lung adenocarcinoma cells by antibodies against epidermal growth factor receptor and by ganglioside GM3: Involvement of receptor-directed protein tyrosine phosphatase(s). *Br. J. Cancer* **75**, 213–220.

van Echten, G., and Sandhoff, K. (1993). Ganglioside metabolism. Enzymology, topology, and regulation. *J. Biol. Chem.* **268,** 5341–5344.

Zeng, G., Gao, L., Birkle, S., and Yu, R. K. (2000). Suppression of ganglioside GD3 expression in a rat F-11 tumor cell line reduces tumor growth, angiogenesis, and vascular endothelial growth factor production. *Cancer Res.* **60,** 6670–6676.

Zurita, A. R., Maccioni, H. J., and Daniotti, J. L. (2001). Modulation of epidermal growth factor receptor phosphorylation by endogenously expressed gangliosides. *Biochem. J.* **355,** 465–472.

[14] Activation of Natural Killer T Cells by Glycolipids

By EMMANUEL TUPIN and MITCHELL KRONENBERG

Abstract

Natural killer T (NKT) cells are a distinct T-cell sublineage, originally named because of their coexpression of an $\alpha\beta$ T cell antigen receptor (TCR) characteristic of T lymphocytes, and NK1.1, a C-type lectin expressed by natural killer (NK) cells. NKT cells use their TCR to recognize glycolipids bound to or presented by CD1d. Until recently, most studies used the synthetic glycolipid α-galactosylceramide (αGalCer) to activate these lymphocytes, and very little was known about the natural antigens recognized by NKT cells. Given the pivotal role played by the NKT cells in many immune responses, including antimicrobial responses, tumor rejection, and the development of autoimmune diseases, the identification of the natural antigens recognized by these cells, and analogs that may alter their cytokine production, are goals of primary importance. This chapter discusses methods that can be used to assess the potency of potential glycolipid antigens for this unique population of T lymphocytes, including methods for *in vitro* NKT cell activation and expansion, *in vivo* activation, and measurement of their avidity for different antigens.

Overview

Natural killer T cells are named after their coexpression of an $\alpha\beta$ T cell antigen receptor (TCR) along with surface receptors typical for NK cells, such as NK1.1 in mice (Kronenberg and Gapin, 2002). For several reasons, however, this classification has become inadequate. First, only three commonly used inbred mouse strains, C57BL/6, NZB, and SJL, express an allelic form of NK1.1 that can be recognized by the available NK1.1 monoclonal antibody (mAb). Second, in addition to CD1d reactive cells, some conventional T lymphocytes up regulate NK1.1/CD161 on activation.

METHODS IN ENZYMOLOGY, VOL. 417   0076-6879/06 $35.00
     DOI: 10.1016/S0076-6879(06)17014-7

Moreover, some CD1d-dependent "NKT" cells do not express NK1.1 (Godfrey et al., 2004).

A newer nomenclature, therefore, was established on the basis of the expression of an invariant α chain TCR rearrangement. Contrary to conventional T cells that express highly diverse antigen receptors, most mouse NKT cells express an invariant Vα14-Jα18 TCR α rearrangement coexpressed with a limited number of β chains (Vβ8.2, Vβ7, or Vβ2). These cells, therefore, are sometimes referred to as Vα14 invariant (Vα14i) NKT cells. The homologous population in human expresses an invariant Vα24-Jα18 rearrangement paired with Vβ11 (Dellabona et al., 1994; Porcelli et al., 1993); these are the orthologs of mouse Vα14 and Vβ11. We refer to the human NKT cell population as Vα24i NKT cells, and collectively to these subpopulations in mice and humans as iNKT cells.

On activation, iNKT cells very rapidly produce large amounts of different cytokines, including IFNγ and IL-4, suggesting they could have a potent immunoregulatory function. This has been confirmed in numerous studies demonstrating the pivotal roles played by iNKT cells in many different types of immune responses and cancer (Smyth et al., 2002; Van Kaer, 2004).

Conventional T lymphocytes respond to peptides presented by the polymorphic class I and class II antigen-presenting molecules. iNKT cells, by contrast, are activated on recognition of glycolipids presented by CD1d, a nonclassical or nonpolymorphic antigen-presenting molecule. CD1d is distantly related to both the class I and class II molecules, and like class I molecules, it consists of a heavy chain noncovalently associated with β2-microglobulin (Porcelli, 1995). When bound to CD1d, the lipid tails of glycolipid are buried in the hydrophobic CD1d groove, leaving predominantly the carbohydrate portion available for TCR recognition together with CD1d (Zajonc et al., 2005). αGalCer, a synthetic glycosphingolipid, originally extracted from a marine sponge in a screen for compounds that could prevent tumor metastases to the liver of mice (Kobayashi et al., 1995), was the first antigen described that activates iNKT cells (Kawano et al., 1997). αGalCer was not thought to be the natural antigen for iNKT cells, however, because of its unusual α linkage of the 1' carbon of the sugar to the 1 carbon of the sphingosine base, whereas in nearly all other glycosphingolipids the bond connecting the sugar to the lipid is in the β anomeric form (Fig. 1). Recent publications, however, showed that Sphingomonas, a common environmental bacteria, have α linked glycosphingolipids, and these compounds are able to bind to CD1d and activate both mouse and human iNKT cells (Kinjo et al., 2005; Mattner et al., 2005; Sriram et al., 2005). iNKT cells are weakly self-reactive for CD1d$^+$ antigen-presenting cells, and isoglobotrihexosyl ceramide (iGb3) was identified recently as an

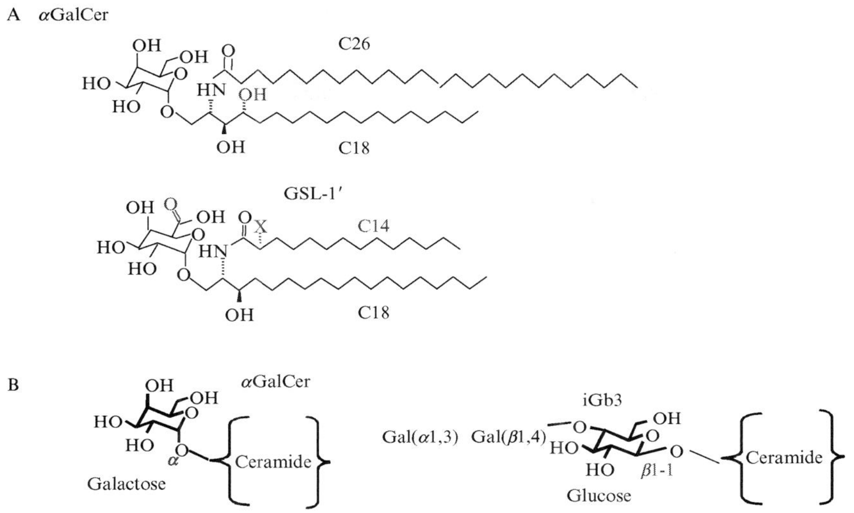

FIG. 1. Structures of glycolipids recognized by *i*NKT cells. (A) The synthetic antigen αGalCer (top) is compared with a *Sphingomonas* bacterial glycosphingolipid (GSL-1'). Differences between the two compounds are noted, including changes in the ceramide lipid and the carboxylate modification of galactose in the bacterial glycosphingolipid. Different *Sphingomonas* species have glycolipids with differences in the carbohydrate and the ceramide lipid. (B) Comparison of αGalCer to the autologous antigeni Gb3. The ceramide is indicated schematically, and the anomeric linkage of the sugar to the lipid is in red.

autoantigen likely to be required for the development of these lymphocytes (Zhou *et al.*, 2004).

In this chapter, we discuss methods used to identify, expand, and assay Vα14*i* NKT cells, and methods to test the potency of different glycolipid antigens.

Vα14i NKT Cell Isolation

Vα14*i* NKT cells can be found nearly everywhere conventional T cells are found, but with the exception of the liver, they are a small minority T-cell population, and, therefore, an important consideration is their absolute number, as well as their frequency. For example, although the total number of Vα14*i* NKT cells in the spleen is equal to or slightly larger than in other organs, they represent less than 1% of the total spleen cells, whereas they represent up to 40% of the liver mononuclear cells. Several methods of liver mononuclear cell preparation can be found, but all are based on the formation of a density gradient. In this chapter, we will present two different methods of liver mononuclear isolation. The choice of the method will depend on the need for a higher total number of mononuclear cells versus a need for a higher degree of cell purity.

Liver Mononuclear Cell Isolation

Solutions for Method 1

- 37.5% isotonic Percoll: for 1 l, mix:
  337.5 ml of Percoll (Amersham-Pharmacia Biotech, #17-0891-01)
  100 ml of 10× phosphate-buffered saline (PBS)
  562.5 ml of double distilled (dd) H_2O
- Red Blood Cell (RBC) Lysing Buffer (Sigma-Aldrich, #R7757)
- PBS/FBS/Az solution: PBS supplemented with 2% fetal bovine serum (FBS) and 0.02% sodium azide (NaN_3)

Method 1 (Rapid Method)

1. After euthanasia of the mouse, the liver is perfused *in situ* by the hepatic portal vein with room temperature (RT) PBS until the organ becomes pale. The liver is then harvested and placed in PBS/FBS/Az solution.
2. The liver is cut into small pieces, pressed gently through a 70-μm cell-strainer (BD Falcon, #352350), and suspended in 40 ml of cold PBS/FBS/Az solution.
3. After centrifugation at 500g for 7 min at 4°, the supernatant is discarded and the cells washed one more time in cold PBS/FBS/Az.
4. The pellet is then resuspended in 25 ml of a *room temperature* Percoll solution and centrifuged at 680–700g for 12 min at *room temperature*. The cells of interest will form a pellet; therefore, the brake can be left on.
5. The supernatant is discarded, and the pellet containing RBC and lymphocytes is washed in 10 ml of cold PBS/FBS/Az solution.
6. The RBCs are lysed by resuspending the pellet in 2 ml of red blood cell lysing buffer for 4–5 min at RT.
7. The reaction is stopped by addition of 10–20 ml of cold PBS/FBS/Az solution and the cells pelleted by centrifugation.
8. The cells are resuspended in 5 ml PBS/FBS/Az and filtered through 70-μm cell-strainer to remove the debris and washed one more time.
9. Finally, the mononuclear cells (MNC) are resuspended in 1 ml of PBS/FBS/Az solution, tissue culture medium, or staining buffer, according to the downstream experiment. When purified from an untreated mouse, an average of 4–5 × 10^6 MNC (60–70% of them being lymphocytes) can be recovered with this method.

Solutions for Method 2

- "100%" isotonic Percoll: 1 part 10× PBS + 9 parts Percoll
- 70% Percoll: 7 ml "100%" isotonic Percoll + 3 ml RPMI plus 5% FBS
- Medium: RPMI-1640 medium (Gibco, #11875-093) supplemented with 5% heat-inactivated FBS

Method 2 (High-purity Method). The liver is harvesting as described in "Method 1."

1. The liver is mashed through a 70-μm cell-strainer into a 50-ml Falcon tube, washed with 40 ml of media, and centrifuged at 500*g* for 5 min at 4°.
2. The cells are resuspended in 18 ml of media and 12 ml of "100%" isotonic Percoll, resulting in a 40% Percoll cell suspension.
3. 10 ml of the 70% isotonic Percoll is carefully lain under the 40% Percoll cell suspension.
4. Centrifugation at *900g* at *room temperature* for 30 min, brake *off.*
5. The mononuclear cells are harvested at the interface of the two layers of Percoll and washed once.
6. The cells are resuspended in 1 ml of the desired solution, depending on the downstream experiment. With this technique, an average of 1 to 2×10^6 MNC could be expected and with up to 80% lymphocytes.

Spleen or Thymus Cell Preparation

There are several reasons for harvesting Vα14*i* NKT cells from organs other than the liver. As already mentioned, although Vα14*i* NKT cells represent only 1% of the spleen cells, their absolute number is usually higher than in the liver and the spleen is easily harvested. Moreover, Vα14*i* NKT from different organs might display different functional properties (Crowe *et al.*, 2005).

Solutions

- Complete medium (CM): RPMI-1640 medium (Gibco, #11875-093) supplemented with 10% heat-inactivated FBS, penicillin G (100 U/ml), streptomycin sulfate (100 μg/ml), L-glutamine (0.292 mg/ml), and 2-mercaptoethanol (48.49 μM)

Method

1. The spleen is harvested, pressed gently through 70-μm cell-strainers, suspended in 20 ml of cold CM, and centrifuged at 500*g* for 5 min at 4°.
2. The supernatant is discarded, and the RBCs are lysed by resuspending the pellet in 2 ml of RBC lysing buffer for 4–5 min at RT.

3. The reaction is stopped by addition of 18 ml of cold PBS/FBS/Az and the cells pelleted by centrifugation.
4. The cells are resuspended in 10 ml CM, filtered through a 70-μm cell strainer to eliminate the debris and washed once more.
5. Finally, the cells are resuspended in 1 ml and counted. Depending on the age of the mouse, 6–10 10^7 MNC will be obtained.

Purification of Vα14i NKT Cells by Cell Sorting

Even if prepared from the liver, the highest percentages of Vα14i NKT cells among total lymphocytes that can be obtained is less than 40%, and in some cases, it might be necessary to further purify them. Positive selection using CD1d-αGalCer tetramers will achieve the highest purity and specificity. Protocols for the production of soluble mCD1d proteins and generation of CD1d-αGalCer tetramers have been described in detail in several articles (Benlagha et al., 2000; Matsuda et al., 2000; Naidenko et al., 1999; Sidobre and Kronenberg, 2002) and will, therefore, not be repeated here. Moreover, CD1d tetramers loaded with a compound closely related to αGalCer are also available from the NIH Tetramer Core Facility. (http://www.yerkes.emory.edu/TETRAMER/CD1d_Tetramers.html).

For sorting with CD1d-αGalCer tetramers, spleen cells are first depleted of B cells and CD8$^+$ T cells using anti-CD19 and anti-CD8α conjugated magnetic beads and run through MACS columns (Miltenyi), according to the manufacturer's protocol. The CD19$^-$CD8$^-$ cell suspension is labeled with CD1d-αGalCer tetramers and anti-TCRβ, washed, and the double positive cells are sorted on a FACSDiva (BD BioSciences). This technique is the most specific one, but it has the disadvantage of partially activating the cells. Indeed, production of a small amount of the cytokine IL-4 by the sorted Vα14i NKT cells can be detected after 48 h of culture, without addition of antigen.

An alternative positive selection method for obtaining Vα14i NKT cells uses anti-TCRβ and anti-NK1.1 mAbs. Despite being less specific than tetramer-based sorting, this method does not cause a high degree of activation. NK1.1$^+$ cells are enriched using anti-phycoerythrin (PE) conjugated magnetic beads (Miltenyi) after staining with anti-NK1.1-PE antibody. The NK1.1$^+$ cell-enriched population is then stained with anti-TCRβ, anti-CD8 and anti-CD19, before washing and sorting. CD8$^+$ and CD19$^+$ cells are gated out, whereas NK1.1$^+$TCRβ^+ cells are collected.

In Vitro Antigen Presentation Assay

CD1d-coated Plates

A simple assay for antigenic potency requires Vα14*i* NKT cells but does not require antigen-presenting cells (APC). This assay works very well with immortalized cells, Vα14*i* NKT cell hybridomas. The derivation and characterization of the mouse CD1d (mCD1d)–autoreactive hybridomas that we use have been described previously (Brossay *et al.*, 1998b; Cardell *et al.*, 1995). This method is probably the most sensitive for glycolipid antigen detection, but it cannot detect glycolipid antigens that require antigen processing (catabolism) or the activity of lysosomal lipid transfer proteins for CD1d loading.

Method

1. 100 μl of soluble mCD1d protein (10 μg/ml in PBS) is coated in 96-well flat bottom plate by 1 h incubation at 37°.
2. The plates are washed four times with 200 μl PBS and nonspecific binding blocked by adding 200 μl of PBS-10% FBS for 1 h.
3. The blocking solution is discarded and 100 μl of the glycolipid of interest is added to the well and incubated for 2–24 h at 37°. Because of the relative insolubility of most glycolipids, it is recommended that the compounds are sonicated for 15 min in warm water (40–50°) bath before adding the glycolipid to the plates. Most compounds are diluted in a vehicle containing 0.5% polysorbate-20 and 0.9% NaCl. Some very insoluble glycolipids are diluted in dimethyl sulfoxide (DMSO), but the soluble recombinant CD1d is very sensitive to the DMSO concentration, and, therefore, compounds dissolved in DMSO do not work with this method.
4. The plates are washed three times with PBS and then twice with CM. Be careful to remove any air bubbles that might have been created by the pipetting.
5. 5×10^4 hybridoma cells in 200 μl are then immediately added to each well. The hybridoma DN3A4-1.2 is useful because of its stable TCR expression and high αGalCer reactivity. It is, however, recommended to confirm the data using additional hybridomas; several Vα14*i* NKT cell hybridomas are available on request from the authors.
6. IL-2 release in the supernatant is measured after 16 h of culture in a sandwich ELISA using a pair of rat anti-mouse IL-2 mAbs for capture (cat. #554424) and detection (cat. #554426) and a recombinant IL-2 standard (cat. #550069) (BD BioSciences, PharMingen).

In Vitro Vα14i NKT Cell Expansion

The antigenic potency of a glycolipid can be assessed by its capacity to induce the *in vitro* proliferation of splenic Vα14i NKT cells. Proliferation can be detected by an increase in the percentage of Vα14i NKT cells and by their uptake of bromo-deoxyuridine (BrdU).

Culture Protocol

1. Mononuclear cells from the spleen are prepared as described, resuspended at a concentration of 2×10^6/ml in complete media (CM), seeded at 4×10^5 cells/well (200 μl/well) in a 96-well round bottom plate.
2. The glycolipid of interest, diluted in vehicle or DMSO and previously sonicated, is then added to the culture.
3. The cells are incubated at 37° and 5% CO_2 for 3–7 days.
4. 14–16 h before the culture end point, 10 μM of BrdU is added to the cells.
5. The cells are harvested and stained for flow cytometry analysis.

Flow Cytometry Staining. The readout consists of an extracellular staining of the cultured cells with a αβTCR mAb and CD1d-αGalCer tetramers, to quantify the proportion of Vα14i NKT cells, and an mAb specific for the B-cell antigen B220 to eliminate false-positive cells. Background is further reduced using an unlabeled antibody specific for mouse IgG Fc receptors. This is coupled with an intracellular staining for the incorporated BrdU. Representative results are shown in Fig. 2.

SOLUTIONS AND MATERIALS

- Staining buffer (SB): $1 \times$ PBS, 2% bovine serum albumin (BSA), 0.1% NaN₃, 10 m*M* ethylenediamine tetraacetic acid (EDTA).
- Fixing buffer (FB): $1 \times$ PBS, 1% paraformaldehyde, 0.1% NaN₃
- Cytofix/cytosperm Plus kit (BD PharMingen #2076KK)
- Unlabeled anti-CD16/32 (IgG Fc receptor) blocking antibody
- Antibody mix is: phycoerythrin (PE)-CD1d/αGalCer tetramer (diluted at 1:150 in SB), R-Phycoerythrin-Cy7 (PeCy7)-B220 (clone RA3-6B2, BD PharMingen #552772, diluted at 1:200 in SB) and allophycocyanin-βTCR (clone H57-597, BD PharMingen #553174, diluted at 1:100 in SB).
- Fluorescein isothiocyanate (FITC) BrdU Flow kit (BD PharMingen, #559619)

METHOD

1. 10^6 cells are added in 200 μl/well in a 96 well round-bottomed plate.

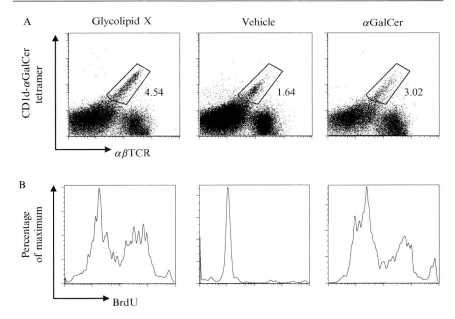

FIG. 2. *In vitro* *i*NKT cell expansion. (A) Percentage of *i*NKT cells (CD1d-αGalCer tetramer[+] and αβTCR[+] cells) among the gated mouse spleen lymphocytes after 5 days of culture with glycolipid X, a monogalactosyl diacylglycerol. (B) Level of BrdU incorporation in the gated *i*NKT cell population. An increase in Brdu incorporation into *i*NKT cell DNA confirms that the increase in the percentage of *i*NKT cells is due to cell proliferation.

2. The cells are centrifuged at 500g for 3 min at 4°.
3. The plate is flicked to remove the media.
4. The cells are resuspended in 25 μl of CD16/32 antibody diluted at 1:200 in SB to block nonspecific binding.
5. After 20 min of incubation at 4° in the dark, 25 μl of the antibody mix is added and the cells incubated 30 min at 4° in the dark.
6. After two washings with 200 μl of SB, the cells are fixed and permeabilized in 100 μl of BD Cytofix/Cytoperm buffer for 20 min at RT.
7. After one washing with 1× BD Perm/Wash buffer, the cells are resuspended for 10 min on ice in 100 μl of BD Cytoperm Plus buffer.
8. The cells are then washed in 1× BD Perm/Wash buffer and resuspended one more time in BD Cytofix/Cytoperm buffer (5 min, RT).
9. After washing, the cells are treated 1 h at 37° with 100 μl of DNase (300 μg/ml) to expose the incorporated BrdU.

10. After the incubation and one washing step, 50 μl of FITC-anti BrdU antibody (diluted at 1:100 in 1× BD Perm/Wash buffer) is added to the cells for 20 min at 4°.
11. After two final washings, the cells are resuspended in FB before flow cytometry analysis in a dual laser FACSCalibur (BD Biosciences) or similar instrument.

This is a sensitive and relatively simple method for antigen detection, but because Vα14i NKT cells can be induced to proliferate by cytokines such as IL-12 released by activated dendritic cells (DC) and macrophages in the absence of antigen, a positive result must be verified by additional experiments. Addition of 10 μg/ml of the anti-CD1d blocking mAb 1B1 (BD PharMingen or eBiosciences) should inhibit TCR-dependent stimulation. Moreover, the use of spleen cells from *IL-12$^{-/-}$* mice (Jackson laboratory, cat number 002693) eliminates the possibility of IL-12–driven proliferation.

Co-culture Assay

This assay uses purified APC and Vα14i NKT cell populations to more precisely dissect the cell–cell interactions required for Vα14i NKT cell activation. We mainly use bone marrow–derived DC, because these are the most potent cells for antigen presentation.

Bone Marrow–derived DC Preparation

MATERIAL AND SOLUTIONS

• Complete Media (CM): as already described
• RBC lysing buffer
• Mouse granulocyte-macrophage colony-stimulating factor (GM-CSF): Peprotech, #315-03
• Bacteriological petri dishes: Fisher brand, #08-757-12

METHOD

1. Femurs and tibiae of mice are removed and purified from the surrounding muscles by rubbing with towel papers. The intact bones are left in 70% ethanol for 2–5 min for disinfection.
2. Both ends of the bones are cut with scissors and the marrow flushed with CM using a 10-ml syringe with 25–26-gauge needle.
3. The cells are pelleted by centrifugation at 500g for 7 min at 4°.
4. The supernatant is discarded, and the RBCs are lysed by resuspending the pellet in 3 ml of RBC lysing Buffer for 4–5 min at RT.
5. The reaction is stopped by addition of 20 ml CM and the cells washed twice with CM before the debris are removed with a 70-μm cell-strainer.

6. Bone marrow cells are seeded at $3–4 \times 10^6$ cells per bacteriological petri dish in 10 ml of CM containing 10 ng/ml of mouse GM-CSF. Bacterial petri dishes should be chosen over culture dishes, because DC are adherent only to culture dishes but not to bacteriological petri dishes. They, therefore, can be more easily harvested.
7. At day 3, 6.5 ml of supernatant (including cells) are collected and discarded and replaced with 7.5 ml of fresh CM + 10 ng/ml of GM-CSF.
8. At day 5, 5 ml of fresh CM + 10 ng/ml of GM-CSF are added to the plate.
9. The bone marrow–derived DC are ready by day 6 and can be collected by vigorous pipetting and counted.

DC Loading with Antigen

10. 10^6/ml DC are cultured in 6-well cell culture plate with the compound of interest (previously sonicated) in CM + 10 ng/ml of GM-CSF.

11. 24 h later, the bone marrow DC are collected in cold PBS or CM by vigorous pipetting and counted.

12. Pulsed DC and *i*NKT cells are then seeded in 96-well round-bottomed plate. If the *i*NKT cells have been obtained after sorting, the ratio DC/*i*NKT cells should be 5:1. If a whole liver MNC preparation is used as source for *i*NKT cells, the ratio is 1:2.

13. 24 h later, the supernatant is harvested and ELISA for IFNγ and IL-4 performed.

In addition to monitoring the status of the *i*NKT cells by measuring their effector functions (cytokine production), this assay can be used to study the difference in the ability of different APC to present/process the glycolipid. In addition to bone marrow–derived DC, the source of the antigen presentation cells could be other cell types, including CD1d-transfected cell lines (Brossay *et al.*, 1998a) or Flt3-derived DC (Gilliet *et al.*, 2002).

In Vivo iNKT Cell Activation Assay

Direct intraperitoneal (*i.p.*) or intravenous (*i.v.*) injection of glycolipid antigens is possible, although this causes the *in vivo* activation of the *i*NKT cells only when very strong agonists such as αGalCer are used. It is, therefore, preferable to use DC incubated with antigen (antigen pulsed) as a way to deliver glycolipids. Indeed, it has been shown that *i*NKT cells respond more vigorously to αGalCer-pulsed DC than antigen given in the free form (Fujii *et al.*, 2002).

Method

1. DC are prepared and antigen pulsed as previously described
2. 5–6 × 10^5 of compound-pulsed DC are injected *i.v.* into mice.
3. Twelve to 14 h later, the mice are killed and serum, liver, and spleen are harvested.

Serum ELISA. Detection of serum IFNγ and IL-4 by ELISA will reflect the systemic effect of *in vivo* *i*NKT activation.

Flow Cytometry Staining and Analysis

On activation, *i*NKT cells upregulate their expression of both the lectin CD69 and CD25, the α chain of the IL-2 receptor. As shown in Fig. 3, extracellular staining for these surface markers, therefore, tracks their activation status. *i*NKT cells respond robustly to antigen-induced activation, and cytokine production can usually be detected by intracellular cytokine staining (ICCS). *i*NKT cells do not require some of the manipulations carried out on conventional T cells before ICCS, including the use of agents such as Brefeldin-A, to trap cytokine inside the producing cells.

SOLUTIONS

- Staining buffer (SB): 1 × PBS, 2% BSA, 0.1% NaN$_3$, 10 mM EDTA
- Fixing buffer (FB): 1 × PBS, 1% paraformaldehyde, 0.1% NaN$_3$
- Antibody mixtures (diluted in SB):
  Mix 1: FITC-CD69 (1:100), PE-CD1d-αGalCer tetramer (1:150), PeCy7-CD25 (1:150) and allophycocyanin -TCR (1:100).
  Mix 2, 3 and 4: PE-CD1d-αGalCer tetramer (1:150), PeCy7-NK1.1 (1:75), allophycocyanin -TCR (1:100) and FITC-IFNγ, IL-4 or isotype control respectively.
- Cytofix/Cytoperm Plus kit (BD PharMingen, #554715)

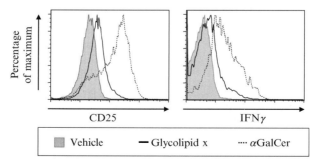

FIG. 3. *In vivo* activation of iNKT cells by glycolipid-pulsed bone marrow–derived DC. Cell surface expression of CD25 (IL-2 receptor α chain) on *i*NKT cells (left) and intracellular IFNγ in *i*NKT cells.

METHOD

1. MNC from the liver and spleen are prepared as previously described.
2. 10^6 cells/200 μl are seeded in a 96-well round bottomed plate and centrifuged.
3. The supernatant is removed, the cells resuspended in 25 μl of CD16/32 Ab diluted at 1:200 in SB and incubated 20 min at 4° in the dark.
4. After the incubation, 25 μl of the different extracellular Abs (CD69, tetramer, CD25 and TCR for mixture 1, tetramer, B220 and TCR for mixture 2, 3 and 4) are added to the different wells.
5. At the end of a 30-min incubation at 4°, the cells are washed twice with 200 μl SB.
6. The cells are then permeabilized in 100 μl of Cytofix/Cytosperm buffer solution and incubated either 20 min or overnight at 4° in the dark.
7. After the incubation, the cells are washed twice with 1× Perm/Wash buffer. During these washes the cells are pelleted at 600–650g because they are now lighter.
8. Cells of mixture 1 are resuspended in 50 μl of Perm/wash.
9. Cells of mixtures 2, 3, and 4 are resuspended in 50 μl of FITC-IFNγ (1:100), FITC-IL-4 (1:200), or isotype control (1:100) Abs, respectively, diluted in 1× Perm/wash solution.
10. After 30 min of incubation at 4° in the dark, the cells are washed twice with 1× Perm/Wash solution before being resuspended in 200 μl of FB for the flow cytometry analysis.

Biochemical Assays

Biochemical assays for measuring the interaction between glycolipids bound to CD1d and the iNKT cell TCR are more direct than immune activation assays, and they can measure the association constant (K_D) and the half-life ($t_{1/2}$) of the binding of CD1d-loaded glycolipid antigens antigen with the TCR of iNKT cells. This can be done using surface plasmon resonance to measure directly the affinity of the interaction between soluble TCR molecules and CD1d loaded with the appropriate antigen. Flow cytometry also can be used, however, to measure the avidity and $t_{1/2}$ using fluorescent-labeled CD1d tetramers, in equilibrium binding and decay of binding measurements. These assays are depicted schematically in Fig. 4. The flow cytometry–based methods are convenient, and they have the advantage that the binding measurements can be done on living cells to TCR molecules in cell membranes. The complete procedures were published by Sidobre $et\ al.$ (2002) and will, therefore, be described only briefly here:

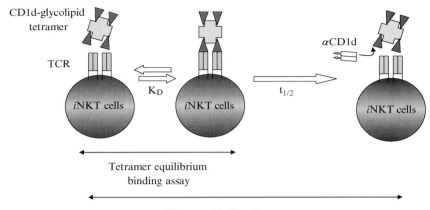

FIG. 4. Schematic depiction of the tetramer binding assays, including equilibrium binding and the decay of tetramer binding assays. Rebinding of the CD1d tetramer is prevented in the decay of tetramer binding assay by the addition of an anti-CD1d (αCD1d) monoclonal antibody.

Tetramer Equilibrium Binding

Solutions

- Staining buffer (SB): $1 \times$ PBS, 2% BSA, 0.1% NaN$_3$, 10 mM EDTA
- Fixing buffer (FB): $1 \times$ PBS, 1% paraformaldehyde, 0.1% NaN$_3$

Method

1. 2×10^5 hybridoma cells are incubated at RT with various concentrations (1.2×10^{-10} to 5.8×10^{-8} M) of CD1d-glycolipid tetramers or CD1d-vehicle tetramer (negative control) for 3 h. The same protocol can be carried out with primary iNKT cells from liver, thymus, or spleen. In this case, 10 μg/ml anti-TCRβ antibody (H57-597, BD PharMingen) is also added, and for flow cytometry analysis, the TCR negative cells are gated out. Note that the β chain–specific antibody does not compete for binding to the $\alpha\beta$ TCR by the CD1d tetramer.
2. After the incubation, the cells are washed twice in SB and fixed with FB before being analyzed by flow cytometry.
3. The mean fluorescence intensity (MFI) of the glycolipid tetramer minus MFI of the vehicle tetramer is plotted against concentrations of the tetramer and the K$_D$ can be assessed by Scatchard analysis.

Decay of Tetramer Binding

Solutions

- Tetramer decay media (TDM): RPMI1640 supplemented with 10% FBS, penicillin, streptomycin, L-glutamine, 2-mercaptoethanol, and 0.05% NaN$_3$.
- Paraformaldehyde: a 4% solution in PBS
- CD1d antibody: BD PharMingen

Method

1. 2.5×10^6 hybridoma cells or primary *i*NKT cells are stained with CD1d-glycolipid tetramer (diluted at 1:10) for 1.5 h at RT in TDM in 96-well plates.
2. After the incubation, the cells are washed twice with ice-cold TDM and resuspended in 2 ml of TDM
3. 100 μg/ml of anti-CD1d antibody is added to the cultures to prevent tetramer rebinding, and cells are equilibrated for 15 min at the desired temperature (4°, 25°, or 37°).
4. Set up a plate on ice with 50 μl/well of TDM.
5. At consecutive time points (determined by the experiment but usually every 5 min), 200 μl of cells is removed and added to a plate on ice to prevent further tetramer decay.
6. At the end of the time period (60 min), cells are washed once and incubated with anti-TCRβ for 30 min on ice if the experiment is done with primary *i*NKT cells.
7. After the incubation, cells are washed twice, fixed with paraformaldehyde, and analyzed by flow cytometry.

The MFI of staining is then determined for at each time point and is graphed as follows:

> X-axis: time in minutes
> Y-axis: log[MFI of bound tetramer/MFI of tetramer at zero time point].

A line of best fit is then determined and the half-life of interaction (t$_{1/2}$) is determined from the slope of the line by the formula: $t_{1/2} = \log(2)/\text{slope}$.

References

Benlagha, K., Weiss, A., Beavis, A., Teyton, L., and Bendelac, A. (2000). *In vivo* identification of glycolipid antigen-specific T cells using fluorescent CD1d tetramers. *J. Exp. Med.* **191,** 1895–1903.

Brossay, L., Chioda, M., Burdin, N., Koezuka, Y., Casorati, G., Dellabona, P., and Kronenberg, M. (1998a). CD1d-mediated recognition of an alpha-galactosylceramide by natural killer T cells is highly conserved through mammalian evolution. *J. Exp. Med.* **188,** 1521–1528.

Brossay, L., Tangri, S., Bix, M., Cardell, S., Locksley, R., and Kronenberg, M. (1998b). Mouse CD1-autoreactive T cells have diverse patterns of reactivity to CD1+ targets. *J. Immunol.* **160,** 3681–3688.

Cardell, S., Tangri, S., Chan, S., Kronenberg, M., Benoist, C., and Mathis, D. (1995). CD1-restricted CD4+ T cells in major histocompatibility complex class II-deficient mice. *J. Exp. Med.* **182,** 993–1004.

Crowe, N. Y., Coquet, J. M., Berzins, S. P., Kyparissoudis, K., Keating, R., Pellicci, D. G., Hayakawa, Y., Godfrey, D. I., and Smyth, M. J. (2005). Differential antitumor immunity mediated by NKT cell subsets *in vivo. J. Exp. Med.* **202,** 1279–1288.

Dellabona, P., Padovan, E., Casorati, G., Brockhaus, M., and Lanzavecchia, A. (1994). An invariant V alpha 24-J alpha Q/V beta 11 T cell receptor is expressed in all individuals by clonally expanded CD4-8-T cells. *J. Exp. Med.* **180,** 1171–1176.

Fujii, S., Shimizu, K., Kronenberg, M., and Steinman, R. M. (2002). Prolonged IFN-gamma-producing NKT response induced with alpha-galactosylceramide-loaded DCs. *Nat. Immunol.* **3,** 867–874.

Gilliet, M., Boonstra, A., Paturel, C., Antonenko, S., Xu, X. L., Trinchieri, G., O'Garra, A., and Liu, Y. J. (2002). The development of murine plasmacytoid dendritic cell precursors is differentially regulated by FLT3-ligand and granulocyte/macrophage colony-stimulating factor. *J. Exp. Med.* **195,** 953–958.

Godfrey, D. I., MacDonald, H. R., Kronenberg, M., Smyth, M. J., and Van Kaer, L. (2004). NKT cells: What's in a name? *Nat. Rev. Immunol.* **4,** 231–237.

Kawano, T., Cui, J., Koezuka, Y., Toura, I., Kaneko, Y., Motoki, K., Ueno, H., Nakagawa, R., Sato, H., Kondo, E., Koseki, H., and Taniguchi, M. (1997). CD1d-restricted and TCR-mediated activation of valpha14 NKT cells by glycosylceramides. *Science* **278,** 1626–1629.

Kinjo, Y., Wu, D., Kim, G., Xing, G. W., Poles, M. A., Ho, D. D., Tsuji, M., Kawahara, K., Wong, C. H., and Kronenberg, M. (2005). Recognition of bacterial glycosphingolipids by natural killer T cells. *Nature* **434,** 520–525.

Kobayashi, E., Motoki, K., Uchida, T., Fukushima, H., and Koezuka, Y. (1995). KRN7000, a novel immunomodulator, and its antitumor activities. *Oncol. Res.* **7,** 529–534.

Kronenberg, M., and Gapin, L. (2002). The unconventional lifestyle of NKT cells. *Nat. Rev. Immunol.* **2,** 557–568.

Matsuda, J. L., Naidenko, O. V., Gapin, L., Nakayama, T., Taniguchi, M., Wang, C. R., Koezuka, Y., and Kronenberg, M. (2000). Tracking the response of natural killer T cells to a glycolipid antigen using CD1d tetramers. *J. Exp. Med.* **192,** 741–754.

Mattner, J., Debord, K. L., Ismail, N., Goff, R. D., Cantu, C., 3rd, Zhou, D., Saint-Mezard, P., Wang, V., Gao, Y., Yin, N., Hoebe, K., Schneewind, O., Walker, D., Beutler, B., Teyton, L., Savage, P. B., and Bendelac, A. (2005). Exogenous and endogenous glycolipid antigens activate NKT cells during microbial infections. *Nature* **434,** 525–529.

Naidenko, O. V., Maher, J. K., Ernst, W. A., Sakai, T., Modlin, R. L., and Kronenberg, M. (1999). Binding and antigen presentation of ceramide-containing glycolipids by soluble mouse and human CD1d molecules. *J. Exp. Med.* **190,** 1069–1080.

Porcelli, S., Yockey, C. E., Brenner, M. B., and Balk, S. P. (1993). Analysis of T cell antigen receptor (TCR) expression by human peripheral blood CD4-8-alpha/beta T cells demonstrates preferential use of several V beta genes and an invariant TCR alpha chain. *J. Exp. Med.* **178,** 1–16.

Porcelli, S. A. (1995). The CD1 family: A third lineage of antigen-presenting molecules. *Adv. Immunol.* **59,** 1–98.

Sidobre, S., and Kronenberg, M. (2002). CD1 tetramers: A powerful tool for the analysis of glycolipid-reactive T cells. *J. Immunol. Methods* **268,** 107–121.

Sidobre, S., Naidenko, O. V., Sim, B. C., Gascoigne, N. R., Garcia, K. C., and Kronenberg, M. (2002). The V alpha 14 NKT cell TCR exhibits high-affinity binding to a glycolipid/CD1d complex. *J. Immunol.* **169,** 1340–1348.

Smyth, M. J., Crowe, N. Y., Hayakawa, Y., Takeda, K., Yagita, H., and Godfrey, D. I. (2002). NKT cells – conductors of tumor immunity? *Curr. Opin. Immunol.* **14,** 165–171.

Sriram, V., Du, W., Gervay-Hague, J., and Brutkiewicz, R. R. (2005). Cell wall glyco-sphingolipids of Sphingomonas paucimobilis are CD1d-specific ligands for NKT cells. *Eur. J. Immunol.* **35,** 1692–1701.

Van Kaer, L. (2004). Natural killer T cells as targets for immunotherapy of autoimmune diseases. *Immunol. Cell. Biol.* **82,** 315–322.

Zajonc, D. M., Cantu, C., 3rd, Mattner, J., Zhou, D., Savage, P. B., Bendelac, A., Wilson, I. A., and Teyton, L. (2005). Structure and function of a potent agonist for the semi-invariant natural killer T cell receptor. *Nat. Immunol.* **6,** 810–818.

Zhou, D., Mattner, J., Cantu, C., 3rd, Schrantz, N., Yin, N., Gao, Y., Sagiv, Y., Hudspeth, K., Wu, Y. P., Yamashita, T., Teneberg, S., Wang, D., Proia, R. L., Levery, S. B., Savage, P. B., Teyton, L., and Bendelac, A. (2004). Lysosomal glycosphingolipid recognition by NKT cells. *Science* **306,** 1786–1789.

Section V

Glycolipid Interaction

[15]   Determination of Glycolipid–Protein
Interaction Specificity

By PABLO H. H. LOPEZ and RONALD L. SCHNAAR

Abstract

Glycolipids are found on all eukaryotic cells. Their expression varies among tissues, with the highest density found in the brain, where glycolipids are the most abundant of all glycoconjugate classes. In addition to playing roles in membrane structure, glycolipids also act as cell surface recognition molecules, mediating cell–cell interactions, as well as binding certain pathogens and toxins. Because of their amphipathic nature, underivatized glycolipids are amenable to immobilization on hydrophobic surfaces, where they can be probed with lectins, antibodies, pathogens, toxins, and intact cells to reveal their binding specificities and affinities. Three particularly useful methods to probe specific glycolipid-mediated recognition events are microwell adsorption (ELISA), thin layer chromatography overlay, and surface plasmon resonance (SPR) spectroscopy.

Overview

Glycolipids are amphipathic molecules consisting of a hydrophilic oligosaccharide chain linked to a hydrophobic lipid that anchors the glycolipid in cell membranes (Hakomori, 2002). Two major classes of glycolipids, glycoglycerolipids and glycosphingolipids, are distinguished by their lipid moieties, diacylglycerol (or alkylacylglycerol) and ceramide, respectively. Whereas glycerol-based glycolipids are most abundant in microbes and plants, glycosphingolipids are by far the most abundant and diverse class of glycolipids in animals. Animal glycosphingolipids are classified into families according to their core glycan sequences, on which many different terminal variations are elaborated (Stults *et al.*, 1989). Glycosphingolipid families are expressed in tissue-specific patterns. For example, ganglio-series glycosphingolipids, based on the core structure $Gal\beta1\text{-}3GalNAc\beta1\text{-}4Gal\beta1\text{-}4Glc\beta1\text{-}1'ceramide$, predominate in the brain (Schnaar, 2000), whereas neolacto-series glycolipids, based on the core structure $Gal\beta1\text{-}4GlcNAc\beta1\text{-}3Gal\beta1\text{-}4Glc\beta1\text{-}1'ceramide$ are common on leukocytes (Stroud *et al.*, 1996). Glycosphingolipids are further subclassified as neutral, sialylated (having one or more sialic acid residues), or sulfated (Stults *et al.*, 1989). Sialylated glycosphingolipids are also known as gangliosides, which are widely distributed in vertebrates and

METHODS IN ENZYMOLOGY, VOL. 417   0076-6879/06 $35.00
                          DOI: 10.1016/S0076-6879(06)17015-9

are the most abundant glycoconjugates on nerve cells. Hundreds of distinct glycolipid structures have been reported (Stults *et al.*, 1989), and their structural diversity underlies glycosphingolipid functions, being responsible for specific cell–cell interactions, pathogen/toxin tropism, and autoimmune pathologies (Hakomori, 2002; Schnaar, 2004; Willison, 2005). Whereas this chapter describes methods to determine glycolipid binding specificities using purified or partially purified natural glycolipids, the methods can also be applied to synthetic neoglycolipids (Feizi and Chai, 2004).

Although the amphipathic nature of glycolipids poses challenges for their isolation and purification, their lipid moieties provide the means to readily immobilize and orient them on surfaces for probing with complementary binding molecules. In this chapter, methods to probe binding specificities of purified glycolipids by means of microplate adsorption and surface plasmon resonance spectroscopy (SPR) are described, as well as methods to identify binding components of glycolipid mixtures using thin layer chromatography (TLC) overlay.

Materials

Glycolipids

Organic solvents are used to solubilize endogenous glycolipids from tissues and cells, where they are typically expressed on plasma membranes. Extraction procedures have been optimized, often using defined chloroform-methanol-water mixtures added in specific solvent sequence and ratio, to maximize precipitation and removal of proteins and nucleic acids while maximizing solubilization of glycosphingolipids along with other lipids (Schnaar, 1994). Because glycosphingolipids aggregate with each other and other lipids in aqueous solution, organic solvents are used throughout subsequent purification steps, which typically involve solvent partition, ion exchange chromatography, and/or silicic acid chromatography. Details for glycolipid purifications, which vary depending on the source and specific nature of the target molecules, have been published elsewhere (Schnaar, 1994), and are beyond the scope of this chapter. Purified glycolipids dissolved in organic solvents (e.g., chloroform-methanol-water [4:8:3]) are typically stable when stored at $-20°$ for years or even decades.

Microplate (ELISA)-Based Recognition

1. 96-Well EIA/RIA clear flat-bottom untreated polystyrene microplate (Corning Life Sciences, Acton, MA, product #9017) (*see* Note 1)
2. Reagent grade butanol (for prewashing microplates)

3. Reagent grade 100% ethanol

4. A 100-fold stock of a lipid mixture consisting of 0.1 mM phosphatidyl-choline (PC) and 0.4 mM cholesterol is prepared in ethanol. To 4.8 ml of 100% ethanol add 0.15 ml of a 3.3 mM stock of phosphatidylcholine in chloroform (Avanti Polar Lipids, Alabaster, AL, product 840053) and 50 μl of 40 mM stock of cholesterol in chloroform (prepared from dry powder, Avanti product 700000). Store the lipid stock tightly capped in a glass tube with Teflon-lined seal at $-20°$.

5. Dulbecco's phosphate-buffered saline (PBS) contains (g/l): NaCl (8.0), KCl (0.20), $CaCl_2$ (0.10), $MgCl_2 \cdot 6H_2O$ (0.10), Na_2HPO_4 (2.3), and KH_2PO_4 (0.20) (Bashor, 1979).

6. Microplate blocking buffer is prepared by dissolving bovine serum albumin (BSA, e.g., Sigma-Aldrich, St. Louis, MO, product A-7030) in PBS at 1 mg/ml

TLC Overlay

1. Aluminum-backed silica gel high-performance HPTLC plates (20 × 20 cm, E. Merck, Darmstadt, Germany, product 5547) (*see* Note 2)

2. TLC running solvent. A wide-spectrum running solvent suitable for many applications is chloroform-methanol-aqueous 0.25% w/v KCl (60:35:8) (*see* Note 3)

3. Glass TLC chromatography developing tank (e.g., 12 × 5 × 12 cm, rectangular, Camag Scientific, Muttenz, Switzerland, product 022.5515).

4. Microliter TLC spotting syringe (Hamilton Co., Reno, NV, product 701N)

5. Glass petri dish

6. Poly(isobutyl methacrylate) (PIBM) solution. A stock solution of 5% w/v of PIBM (e.g., Sigma-Aldrich, product 181544) dissolved in chloroform is prepared. The stock solution is diluted to working concentration (e.g., 0.1% w/v) in *n*-hexane immediately before use (*see* Note 4)

7. Binding buffer is prepared by adding 1 mg/ml BSA and 0.05% (w/v) Tween-20 (e.g., Pierce Chemical Co., Rockford, IL, product 28320) to PBS

Surface Plasmon Resonance Spectroscopy

1. SPR-based biosensor (e.g., Biacore 3000, Biacore AB, Uppsala, Sweden).

2. Lipid vesicle extrusion apparatus (Avanti Polar Lipids, product 610000) and 50-nm polycarbonate filters (product 610003)

3. 1,2-Dimyristoyl-*sn*-glycero-3-phosphocholine (DMPC, Avanti Polar Lipids product 850345)
4. SPR blocking buffer is prepared by dissolving 0.1 mg/ml BSA in 10 mM sodium phosphate, 150 mM sodium chloride, and adjusting to pH 7.4 (degassed and filtered daily)
5. Running buffer consists of 10 mM phosphate buffer, pH 7.4, 150 mM sodium chloride, pH 7.4 (degassed and filtered daily)
6. Sensor chip L1 (product BR-1005-58), Biacore AB
7. Washing solution consists of 20 mM CHAPS detergent (e.g., Sigma C-5070) in water
8. Regenerating solution consists of 20 mM NaOH in water (degassed and filtered daily)

Methods

The three methods for determining glycolipid–protein interactions described here—microplate, TLC, and SPR—provide a wide range of capabilities applicable to different experimental goals. In each case, glycolipids are non-covalently adsorbed to a solid support (microwell, TLC plate or SPR chip), where they remain stably attached in aqueous solutions. Exposure of the adsorbed glycolipids to potential binding proteins (lectins, antibodies, toxins) or biological entities (viruses, bacteria, intact cells) results in specific interactions that are detected using direct or indirect methods. Microplate adsorbed purified glycolipids can be used to determine relative binding specificities of soluble lectins, antibodies, toxins, pathogens, or cells, or for high-throughput screening, for example, of sera or hybridoma supernatants. TLC overlay is most useful for detecting and identifying binding species within mixtures of partially purified glycolipids from natural sources. SPR using purified glycolipids and unlabeled soluble glycan binding proteins (GBPs) provides real-time binding kinetics for determination of affinities and specificities of glycolipid–protein interactions.

Microplate (ELISA)-Based Recognition

Because of their amphipathic nature, glycolipids spontaneously and stably adsorb to hydrophobic surfaces under the appropriate conditions. Commercially available polystyrene microplates are excellent substrates for glycolipid immobilization and are amenable to subsequent incubation with GBPs and their detection by direct or indirect methods (Collins *et al.*, 2000; Foxall *et al.*, 1992; Schnaar *et al.*, 2002).

1. To remove mold release agents used in their manufacture, microplates are prewashed (batchwise) with butanol and ethanol and then stored for

subsequent use. Microplates are immersed upright in a shallow glass dish (e.g., a glass baking dish) of butanol and allowed to incubate $\geq$15 min. Plates are drained and immersed in three sequential dishes of ethanol, draining between transfers. Finally, plates are drained, allowed to air dry, and are stored in sealed plastic bags.

2. Desired concentrations of glycolipid (typically 2–200 pmol/well) are adsorbed, along with a constant amount of PC (25 pmol/well) and cholesterol (100 pmol/well), as follows. PC/cholesterol working solution is prepared by diluting the stock 100-fold in 100% ethanol (final concentration 1 μM PC, 4 μM cholesterol). The total amount of glycolipid required is calculated, and an aliquot containing a 10–20% excess (to account for handling losses) is transferred to a glass test tube and evaporated to dryness under a stream of nitrogen or in an evaporator (e.g., SpeedVac Concentrator, Thermo Electron Corp., Waltham, MA). The dried glycolipid is redissolved at the desired concentration(s) in the PC/cholesterol working solution. A bath sonicator (brief sonication) and/or vigorous vortex mixing are used to ensure that the glycolipids are efficiently redissolved. Glycolipid solutions are stored in screw-capped glass tubes until ready for addition to microplate wells (*see* Note 5).

3. Immediately before addition to microplate wells, glycolipid solutions (from Step 2) are diluted with an equal volume of water and mixed vigorously. A 50-μl aliquot is then pipetted into each well of a 96-well microplate using standard single or multichannel micropipettors with standard plastic disposable tips. The plate is then placed on a well-ventilated bench or chemical fume hood and left *uncovered* at ambient temperature for 90 min to allow partial evaporation. After incubation, any remaining solvent is removed by inversion and vigorous shaking of the microplate, and the wells are washed three times by repeated immersion in water, inversion, and shaking. The plate is righted and water (200 μl/well) added. Glycolipid-adsorbed plates may be used immediately or stored in water for several hours (e.g., overnight) at ambient temperature without measurable loss of adsorbed lipids (*see* Note 6).

4. Water is removed from the glycolipid-adsorbed wells by inverting and shaking the microplate, then is replaced with 200 μl/well of microplate blocking buffer. The plate is incubated covered for 30 min at 37°, then the blocking buffer is removed by inversion and shaking, and the wells washed three times with PBS.

5. Binding and detection. Microplate-adsorbed glycolipids are amenable to probing with lectins, proteins, and other biological entities that are either directly labeled or that can be detected with secondary reagents. Any standard ELISA-type protocol can be applied to the microplates. One protocol for determining anti-glycolipid antibody binding specificity follows (Schnaar *et al.*, 2002). Primary antibodies (antisera, hybridoma

supernatants, monoclonal antibodies, etc.) are diluted (typically $\sim$1 μg/ml antibody) in microplate blocking buffer and 50 μl is added to each glycolipid-adsorbed well. After 90 min at ambient temperature, the microplate is washed three times by immersion in PBS, inversion, and shaking. Secondary antibody (e.g., alkaline phosphatase (AP)–conjugated anti-human or anti-mouse IgG, as appropriate) diluted (typically to $\sim$1 μg/ml) in microplate blocking buffer, is then added at 50 μl/well and the microplate incubated 45 min at ambient temperature. Finally, the microplate is washed by immersion in PBS, then water (twice) and AP substrate, consisting of 2 mg/ml p-nitrophenylphosphate in developing buffer (100 mM Tris, 100 mM NaCl, 5 mM MgCl$_2$) is added at 100 μl/well. Color development is determined with a microplate reader. An example of the use of microplate-adsorbed gangliosides to determine binding specificities of antiganglioside monoclonal antibodies is shown in Fig. 1 (Schnaar et al., 2002) (see Note 7).

TLC Overlay

TLC overlay is performed in two steps (Karlsson and Stromberg, 1987; Magnani et al., 1980). First, a glycolipid sample is applied to a silica gel TLC plate and developed to resolve the different species. Second, the developed plate is coated with a thin plastic film and overlain with aqueous solution containing soluble GBPs or other biological binding entities to be analyzed. After washing to remove unbound material, bound material is detected either directly or using secondary reagents. When combined with enzymatic and/or chemical modification *in situ*, TLC overlay is a powerful tool for the analysis of glycolipid–protein interactions (Schnaar and Needham, 1994).

Thin Layer Chromatography

1. The chromatography tank is pre-equilibrated by adding developing solvent (typically chloroform-methanol-aqueous mixtures, see "Materials") to a depth of $\sim$0.5 cm in the bottom of the chromatography developing tank, which is then covered with a grease-free lid (glass or metal) and incubated for at least 30 min before introducing the spotted TLC plate. The tank is placed in an area of uniform temperature protected from drafts or heat sources.

2. TLC plates are prepared for spotting. Aluminum-backed HPTLC plates are cut to desired sizes with a scissors, using care to avoid folding or flaking of the silica gel sorbent. The width of the plate should be 3 cm plus 1 cm per sample to be spotted. The silica gel sorbent along cut edges is smoothed using the dry tip of a gloved finger. Using a ruler and pencil,

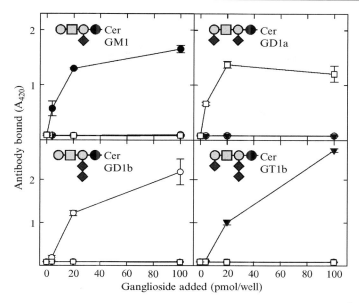

FIG. 1. Specific binding of four different antiganglioside monoclonal antibodies[1] to major brain gangliosides (Schnaar *et al.,* 2002). The four major brain gangliosides GM1 (—•—), GD1a (—□—), GD1b (—○—), and GT1b (—▾—) were adsorbed to microplate wells, along with phosphatidylcholine and cholesterol, at the indicated ganglioside concentrations (pmol/well added). Highly specific binding of four different IgG1-class antibodies (one for each of the gangliosides, as indicated) was demonstrated using the methods described. Data are presented as the mean and range for duplicate determinations. Sugar structures are represented using the symbol nomenclature of the Consortium for Functional Glycomics (http://functionalglycomics. org). The four major brain gangliosides shown have the same neutral glycan backbone (Galβ1-3GalNAcβ1-4Galβ1-4Glcβ1-1$'$ceramide) with different numbers and positions of sialic acids (diamonds) as shown. Modified from Schnaar *et al.* (2002), with permission.

light marks for sample application are drawn along a line 1 cm above and parallel to the bottom of the TLC plate. Each sample application line should be 0.5 cm long, leaving 1.5 cm unused on the lateral edges of the plate (where distortion may occur during development) and 0.5 cm between samples (*see* Note 8).

3. Glycolipid samples are dissolved in organic solvents for spotting. For pure samples, 100 pmol/lane (or less for high-affinity binding proteins) is

---

[1] Under a licensing agreement between Seikagaku America and the Johns Hopkins University, Dr. Schnaar is entitled to a share of royalty received by the University on sales of the monoclonal antibodies described in Figs. 1 and 2. The terms of this arrangement are being managed by the Johns Hopkins University in accordance with its conflict of interest policies.

sufficient. For mixtures, 1 nmol (or more) may be spotted per lane (*see* Note 9). Optimally, each sample should be spotted in a total volume of 1 μl, although larger volumes (up to 10 μl) may be spotted with sufficient care and patience. Aliquots of glycolipid sample in organic storage solutions are evaporated to dryness under a nitrogen stream or in an evaporator then redissolved in a small volume of methanol or chloroform-methanol-water (4:8:3) for spotting.

4. Using a Hamilton syringe with the beveled edge of the needle closely apposed but not touching the TLC plate surface, samples are applied along the premarked pencil lines. For each lane, ~0.5 μl is applied, allowed to dry, then application and drying are repeated until the total desired volume has been spotted. Sufficient time is allowed after the last sample is applied for the spotting solvent to completely evaporate (*see* Note 10).

5. The TLC plate, with samples applied, is placed in the pre-equilibrated developing tank, with the top of the plate resting against the tank wall. The tank is kept covered at uniform temperature away from drafts or heat sources until the solvent moves up the plate by capillary action to within a few millimeters from the top of the plate. To ensure reproducibility of TLC runs, used developing solvent is discarded after each run (*see* Note 11).

TLC Overlay

1. After the solvent has reached the top of the TLC plate, it is removed from the developing tank and allowed to dry completely. Drying may be aided by unheated forced air (*see* Note 10). One corner of the aluminum-backed TLC plate is bent toward the sorbent-adsorbed surface using a forceps to facilitate easier handling during subsequent steps. A glass petri dish is filled to ~0.5-cm depth with 0.1% PIBM in hexane. The TLC plate is slowly inserted, at an angle into the solution until it is completely immersed. The petri dish is covered and incubated for 90 sec. The TLC plate is removed and held vertically over adsorbent paper to allow excess PIBM solution to drain away from the surface. The treated TLC plate is dried completely, using unheated forced air to speed drying if desired (*see* Note 12).

2. A dilution of GBP (typically 1 μg/ml) in binding buffer is placed in a small plastic tray whose size is close to that of the TLC plate. The TLC plate is immersed in the solution and pressed to the bottom of the container with a forceps to remove bubbles. The plate is incubated with GBP solution overnight at 4° with gentle shaking in a humidified atmosphere (*see* Note 13).

3. Bound proteins may be detected directly or with secondary reagents. For antibody binding (using sera or purified antibodies) after incubation of the TLC plate in GBP solution, the plate is washed three times for 1 min by immersion in PBS, then the plate is overlain with 0.5 μg/ml of horseradish peroxidase–conjugated secondary antibody in binding buffer for 90 min at

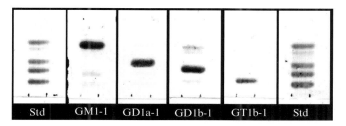

FIG. 2. TLC immuno-overlay of major brain gangliosides with antiganglioside monoclonal antibodies[1] (Schnaar *et al.*, 2002). A mixture of the four major brain gangliosides (GM1, GD1a, GD1b, GT1b) was spotted at the origin of replicate TLC lanes. After development, the plate was cut into sections, which were either stained with a chemical reagent to reveal the gangliosides ("Std" lanes, top to bottom: GM1, GD1a, GD1b, GT1b) or immunostained with the indicated anti-ganglioside monoclonal antibody, designated by its ganglioside binding specificity and IgG subclass (IgG1). Microplate binding data for the same four monoclonal antibodies is shown in Fig. 1. Modified from Schnaar *et al.* (2002), with permission.

ambient temperature. The plate is then washed three times for 1 min in PBS. Bound secondary antibody is detected by dipping the plate in substrate solution containing 2.8 mM 4-chloro-1-naphthol, 0.01% H_2O_2 and 3.3% methanol in 20 mM Tris–HCl buffer, pH 7.4. A positive reaction results in a purple band. The reaction is stopped by washing the plate in PBS; then the plate is allowed to air dry. An example of TLC immunooverlay is shown in Fig. 2 (*see* Note 14).

TLC Blotting. An alternative to direct TLC overlay is TLC blotting, in which TLC-resolved glycolipids are transferred quantitatively to polyvinylidene fluoride (PVDF) membranes (Taki *et al.*, 1994). Once transferred, glycolipids remain stably adsorbed to the hydrophobic membrane and are amenable to the same procedures used in Western blotting of PVDF-transferred proteins.

1. Glycolipids are resolved using glass-backed silica gel HPTLC plates using the protocols described above.
2. After developing and drying, the plate is dipped in blotting solvent (isopropanol-0.2% aqueous $CaCl_2$-methanol [40:20:7]) for 20 sec, placed sorbent face up, and covered with a PVDF membrane sheet (Immobilon-P, Millipore, Billerica, MA) cut to the same size. A glass fiber filter sheet (type GF/A, Whatman PLC, Middlesex, UK) is placed on top of the PVDF.
3. An iron preheated to 180° is pressed with even pressure on top of the glass fiber filter for 30 sec. The PVDF membrane is removed, washed with PBS, and probed using standard Western-blotting techniques (*see* Note 15).

Surface Plasmon Resonance (SPR) Spectroscopy

SPR is performed by immobilizing glycolipids on a commercial SPR sensor chip, followed by exposure to unlabeled soluble GBPs (Kuziemko *et al.*, 1996; MacKenzie *et al.*, 1997; MacKenzie and Hirama, 2000; Nakajima *et al.*, 2001). Binding is performed in a biosensor designed to detect mass bound to the sensor chip surface. Two biosensor chip surface chemistries are typically used to study lipid-mediated binding events: a gold chip derivatized with a long-chain thioalkanes to generate a flat hydrophobic monolayer on which lipids are immobilized (e.g., HPA chip, Biacore AB) and a chip derivatized with alkane-derivatized dextran (e.g., L1 chip, Biacore AB), to which preformed glycolipid-containing liposomes (bilayer lipid vesicles) are anchored. The later method is described here because of its high capacity and design flexibility (*see* Note 16).

Glycolipid-Containing Liposomes

1. The glycolipid of interest and carrier phospholipid (DMPC) are combined in organic solvent in a glass tube and evaporated to dryness under a stream of nitrogen or using a rotary evaporator. For example, 0.04 mg of glycolipid in storage solvent (chloroform-methanol-water [4:8:3]) is added to a glass tube, dried under nitrogen, then redissolved in 200 μl of chloroform-methanol (2:1) containing 2 mg of DMPC. The lipid mixture is again dried under nitrogen, then placed under vacuum for 2 h. As a control, identical liposomes are prepared without added glycolipid.

2. Thoroughly dried lipids are rehydrated by suspension in running buffer prewarmed to 37° at a final lipid concentration of ~10 mM (300 μl for the previous example). The suspension is mixed vigorously for 30 min, avoiding the generation of bubbles and keeping the solution at $\geq$25° to create a white suspension.

3. To prepare unilamellar vesicles, the suspension is passed 15 times through a 50-nm pore polycarbonate filter using an extrusion apparatus per the manufacturer's instructions. The turbid suspension clarifies after extrusion. To remove large vesicles, the extruded preparation may be centrifuged at 100,000g for 30 min at 15°. The supernatant, containing unilamellar vesicles, is stable for up to 3–4 days at 4° under nitrogen.

Liposome Immobilization and SPR. The following procedure is for use in a Biacore 3000 (Biacore AB) or similar biosensor.

1. An L1 sensor chip is docked in the biosensor and cleaned for 10 min with washing solution at a flow rate of 10 μl/min. The injection needle is cleaned with water, and the system primed with running buffer.

2. Control and glycolipid-containing liposome preparations are injected into different channels (flow cells) of the sensor chip at a concentration of 500 μM total lipid. Liposome preparations are injected at a flow rate of 2 μl/min and stopped when mass adsorption reaches 2500 RU.

3. Lipid-adsorbed chips are washed for 12 sec with regenerating solution at 100 μl/min, then immediately with SPR blocking buffer for 5 min at 10 μl/min. The cells are then washed again with regeneration solution for 12 sec followed immediately with running buffer.

4. When the baseline stabilizes, running buffer containing GBP is injected at a rate of 40 μl/min. Typically, concentrations of GBP used range from 10 nM–1 μM, depending on affinity. If nonspecific binding is unacceptably high, GBP can be diluted in blocking buffer. The GBP is diluted to produce a specific response of ~100 RU (maximum) (see Note 17).

5. Binding curves are subsequently performed at multiple flow rates (e.g., 20–100 μl/min) using the optimal GBP concentration. In the absence of mass transport or rebinding limitations, the binding and dissociation rates should be independent of the flow rate. If the binding or dissociation rates increase with increasing flow rate, subsequent studies should be performed using the lowest flow rate that results in the maximum binding and dissociation rates for a fixed concentration of GBP.

6. Using the optimal flow rate, binding determinations are performed using different concentrations of GBP to provide a range of binding and dissociation rates (see example in Fig. 3). Typically, binding curves include ~3 min of association and up to 15 min of dissociation. For high-affinity ligands, longer dissociation times may be required. Between runs, the cells are washed with regeneration solution (12 sec, 100 μl/min) and running buffer to re-establish the initial baseline (see Note 18).

7. Data are fitted to appropriate models of receptor-ligand binding, using the biosensor software or independent curve fitting programs. Most such programs allow for consideration of multivalent binding, which may be appropriate for GBPs such as multivalent lectins and antibodies.

Notes

1. The example given is for 96-well microplates but can be applied to other microplate formats. Plain (untreated) polystyrene plates are required, because "tissue culture"–treated plates are modified to increase the hydrophilic nature of the plastic surface, making them less suitable for immobilization of amphipathic molecules.

2. Aluminum-backed plates are convenient, in that they can be easily cut to size with a scissors. For TLC blotting and centrifuge-based techniques

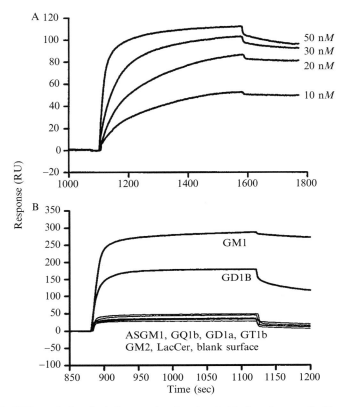

FIG. 3. SPR analysis of cholera toxin A subunit binding to ganglioside-containing liposomes immobilized on sensor chips using the method described by MacKenzie *et al.* (1997) (*see* Note 16). (A) Effect of cholera toxin concentration on binding to immobilized GM1. Kinetic plots of binding of different toxin concentrations (as indicated) to GM1-containing liposomes on sensor chips are shown; (B) Plots of binding of 100 nM toxin to liposomes containing different glycolipids (as indicated) are shown. Note the preferential binding to GM1. Modified from MacKenzie *et al.* (1997), with permission.

(eukaryotic cell adhesion), nonpliable glass-backed plates (E. Merck 5635) are more typically used.

　　3. Volume ratios of chloroform-methanol-0.25% aqueous KCl (or CaCl$_2$) may be altered to 70:30:5 to enhance separation of small or nonpolar glycolipids, or 45:45:10 for large and/or highly charged glycolipids. The efficacy of various solvent systems to separate neutral and charged glycolipids are published elsewhere (Schnaar and Needham, 1994).

4. PIBM working solutions should be prepared immediately before use to avoid changes in concentration because of solvent evaporation. Do not attempt to prepare PIBM working solution directly from powder into *n*-hexane, because the plastic is difficult to dissolve in hexanes.

5. Glycolipid added at 100 pmol/well to 96-well microplates typically provides ample glycolipid for binding studies. Adding more than 200 pmol/well is not productive, because the surface area of each well limits maximum adsorption (Blackburn *et al.*, 1986). Using lower concentrations, or concentration curves, can provide insights on relative binding affinities (Collins *et al.*, 2000). For large-scale screening (e.g., of sera or hybridoma supernatants), 25–50 pmol/well is usually sufficient. Inclusion of PC and cholesterol enhance the stability of adsorbed glycolipids and ensure a uniform background substratum over a large range of glycolipid concentrations. Some laboratories omit co-adsorbed carrier lipids (Foxall *et al.*, 1992). Adsorption efficiency is typically 40–70%, depending on the particular glycolipid used (Blackburn *et al.*, 1986).

6. During the partial evaporation step, the gradual loss of ethanol results in partitioning of the lipids onto the plastic surface, where they remain stably adsorbed in aqueous solutions. Some laboratories use an alternate procedure, dissolving glycolipids in methanol or methanol-water (1:1), adding to microwells, then allowing the solvent to evaporate completely (e.g., overnight) before water washing and use (Foxall *et al.*, 1992; Karlsson and Stromberg, 1987).

7. Whereas binding of soluble GBPs is performed as described, binding of intact cells, especially larger eukaryotic cells, requires a different approach (Collins *et al.*, 2000). Fluid shear inherent in standard microwell plate washing typically disrupts eukaryotic cell adhesion to immobilized glycolipids. Therefore, centrifugal force, which is adjustable and can be much milder than fluid sheer, is preferred. For this purpose, a custom Plexiglas box was developed, into which a fluid-filled microwell plate can be inverted and sealed. The sealed box fits into a table-top centrifuge carrier, where it is subjected to centrifugal forces sufficient to detach nonadherent cells but leave specifically adherent cells attached to the glycolipid-adsorbed surface. After centrifugation, the plate is recovered, and adherent cells are quantified. Complete details are published elsewhere (Collins *et al.*, 2000).

8. It is often valuable to compare chemical staining and protein binding patterns, or binding of different proteins, to the same glycolipid mixtures resolved on the same TLC plate (see Fig. 2). This is accomplished by spotting replicate samples spaced laterally on the plate, developing the TLC, then cutting the plate into sections that are subjected to different chemical or GBP overlay procedures.

9. Partially purified glycolipid extracts from tissue may be used, although contaminating phospholipids may compromise TLC resolution. Partially purified glycolipid extract from the equivalent of 10–20 mg (wet weight) of tissue per lane is applied. This amount can be increased or decreased empirically. Samples spotted too heavily will fail to resolve and will instead produce a smear. Samples spotted too lightly may not provide sufficient glycolipid for detection, especially for less abundant species. When interpreting GBP binding to glycolipid mixtures, the relative abundance of each species must be considered. Low-affinity binding to highly abundant species will be as robust as high-affinity binding to less abundant glycolipids (Suetake and Yu, 2003).

10. An industrial forced air dryer (e.g., product HG-201A, Master Appliance Corp., Racine, WI) may be used, without heat, to speed drying of applied samples and aid in spotting larger volumes. Set the dryer so that the air flows over and parallel to the plate surface.

11. A 5-cm long plate may provide ample resolution for many glycolipid mixtures. Running a longer plate may not improve separation, because development time will be increased, and diffusion of the samples may compromise further resolution.

12. PIBM coating results in lower background binding and greatly enhanced sensitivity. It has been postulated that the hydrophobic plastic coating reorients the glycolipids so that their polar glycans are oriented away from the sorbent layer and outwards into the solvent, where they may interact more efficiently with complementary binding proteins (Karlsson and Stromberg, 1987). Although the recommended PIBM concentration (0.1 % w/v) typically results in low background and high sensitivity for GBPs, it may be adjusted empirically to enhance signal/noise, with lower concentrations optimal for intact cell adhesion.

13. For high-affinity binding proteins, such as bacterial toxins, concentrations as low as 10 ng/ml may be sufficient. Low temperature incubation over longer periods (e.g., 4°, overnight) typically enhances sensitivity. The choice of binding buffer may be modified on the basis of the specific requirements and sensitivities of the GBP to ionic strength, divalent cations, and pH.

14. Higher sensitivity may be obtained using enhanced chemiluminescence for detection (Arnsmeier and Paller, 1995). The procedure described is suitable for soluble binding proteins (antibodies, lectins, toxins), as well as for viruses and bacteria, but not for eukaryotic cells that are susceptible to removal by fluid sheer. Intact eukaryotic cell adhesion to TLC-resolved glycolipids is detected using mild centrifugation in a fluid-filled custom-designed Plexiglas box to remove nonadherent cells (Schnaar, 1994).

15. A commercial device developed to enhance the reproducibility of TLC blotting is available (TLC Thermal Blotter Model AC-5970, Atto Corp., Tokyo, Japan).

16. Another option is to add a lipid antigen unrelated to the binding interaction under study to the glycolipid/phospholipid mixture. The unrelated lipid antigen is then used as a handle to capture the resulting liposomes on an antibody-coated biosensor chip (MacKenzie et al., 1997). Alternately, GBP can be bound to the chip surface and exposed to glycolipid-containing liposomes in solution (Sandhoff et al., 2005).

17. Binding curves should be close to a plateau at the end of the association phase (≤ 3 min). If this is not the case, the ligand density on the sensor chip may be decreased to <2500 RU, and/or the GBP concentration may be increased.

18. If regeneration (return to baseline) is incomplete, the time of treatment with regenerating solution may be increased, or a more stringent regenerating solution (consistent with the stability of the immobilized glycolipid-containing liposomes) may be used.

References

Arnsmeier, S. L., and Paller, A. S. (1995). Chemiluminescence detection of gangliosides by thin-layer chromatography. *J. Lipid Res.* **36,** 911–915.

Bashor, M. M. (1979). Dispersion and disruption of tissues. *Methods Enzymol.* **58,** 119–131.

Blackburn, C. C., Swank-Hill, P., and Schnaar, R. L. (1986). Gangliosides support neural retina cell adhesion. *J. Biol. Chem.* **261,** 2873–2881.

Collins, B. E., Yang, L. J. S., and Schnaar, R. L. (2000). Lectin-mediated cell adhesion to immobilized glycosphingolipids. *Methods Enzymol.* **312,** 438–446.

Feizi, T., and Chai, W. (2004). Oligosaccharide microarrays to decipher the glyco code. *Nat. Rev. Mol. Cell. Biol.* **5,** 582–588.

Foxall, C., Watson, S. R., Dowbenko, D., Fennie, C., Lasky, L. A., Kiso, M., Hasegawa, A., Asa, D., and Brandley, B. K. (1992). The three members of the selectin receptor family recognize a common carbohydrate epitope, the sialyl Lewis[x] oligosaccharide. *J. Cell Biol.* **117,** 895–902.

Hakomori, S. (2002). The glycosynapse. *Proc. Natl. Acad. Sci. USA* **99,** 225–232.

Karlsson, K.-A., and Stromberg, N. (1987). Overlay and solid phase analysis of glycolipid receptors for bacteria and viruses. *Methods Enzymol.* **138,** 220–232.

Kuziemko, G. M., Stroh, M., and Stevens, R. C. (1996). Cholera toxin binding affinity and specificity for gangliosides determined by surface plasmon resonance. *Biochemistry* **35,** 6375–6384.

MacKenzie, C. R., Hirama, T., Lee, K. K., Altman, E., and Young, N. M. (1997). Quantitative analysis of bacterial toxin affinity and specificity for glycolipid receptors by surface plasmon resonance. *J. Biol. Chem.* **272,** 5533–5538.

MacKenzie, C. R., and Hirama, T. (2000). Quantitative analyses of binding affinity and specificity for glycolipid receptors by surface plasmon resonance. *Methods Enzymol.* **312,** 205–216.

Magnani, J. L., Smith, D. F., and Ginsburg, V. (1980). Detection of gangliosides that bind cholera toxin: Direct binding of ^{125}I-labeled toxin to thin-layer chromatography plates. *Anal. Biochem.* **109,** 399–402.

Nakajima, H., Kiyokawa, N., Katagiri, Y. U., Taguchi, T., Suzuki, T., Sekino, T., Mimori, K., Ebata, T., Saito, M., Nakao, H., Takeda, T., and Fujimoto, J. (2001). Kinetic analysis of binding between Shiga toxin and receptor glycolipid Gb3Cer by surface plasmon resonance. *J. Biol. Chem.* **276,** 42915–42922.

Sandhoff, R., Grieshaber, H., Djafarzadeh, R., Sijmonsma, T. P., Proudfoot, A. E., Handel, T. M., Wiegandt, H., Nelson, P. J., and Grone, H. J. (2005). Chemokines bind to sulfatides as revealed by surface plasmon resonance. *Biochim. Biophys. Acta* **1687,** 52–63.

Schnaar, R. L. (1994). Isolation of glycosphingolipids. *Methods Enzymol.* **230,** 348–370.

Schnaar, R. L. (1994). Immobilized Glycoconjugates for Cell Recognition Studies. *In* "Neoglycoconjugates: Preparation and Application" (Y. C. Lee and R. T. Lee, eds.), pp. 425–443. Academic Press, San Diego.

Schnaar, R. L. (2000). Glycobiology of the Nervous System. *In* "Carbohydrates in Chemistry and Biology, Part II:Biology of Saccharides" (B. Ernst, G. W. Hart, and P. Sinaÿ, eds.), pp. 1013–1027. Wiley-VCH, Weinheim, Germany.

Schnaar, R. L. (2004). Glycolipid-mediated cell-cell recognition in inflammation and nerve regeneration. *Arch. Biochem. Biophys.* **426,** 163–172.

Schnaar, R. L., Fromholt, S. E., Gong, Y., Vyas, A. A., Laroy, W., Wayman, D. M., Heffer-Lauc, M., Ito, H., Ishida, H., Kiso, M., Griffin, J. W., and Sheikh, K. A. (2002). Immunoglobulin G-class mouse monoclonal antibodies to major brain gangliosides. *Anal. Biochem.* **302,** 276–284.

Schnaar, R. L., and Needham, L. K. (1994). Thin-layer chromatography of glycosphingolipids. *Methods Enzymol.* **230,** 371–389.

Stroud, M. R., Handa, K., Salyan, M. E. K., Ito, K., Levery, S. B., Hakomori, S., Reinhold, B. B., and Reinhold, V. N. (1996). Monosialogangliosides of human myelogenous leukemia HL60 cells and normal human leukocytes. 1. Separation of E-selectin binding from nonbinding gangliosides, and absence of sialosyl-Lex having tetraosyl to octaosyl core. *Biochemistry* **35,** 758–769.

Stults, C. L. M., Sweeley, C. C., and Macher, B. A. (1989). Glycosphingolipids: Structure, biological source, and properties. *Methods Enzymol.* **179,** 167–214.

Suetake, K., and Yu, R. K. (2003). Thin-layer chromatography; immunostaining of glycolipid antigens; and interpretation of false-positive findings with acidic lipids. *Methods Enzymol.* **363,** 312–319.

Taki, T., Handa, S., and Ishikawa, D. (1994). Blotting of glycolipids and phospholipids from a high-performance thin-layer chromatogram to a polyvinylidene difluoride membrane. *Anal. Biochem.* **221,** 312–316.

Willison, H. J. (2005). The immunobiology of Guillain-Barre syndromes. *J. Peripher. Nerv. Syst.* **10,** 94–112.

[16] Analysis of Carbohydrate–Carbohydrate Interactions Using Gold Glyconanoparticles and Oligosaccharide Self-Assembling Monolayers

By ADRIANA CARVALHO DE SOUZA and JOHANNIS P. KAMERLING

Abstract

Carbohydrates are the most extended structures exposed at the surface of most cells. These carbohydrate chains, when arranged in polyvalent clusters, offer a rich supply of low-affinity binding sites, making them a reliable and flexible system to regulate cell adhesion and recognition. The very first model system for cell–cell recognition by means of carbohydrate–carbohydrate interactions in the animal kingdom came from a primitive invertebrate animal: the marine sponge. During the past 50 years, studies have shown that highly repetitive carbohydrate motives on extracellular proteoglycan supramolecular complexes of marine sponge cells are involved in the species-specific adhesion. In this chapter, some glyconanotechnology procedures are described for the detailed investigation of the role of a carbohydrate epitope in the marine sponge cell recognition. The various protocols are generally applicable in other areas of glycoscience.

Overview

Carbohydrates are widely distributed in nature in the form of polysaccharides or as part of glycoconjugates such as glycoproteins and glycolipids. Their structural diversity exceeds by far that of proteins and nucleic acids, because of the presence of monomers capable of forming different glycosidic linkages, including branching points. This astonishing structural diversity allows carbohydrates to encode information for a large variety of biological functions. Once regarded mainly as energy-yielding molecules and as structural elements, carbohydrates have become known in the past decades for their role in cell-to-cell communication (Bucior and Burger, 2004b; Hakomori, 2004b), inflammation (Kansas, 1996; Rabinovich *et al.*, 2002), signal transduction (Sacchettini *et al.*, 2001), and fertility and development (Tiemeyer and Goodman, 1996; Vacquier and Moy, 1997).

Carbohydrate chains serve as receptors for enzymes (Kleene and Berger, 1993), hormones (Papandreou *et al.*, 1990), infectious bacteria (Lin *et al.*, 2002; Sung *et al.*, 2001), viruses (Mammen *et al.*, 1995), and toxins (St. Hilaire *et al.*, 1994; Wolfhagen *et al.*, 1994), all classified as carbohydrate–protein

METHODS IN ENZYMOLOGY, VOL. 417
0076-6879/06 $35.00
DOI: 10.1016/S0076-6879(06)17016-0

interactions. For a long time, direct carbohydrate–carbohydrate interactions between the glycan chains of glycoproteins, glycolipids, and proteoglycans have been underestimated, if not ignored; however, the number of reports on this subject is growing (Table I; for a recent review, see Hakomori [2004a]).

The strength of the interactions, whereby carbohydrates are involved, is a few orders of magnitude lower than those observed for protein–protein interaction, with carbohydrate–carbohydrate interactions weaker than carbohydrate–protein interactions. Therefore, the binding of carbohydrates to carbohydrate receptors often requires affinity enhancement to attain biologically relevant strength. For such purposes, nature organizes the cell surface glycoconjugates into clusters or superstructures. The affinity enhancement because of these possibilities for multivalent interactions, known as multivalency, is substantially larger than the effect of the increased concentration of the ligand-receptor pair.

For a long time, the need for multivalent carbohydrate arrangements formed a barrier for the investigation of carbohydrate–carbohydrate interactions on the molecular level. However, recent developments in nanotechnology have provided us with the tools to expand our knowledge on the mechanism behind carbohydrate–carbohydrate interactions (Carvalho de Souza *et al.*, 2004, 2005; de Paz *et al.*, 2005; de la Fuente *et al.*, 2001; Hernáiz *et al.*, 2002; Rojo *et al.*, 2004; Tromas *et al.*, 2001). This chapter describes the protocols for the use of gold glyconanoparticles (GNPs) and oligosaccharide self-assembled monolayers (SAMs) as model systems to investigate the carbohydrate-mediated cell recognition events on the marine sponge *Microciona prolifera*. It should be mentioned that these protocols are also suitable for studies in other areas in which carbohydrate–carbohydrate interactions are involved.

Marine Sponges

The species-specific cell aggregation in the marine sponge *Microciona prolifera* involves proteoglycan-like macromolecular complexes otherwise known as aggregation factors. The supramolecular structure of the *M. prolifera* aggregation factor (MAF) was explained by immunochemical and electrophoretic procedures, combined with atomic force microscopy (AFM) imaging (Jarchow *et al.*, 2000). The main proteins of MAF, MAFp3, and MAFp4 are highly polymorphic molecules (Fernàndez-Busquets and Burger, 1997). AFM imaging of purified MAF revealed a sunburst-like molecule in which the central ring is formed by approximately 20 units of MAFp3, and the 20 arms are formed exclusively by the MAFp4 proteoglycan (Fernàndez-Busquets and Burger, 2003). In a Ca^{2+}-independent process, 6-kDa glycans located on MAFp4 arms (known as g-6)

TABLE I
EXAMPLES OF CARBOHYDRATE–CARBOHYDRATE INTERACTIONS

Structures	Function	Studied system	References
Le^x-Le^x	Basis of the first cell adhesion event in preimplantation of mouse embryo and autoaggregation of mouse embryonal carcinoma F9 cells	Compaction of the mouse embryo at the morula stage; undifferentiated F9 mouse embryonal cells; Le^x-GNPs and Le^x gold surfaces by SPR and AFM; Le^x dimers by NMR spectroscopy	de la Fuente and Penadés, 2004; Eggens et al., 1989; de la Fuente et al., 2001; Gege et al., 2002; Geyer et al., 2000; Gourier et al., 2004; Hakomori, 2003; Hernáiz et al., 2002; Kojima et al., 1994; Tromas et al., 2001
g-200-g-200 glycan	Definition of species-specific sponge cell aggregation and recognition	Species-specific marine sponge cell aggregation; proteoglycan-coated beads aggregation; molecular force microscopy with proteoglycan and glycan monolayers; self-recognition of g-200 sulfated disaccharide epitope by SPR and molecular force microscopy; aggregations of GNPs coated with sulfated disaccharide related structures	Bucior and Burger, 2004a; Bucior et al., 2004; Carvalho de Souza et al., 2004, 2005 Fernández-Busquets and Burger, 2003; Jarchow et al., 2000; Misevic and Burger, 1990, 1993; Popescu and Misevic, 1997; Popescu et al., 2003
GM3-Gg3	Interaction between mouse B16 melanoma and T-cell lymphoma and between B16 cells and mouse endothelial cells	Aggregation of melanoma cells with lymphoma or endothelial cells; interactions of GM3 langmuir monolayers with Gg3-polystyrene conjugates by SPR; interactions of GM3 langmuir monolayers with lactosylsphingosine micelles	Kojima and Hakomori, 1989, 1991; Matsuura and Kobayashi, 2004; Matsuura et al., 2000; Santacroce and Basu, 2003, 2004
(KDN)GM3-Gg3	Sperm-to-egg interaction in rainbouw trout fertilization	Adherence of liposomes containing (KDN)GM3 to Gg3 coated plastic plates	Yu et al., 2002
GalCer-Sulfatide	Myelin compaction	Ca^{2+}-mediated interaction by electrospray ionization mass spectrometry	Boggs et al., 2004; Koshy and Boggs, 1996

adhere to cell surface receptors, whereas 200-kDa glycans protruding from the MAFp3 central ring (known as g-200) promote cell adhesion by means of a Ca^{2+}-dependent self-association process (for a detailed review, see Fernàndez-Busquets and Burger [2003]). Two monoclonal antibodies prepared against MAF, called Block 1 and Block 2, were able to inhibit the Ca^{2+}-dependent self-aggregation process without influencing the cell binding process (Misevic and Burger, 1993; Misevic et al., 1987). In the same studies, it was observed that these monoclonal antibodies recognize highly repetitive carbohydrate epitopes (>1000 antigenic sites for Block 1 Fab fragments, >2500 antigenic sites for Block 2 Fab fragments) in the acidic g-200 glycan of MAFp3. Characterization of these epitopes revealed two small oligosaccharide fragments, the sulfated disaccharide GlcpNAc3S (β1-3)Fucp (Block 2; Spillmann et al. [1995]) and the pyruvylated trisaccharide Galp4,6Pyr(β1-4)GlcpNAc(β1-3)Fucp ([Block 1; Spillmann et al. [1993]). Both fragments turned out to be directly involved in the Ca^{2+}-dependent MAF self-aggregation.

Using synthetic β-D-GlcpNAc3S-(1→3)-α-L-Fucp-(1→O), multivalently presented as bovine serum albumin conjugates, we have shown with surface plasmon resonance spectroscopy that the sulfated disaccharide interacts with itself in the presence of Ca^{2+}-ions. This phenomenon does not simply occur through electrostatic interactions, because other sulfate-containing carbohydrates analyzed with the same procedure did not show any self-recognition (Haseley et al., 2001). The replacement of Ca^{2+} ions by Mg^{2+} or Mn^{2+} ions eradicated completely the self-recognition of the sulfated disaccharide, as earlier also demonstrated for the g-200 glycan (Dammer et al., 1995; Popescu and Misevic, 1997).

Recently, transmission electron microscopy (TEM) imaging of GNPs coated with β-D-GlcpNAc3S-(1→3)-α-L-Fucp-(1→O)(CH$_2$)$_3$S(CH$_2$)$_6$SH and related structures, in the presence or absence of Ca^{2+} ions, have given valuable information on the mechanism of this disaccharide self-recognition (Carvalho de Souza et al., 2004, 2005). On the basis of these model studies, it was postulated that after the coordination of Ca^{2+} ions by several functional groups, the sulfated disaccharide would reach the adequate conformation, wherein other interactions, such as hydrophobic interactions, will take place to stabilize the entire complex.

Synthesis of Spacer-Containing Oligosaccharides

The most applied synthetic procedure for the preparation of GNPs is based on the reduction of gold salts in the presence of a thiol-spacer–containing oligosaccharide. The choice of the thiol spacer used to connect the oligosaccharide to the gold nanoparticles is of great importance in the entire strategy. Currently, fully synthetic thiol-spacer–containing oligosac-charides have been mostly used to prepare GNPs. For a summary, see Table II.

The full synthesis of thiol-spacer–containing oligosaccharides can be difficult and time-consuming. To further the use of GNPs in the field of glycobiology, recently we have developed a simple method to prepare thiolated oligosaccharides by reductive amination of isolated glycans with protected cysteamine (Halkes *et al.*, 2005). However, it should be noted that in the latter approach the "reducing-site" saccharide ring of the glycans is permanently opened in the reductive amination step.

Synthesis of Thiol-Spacer–Containing Oligosaccharides

Here, we focus on oligosaccharide allyl glycosides as precursors in the synthesis of thiol-spacer–containing oligosaccharides. The synthesis of the allyl glycosides *7–13* (Scheme 1) involves the combination of diverse protection-deprotection strategies (Carvalho de Souza *et al.*, 2004).

Elongation

A solution of an allyl glycoside (Scheme 1; 50 μmol *7* to *13)* and 1, 6-hexanedithiol (500 μmol) in 1.5 ml MeOH is irradiated in a 3-ml quartz vial, using a VL-50C Vilber Lourmat UV Lamp. After 2 h, TLC on Silica Gel 60 $F_{25}4$ (Merck; compounds are colored on the TLC plates by spraying with orcinol (2 mg/ml) in 20% (v/v) methanolic H_2SO_4 followed by careful heating with a hot-air gun) shows the formation of a new product with higher R_f. Then, the solvent is concentrated under reduced pressure at 40°, and the residue is loaded on a column (8 mm × 100 mm) of Silica Gel 60 (Merck, 0.040–0.063 mm). First, the excess of 1,6-hexanedithiol is eluted with CH_2Cl_2/MeOH 9:1, then a carbohydrate-containing fraction can be collected by elution with MeOH. The solution is concentrated under reduced pressure at 40° and the residue, dissolved in water, is loaded on a C-18 Extract-Clean column (Alltech). After elution of unreacted allyl glycoside with water (3 × 3 ml), the thiol-spacer–containing oligosaccharide (Scheme 1; *1a/b* to *6*) is eluted with MeOH (3 × 3 ml). Finally, the solution is concentrated *in vacuo*, and the desired product is obtained, after lyophilization from water, as a white, amorphous powder.

Synthesis of Gold Nanoparticles

Since Faraday's publication on the synthesis of gold colloids from a two-phase reaction system (Faraday, 1875), several new methods have been developed on the basis of the reduction of gold(III) derivatives. Among these methods, the reduction of gold tetrachloroaurate ($HAuCl_4$) in water

TABLE II

EXAMPLES OF CARBOHYDRATE–CARBOHYDRATE INTERACTIONS USING GNPs

Spacer	GNPs	References
	Le^x-GNPs	de la Fuente and Penadés, 2002; de la Fuente et al., 2001
	Gal-GNPs, Glc-GNPs, Sp-GNPs	Nolting et al., 2003
	Man-GNPs, TF-GNPs	de Paz et al., 2005; Lin et al., 2002; Svarovsky et al., 2005
	Marine sponge GNPs	Carvalho de Souza et al., 2004, 2005
	Lac-GNPs	de la Fuente et al., 2001

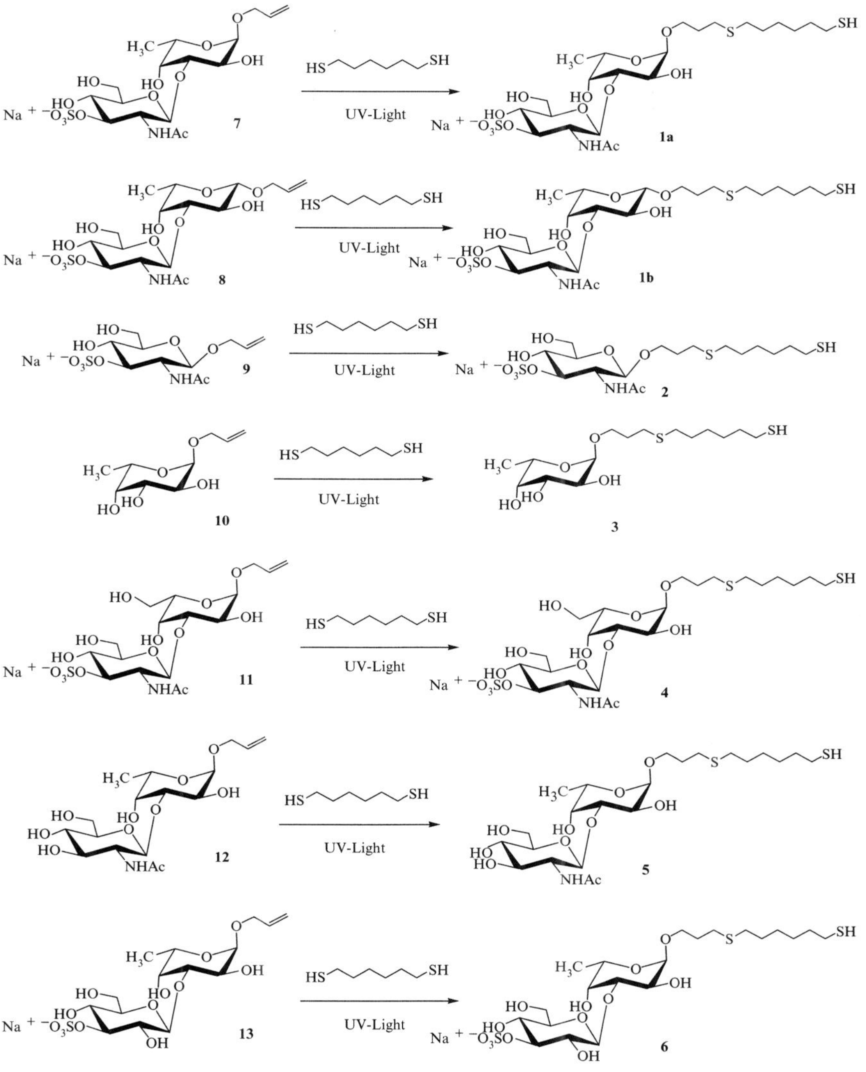

SCHEME 1. Elongation of allyl glycosides with 1,6-hexanedithiol.

by sodium citrate is one of the most popular approaches (Turkevitch *et al.*, 1951). In 1973, Frens published a method to obtain gold nanoparticles with a prechosen size, as a result of the variation of the ratio between reducing/stabilizing agents to gold (Frens, 1973). More recently, the preparation of sodium 3-mercaptopropionate–stabilized gold nanoparticles, by the combined addition of sodium citrate and sodium 3-mercaptopropionate, was published (Yonezawa and Kunitake, 1999). Another recent procedure for the preparation of gold cluster hydrosols is based on the reduction of gold(III) salts by tetrakis(hydroxymethyl)phosphonium chloride (Duff *et al.*, 1993). Since its publication in 1995, the Brust method for the synthesis of thermally stable and air-stable gold nanoparticles has had a large influence on the development of gold nanoparticles research (Brust *et al.*, 1995). The original report describes the reduction of gold tetrachloroaurate by $NaBH_4$ in the presence of dodecanethiol, in a two-phase system. Variations on the Brust method made it possible to prepare gold glyconanoparticles from a methanol solution of a thiol-spacer-containing oligosaccharide (Carvalho de Souza *et al.*, 2004; de la Fuente *et al.*, 2001; Lin *et al.*, 2002).

Gold Glyconanoparticles

A 10-mM methanolic solution of a thiol-spacer–containing oligosaccharide (5 equiv.; *1a/b* to *6*) is added to a 25-mM aqueous solution of tetrachloroauric acid (1 equiv.). Then, an 1-M aqueous solution of $NaBH_4$ (22 equiv.) is slowly added under rigorous stirring. The obtained black suspension is stirred for 2 h at room temperature. After concentration, a solution of the residue in water (10 ml) is loaded on a 30-kDa Nalgene centrifugal filter and washed with water (5 × 15 ml). Finally, a solution of the retentate in water is lyophilized, yielding gold glyconanoparticles (Scheme 2; *Au-1a/b* to *Au-6*) as brown, amorphous powders.

Analysis of Gold Glyconanoparticles

Several analytical methods exist that can be applied to characterize gold nanoparticles. The core dimensions can be measured by scanning tunneling microscopy (STM) (Terrill *et al.*, 1995), atomic force microscopy (AFM) (Terrill *et al.*, 1995), transmission electron microscopy (TEM) (Brust *et al.*, 1995; Hostetler *et al.*, 1998), small-angle X-ray scattering (SAXS) (Hostetler *et al.*, 1998; Terrill *et al.*, 1995), and X-ray diffraction (XRD) (Hostetler *et al.*, 1998; Leff *et al.*, 1995). A combination of TEM analysis of a fractionally crystallized sample, MALDI mass spectrometry, and theoretical calculations has shown that the more likely shape of the gold core is a truncated octahedron, and the preferred core atom populations ("magic number" or completed metal shells) include, for example, clusters of 225, 314, and 459 atoms (Gutiérrez-Wing *et al.*, 1998; Whetten *et al.*, 1996). Knowing and controlling the core dimensions is very important to establish the size and mass of gold nanoparticles.

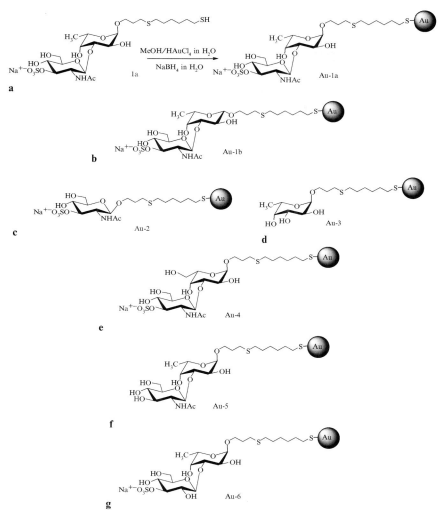

SCHEME 2. **a**, Synthesis of gold glyconanoparticles; **b**, **c**, **d**, **e**, **f**, and **g**, other examples of gold glyconanoparticles.

The protective monolayers in gold nanoparticles have also been examined with multiple techniques (Daniel and Astruc, 2004; Templeton *et al.*, 2000). The average number of ligands per core can be derived from a combination of elemental analysis (de la Fuente *et al.*, 2001; Terrill *et al.*, 1995) and core size analysis. This number can also be deduced from x-ray

photoelectron spectroscopy (XPS) (Hostetler *et al.*, 1998), thermogravi-
metric analysis (TGA) (Hostetler *et al.*, 1998; Templeton *et al.*, 1999; Terrill
et al., 1995), or monosaccharide analysis, in the specific case of gold glyconanoparticles (Carvalho de Souza *et al.*, 2004; Halkes *et al.*, 2005). Further
information about the structure, dynamics, and packing of the nanoparticle
protective monolayers can be obtained with FT-IR spectroscopy (Brust
et al., 1995; Templeton *et al.*, 1999), differential scanning calorimetry
(DSC) (Badia *et al.*, 1997; Brust *et al.*, 1995; Hostetler *et al.*, 1998), or
NMR spectroscopy (Badia *et al.*, 1997; Hostetler *et al.*, 1998; Terrill *et al.*,
1995). Here, we describe a set of analytical procedures selected for the
analysis of gold glyconanoparticles in our laboratory.

Transmission Electron Microscopy

High-resolution TEM is the most common technique for characterization of the gold core. The micrographs, collected with a Philips Tecnai 12
microscope at 120 kV accelerating voltage, and depicted in Fig. 1, were
obtained following a general procedure: Aliquots (1 μl) of aqueous solutions of gold glyconanoparticles (0.1 mg/ml; *Au-1a* to *Au-6*) are placed onto
copper grids coated with carbon film (QUANTIFOIL on a 200-square
mesh copper grid, hole shape R 2/2). The grids are left to dry at room
temperature for several hours.

Inspection of the micrographs shows in all cases a clear dispersity of the
core size. The particle size distribution of the gold glyconanoparticles is
automatically determined from several micrographs of the same sample,
using analySIS 3.2 (Soft Imaging System GmbH). Following the protocol
for the preparation of gold glyconanoparticles as described previously,
the nanoparticles have a mean diameter ranging from 1.5–1.8 nm with a
dispersity varying from 35–44% (Table III).

Monosaccharide Analysis

In an ampoule, the GNPs (0.1–0.5 mg) are mixed with a mannitol
solution (internal standard: 10–100 mmol). After lyophilization and drying
over P_2O_5 in a vacuum desiccator for 18 h, the residue is dissolved in 1 M
methanolic HCl (0.5 ml). Nitrogen is bubbled through the solution for
30 sec, and then the ampoule is sealed. The solution is heated for 24 h at
85°; subsequently, neutralization is carried out by addition of solid silver
carbonate. *N*-(Re)acetylation is performed by addition of acetic anhydride
(10–50 μl). After mixing, the resulting suspension is kept at room temperature for 24 h in the dark. The precipitate is then thoroughly triturated and,
after centrifugation, the supernatant is collected. The residue of silver salts
is washed twice with 0.5 ml dry methanol. The pooled supernatants are

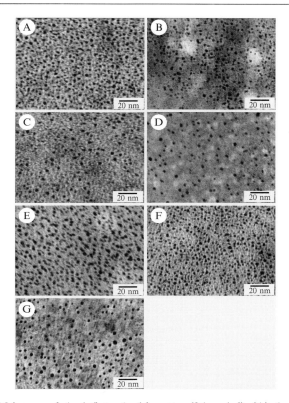

FIG. 1. TEM images of *Au-1a/b* to *Au-6* in water (0.1 mg/ml). (A) *Au-1a*; (B) *Au-1b*; (C) *Au-2*; (D) *Au-3*; (E) *Au-4*; (F) *Au-5*; (G) *Au-6* (see Scheme 2).

evaporated under reduced pressure at 35°. The final residue is dried for 12 h in a vacuum desiccator over P_2O_5. Then, the sample is trimethylsilylated with a mixture of pyridine-hexamethyldisilazane-chlorotrimethylsilane (5:1:1; 100 μl) for 30 min at room temperature (Kamerling and Vliegenthart, 1989).

The trimethylsilylated methyl glycosides are analyzed by GLC on an EC-1 capillary column (30 m × 0.32 mm, Alltech) using a Chrompack CP 9002 gas chromatograph (temperature program, 140–240° at 4°/min). The identification of the monosaccharide derivatives is confirmed by gas chromatography/mass spectrometry on a Fisons Instruments GC 8060/MD 800 system (Interscience), equipped with an AT-1 capillary column (30 m × 0.25 mm, Alltech), using the same temperature program. The monosaccharide analysis data and the size distribution obtained from the TEM analysis

TABLE III

RESULTS OF MONOSACCHARIDE ANALYSIS, MEAN DIAMETER MEASUREMENTS AND CALCULATION
OF THE NUMBER OF LIGANDS PER CORE FOR GOLD GLYCONANOPARTICLES Au-1a/1b to Au-6

GNPs	D_{core} (nm)	Dispersity	%CHO	# Gold atoms (Hostetler et al., 1998)	# CHO ligands[a]
Au-1a	1.82 ± 0.8	44%	36%	201	33
Au-1b	1.71 ± 0.6	35%	40%	201	39
Au-2	1.51 ± 0.6	40%	23%	116	13
Au-3	1.80 ± 0.7	39%	22%	201	31
Au-4	1.80 ± 0.6	33%	39%	201	37
Au-5	1.63 ± 0.6	37%	41%	140	34
Au-6	1.55 ± 0.6	39%	37%	116	22

[a]Calculated according to $n_{CHO} = [(n_{Au} \times MW_{Au}/100 - \%_{CHO}) \times \%_{CHO}]/MW_{CHO}$; where n_{CHO} is the number of sugar molecules on GNPs, n_{Au} is the number of gold atoms in GNPs.

can be combined to determine the average of ligand molecules per core (Table III).

^{1}H-NMR Spectroscopy

The 500-MHz ^{1}H-NMR spectra are recorded at 300 K with a Bruker AMX 500 spectrometer; δH values are given in ppm relative to the signal for internal Me$_4$Si (δ_H = 0, CD$_3$OD) or internal acetone (δ_H = 2.22, D$_2$O). In Fig. 2, the ^{1}H-NMR spectra of two oligosaccharides (1a and 1b) alone and bound to gold nanoparticles are displayed along with some assignments. On binding to the gold, the resonances broaden and some chemical shifts are slightly shifted. However, some important fingerprint resonances of the oligosaccharide ligands can still be recognized in the gold glyconanoparticles spectra.

It is well known that a strong NMR line broadening is observed in spectra of gold nanoparticles (Badia et al., 1996). This spectral broadening is the combined result of several factors: (1) The solid-like appearance of gold nanoparticles—the groups closest to the thiolate/Au interface are more densely packed and thereby experience fast spin relaxation from dipolar interaction, similar to solid-state samples; (2) the distribution of chemical shifts—the site (terraces, edges, vertices) where the thiolate is connected to the gold core has specific influence on the chemical shift of the neighboring groups, thereby causing a substantial broadening of the resonance for the groups closest to the sulfur atom. This effect fades sharply with distance from the metal core; (3) spin-spin relaxation (T_2)—similar to proteins and

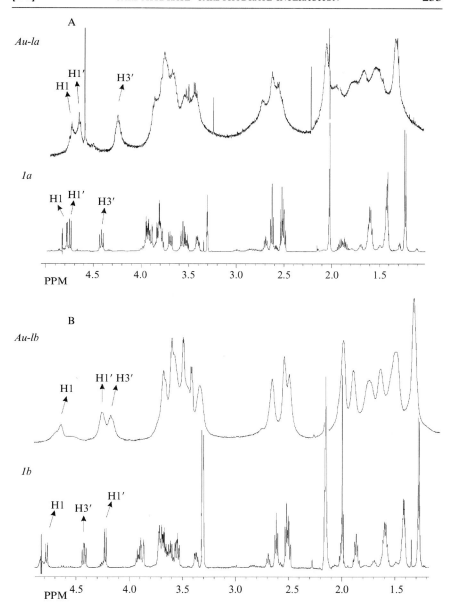

FIG. 2. (A) Comparison of the ¹H-NMR spectra of gold glyconanoparticles *Au-1a* (in D₂O) and thiol-spacer-containing disaccharide *1a* (in CD₃OD). (B) Comparison of the ¹H-NMR spectra of gold glyconanoparticles *Au-1b* (in D₂O) and thiol-spacer-containing disaccharide *1b* (in CD₃OD).

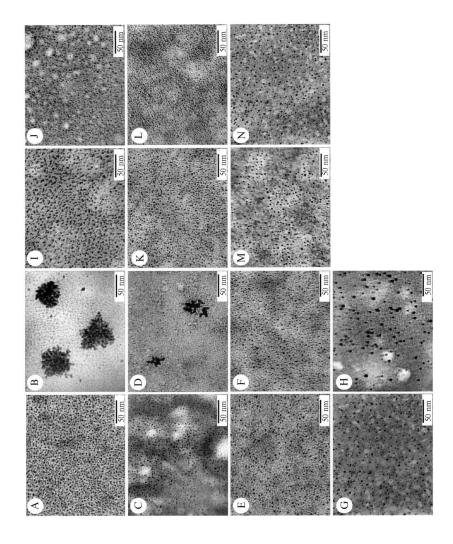

polymers, a NMR line broadening is caused by the slow rotation in solution of the macromolecules (Badia *et al.*, 1997; Hostetler *et al.*, 1998).

Aggregation Experiments Using TEM

For a typical experiment in water, a single drop (1 μl) of an aqueous gold glyconanoparticles solution (0.1 mg/ml incubated for 16 h) is deposited onto a carbon-coated copper grid (QUANTIFOIL on a 200-square mesh copper grid, hole shape R 2/2), and allowed to dry at room temperature for several hours. Experiments whereby water is replaced by either 10 mM CaCl$_2$ or 10 mM MgCl$_2$ are performed following the same procedure. To verify the importance of Ca^{2+} ions for the integrity of the aggregates, a solution of EDTA (final concentration, 50 mM) is added to a sample of *Au-1a* aggregates, then the mixture is loaded on a 30-kDa Nalgene centrifugal filter and washed with water (5 × 15 ml). The residue is dissolved in water (0.1 mg/ml), and an aliquot (1 μl) is examined by TEM.

Figure 3 shows a comparison of the gold nanoparticles *Au-1a/b* to *Au-6*, incubated in water and in 10 mM CaCl$_2$. TEM micrographs of *Au-1a/b* to *Au-6* in water show, in all cases, uniformly dispersed nanodots throughout the grid surface, and no aggregates are observed (Carvalho de Souza *et al.*, 2004, 2005). Of the TEM micrographs of *Au-1a/b* to *Au-6* in aqueous 10 mM CaCl$_2$ (0.1 mg/ml), being the calcium concentration commonly found in sea water, only the *Au-1a* and *Au-1b* nanoparticles present aggregates. The complete dispersion of the *Au-1a* aggregates after treatment with EDTA, confirms the dependence on Ca^{2+} ions of the self-recognition of β-D-Glc$p$NAc3S-$(1 \rightarrow 3)$-α-L-Fuc$p$-$(1 \rightarrow O)$. Furthermore, incubation of *Au-1a* nanoparticles in 10 mM MgCl$_2$ (0.1 mg/ml) does not result in any aggregate formation, in agreement with studies performed with purified MAF (Dammer *et al.*, 1995; Popescu and Misevic, 1997).

Atomic Force Microscopy

Nowadays, atomic force microscopy (AFM) is frequently applied to determine interaction forces between biomolecules (Willemsen *et al.*, 2000).

FIG. 3. TEM images of *Au-1a/b* to *Au-6* after 16 h incubation. (A) *Au-1a* in water (0.1 mg/ml); (B) *Au-1a* in 10 mM CaCl$_2$ (0.1 mg/ml); (C) *Au-1b* in water (0.1 mg/ml); (D) *Au-1b* in 10 mM CaCl$_2$ (0.1 mg/ml); (E) *Au-2* in water (0.1 mg/ml); (F) *Au-2* in 10 mM CaCl$_2$ (0.1 mg/ml); (G) *Au-3* in water (0.1 mg/ml); (H) *Au-3* in 10 mM CaCl$_2$ (0.1 mg/ml); (I) *Au-4* in water (0.1 mg/ml); (J) *Au4* in 10 mM CaCl$_2$ (0.1 mg/ml); (K) *Au-5* in water (0.1 mg/ml); (L) *Au-5* in 10 mM CaCl$_2$ (0.1 mg/ml); (M) *Au-6* in water (0.1 mg/ml); (N) *Au-6* in 10 mM CaCl$_2$ (0.1 mg/ml). The images were obtained with a Philips Tecnai 12 microscope at 120 kV accelerating voltage.

Starting with the detection of the first discrete unbinding forces of single molecular bonds (Florin *et al.*, 1994) by AFM only a few years ago, measurements have become more and more quantitative (Evans, 1999; Merkel *et al.*, 1999).

Recently, AFM was used to quantify the unbinding forces of different carbohydrate–carbohydrate interactions (Dammer *et al.*, 1995; Fritz *et al.*, 1997; Tromas *et al.*, 2001).

Here, we report the use of AFM to determine the binding strength between the sulfated disaccharide epitopes involved in the self-association of the aggregation factor of the marine sponge *M. prolifera*. To this end, a series of approach-retract cycles of tips and substrates, functionalized with self-assembled monolayers (Ulman, 1996) of thiol-spacer-containing disaccharide *1a* (Scheme 2), was performed in both water and aqueous 10 mM $CaCl_2$. Force-distance curves typical for biomolecules were obtained with the functionalized tip and substrate in the presence of Ca^{2+} ions.

Gold/Mica Substrate

To freshly cleaved flat mica plates a chromium layer (2.5 nm) followed by a gold layer (200 nm) are deposited by evaporation in high vacuum. Before use, gold substrates are annealed for 2 min in a gas flame to obtain Au(III) terraces (Casero *et al.*, 2002). Figure 4A shows a 300-nm-wide image of a bare gold substrate. The image reveals grains in the range of 50–100 nm with deep boundaries, but after annealing the grains are smoothed (Fig. 4B).

For the functionalization of the gold substrate, it is immersed for 2 h in a 1 mM methanolic solution of *1a*. Then, the substrate is rinsed several times with water and methanol, and dried under a stream of argon.

Gold Tip Functionalization

Commercial silicon nitride rectangular cantilevers with V-shaped gold-coated tips (Olympus, Bio-Lever series; gold coating at both sides of the tip varying from 20–30 nm), and nominal spring constant 0.03 N/m are used. The cantilevers are functionalized with disaccharide *1a* by immersion for 2 h in a 1 mM methanolic solution of *1a*, then rinsed several times with methanol and dried in air. Because of the low spring constant, some tips can bend completely during removal from the *1a* solution. Bended tips are not used for force-distance measurements.

Force Curve Measurements

A Nanoscope III multimode AFM (Digital Instruments, Santa Barbara, CA) instrument is used for the force-distance measurements. The substrates

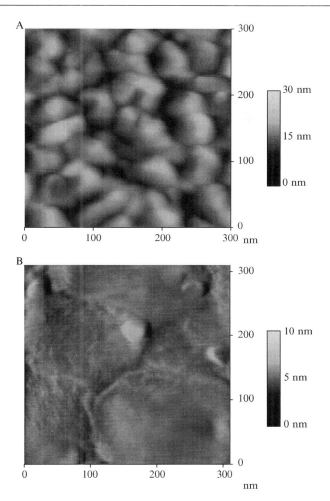

FIG. 4. (A) 300 nm × 300 nm contact-mode AFM image of a gold/mica substrate; (B) 311 nm × 311 nm contact-mode AFM image of a 2-min annealed gold/mica substrate.

are mounted on an E-scanner, which is calibrated on a standard grid of a Nanoscope III AFM. Interaction measurements are performed in water or aqueous 10 mM CaCl$_2$ with a fluid cell without O-ring. Unloading rates of 400 nm/sec and ramp sizes of 200 nm are used.

As visualized in Fig. 5A and B, the force-distance curves, obtained with the tip and gold-substrate, both functionalized with SAM of disaccharide *1a* in the presence of 10 mM Ca^{2+} ions, present typical features for interactions

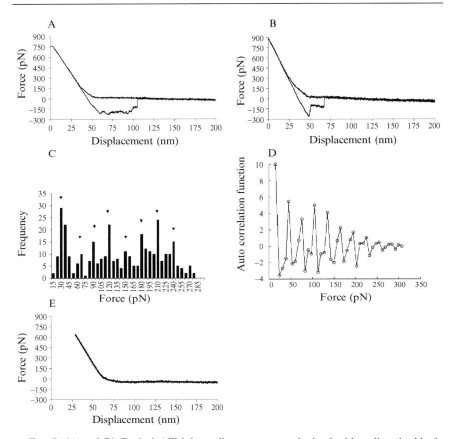

FIG. 5. (A and B) Typical AFM force-distance curves obtained with a disaccharide *1a* functionalized gold tip and substrate in 10 mM CaCl$_2$. (C) Histogram of the adhesion forces between the disaccharide *1a* functionalized tip and substrate in 10 mM CaCl$_2$. (D) Autocorrelation analysis of the histogram in (C). (E) Typical AFM force-distance curve obtained between the disaccharide *1a* functionalized tip and sample in water.

between biomolecules, with, in some cases, multiple, stepwise break processes. A histogram derived from more than 300 measured final rupture forces (Florin *et al.*, 1994) reveals eight maxima that represent an integer number of force quantum for the unbinding of a pair of sulfated disaccharide molecules (Fig. 5C). Only the last step of unbinding is considered for analysis, because it is conceivable that the other steps are nonlinear convolutions of multiple unbinding processes. The autocorrelation analysis of the histogram (Fig. 5D) shows pronounced periodicity of the autocorrelation function, indicating that the measured adhesion forces are composed of integer multiples. The autocorrelation function is calculated from the histogram after subtraction of its envelope function (Marchand and Marmet, 1983). The force quantum of 30 ± 6 pN is attributed to the interaction of a single disaccharide $1a$ pair.

The control experiment in water confirms that the adhesion force of 30 pN corresponds to the Ca^{2+}-dependent sulfated disaccharide self-recognition (Fig. 5E). No interaction is observed in 300 approach-retract cycles performed in different areas of the disaccharide $1a$ sample.

References

Badia, A., Cuccia, L., Demers, L., Morin, F., and Lennox, R. B. (1997). Structure and dynamics in alkanethiolate monolayers self-assembled on gold nanoparticles: A DSC, FT-IR, and deuterium NMR study. *J. Am. Chem. Soc.* **119**, 2682–2692.

Badia, A., Gao, W., Singh, S., Demers, L., Cuccia, L., and Reven, L. (1996). Structure and chain dynamics of alkanethiol-capped gold colloids. *Langmuir* **12**, 1262–1269.

Boggs, J. M., Wang, H., Gao, W., Arvanitis, D. N., Gong, Y., and Min, W. (2004). A glyco-synapse in myelin? *Glycoconjugate J.* **21**, 97–110.

Brust, M., Fink, J., Bethell, D., Schiffrin, D. J., and Kiely, C. (1995). Synthesis and reactions of functionalized gold nanoparticles. *J. Chem. Soc. Chem. Commun.* 1655–1656.

Bucior, I., and Burger, M. M. (2004a). Carbohydrate-carbohydrate interaction as a major force initiating cell-cell recognition. *Glycoconjugate J.* **21**, 111–123.

Bucior, I., and Burger, M. M. (2004b). Carbohydrate-carbohydrate interactions in cell recognition. *Curr. Opin. Struct. Biol.* **14**, 631–637.

Bucior, I., Scheuring, S., Engel, A., and Burger, M. M. (2004). Carbohydrate-carbohydrate interaction provides adhesion force and specificity for cellular recognition. *J. Cell Biol.* **165**, 529–537.

Carvalho de Souza, A., Halkes, K. M., Meeldijk, J. D., Verkleij, A. J., Vliegenthart, J. F. G., and Kamerling, J. P. (2004). Synthesis of gold glyconanoparticles: Possible probes for the exploration of carbohydrate-mediated self-recognition of marine sponge cells. *Eur. J. Org. Chem.* 4323–4339.

Carvalho de Souza, A., Halkes, K. M., Meeldijk, J. D., Verkleij, A. J., Vliegenthart, J. F. G., and Kamerling, J. P. (2005). Gold glyconanoparticles as probes to explore the carbohydrate-mediated self-recognition of marine sponge cells. *ChemBioChem* **6**, 828–831.

Casero, E., Vázquez, L., Martín-Benito, J., Morcillo, M. A., Lorenzo, E., and Pariente, F. (2002). Immobilization of metallothionein on gold/mica surfaces: Relationship between surface morphology and protein-substrate interaction. *Langmuir* **18**, 5909–5920.

Dammer, U., Popescu, O., Wagner, P., Anselmetti, D., Güntherodt, H.-J., and Misevic, G. N. (1995). Binding strength between cell adhesion proteoglycans measured by atomic force microscopy. *Science* **267**, 1173–1175.

Daniel, M.-C., and Astruc, D. (2004). Gold nanoparticles: Assembly, supramolecular chemistry, quantum-size-related properties, and applications toward biology, catalysis, and nanotechnology. *Chem. Rev.* **104**, 293–346.

de la Fuente, J. M., Barrientos, A. G., Rojas, T. C., Rojo, J., Canada, J., Fernández, A., and Penadés, S. (2001). Gold glyconanoparticles as water-soluble polyvalent models to study carbohydrate interactions. *Angew. Chem. Int. Ed.* **40**, 2258–2261.

de la Fuente, J. M., and Penadés, S. (2002). Synthesis of Le^x-neoglycoconjugate to study carbohydrate-carbohydrate associations and its intramolecular interaction. *Tetrahedron: Asymmetry* **13,** 1879–1888.

de la Fuente, J. M., and Penadés, S. (2004). Understanding carbohydrate-carbohydrate interactions by means of glyconanotechnology. *Glycoconjugate J.* **21,** 149–163.

de Paz, J.-L., Ojeda, R., Barrientos, A. G., Penadés, S., and Martín-Lomas, M. (2005). Synthesis of a Le^y neoglycoconjugate and Le^y-functionalized gold glyconanoparticles. *Tetrahedron: Asymmetry* **16,** 149–158.

Duff, D. G., Baiker, A., and Edwards, P. P. (1993). A new hydrosol of gold clusters. 1. Formation and particle size variation. *Langmuir* **9,** 2301–2309.

Eggens, I., Fenderson, B. A., Toyokuni, T., Dean, B., Stroud, M. R., and Hakomori, S. (1989). Specific interaction between Le^x and Le^x determinants - A possible basis for cell recognition in preimplantation embryos and in embryonal carcinoma-cells. *J. Biol. Chem.* **264,** 9976–9984.

Evans, E. (1999). Looking inside molecular bonds at biological interfaces with dynamic force spectroscopy. *Biophys. Chem.* **82,** 83–97.

Faraday, M. (1875). The Bakerian lecture: Experimental reaction of gold (and other metals) to light. *Philos. Trans.* **147,** 145–181.

Fernàndez-Busquets, X., and Burger, M. M. (1997). The main protein of the aggregation factor responsible for species-specific cell adhesion in the marine sponge *Microciona prolifera* is highly polymorphic. *J. Biol. Chem.* **272,** 27839–27847.

Fernàndez-Busquets, X., and Burger, M. M. (2003). Circular proteoglycans from sponges: First members of the spongican family. *Cell. Mol. Life Sci.* **60,** 88–112.

Florin, E. L., Moy, V. T., and Gaub, H. E. (1994). Adhesion forces between individual ligand-receptor pairs. *Science* **264,** 415–417.

Frens, G. (1973). Controlled nucleation for the regulation of the particle size in monodisperse gold suspensions. *Nature: Phys. Sci.* **241,** 20–22.

Fritz, J., Anselmetti, D., Jarchow, J., and Fernàndez-Busquets, X. (1997). Probing single biomolecules with atomic force microscopy. *J. Struct. Biol.* **119,** 165–171.

Gege, C., Geyer, A., and Schmidt, R. R. (2002). Synthesis and molecular tumbling properties of sialyl Lewis X and derived neoglycolipids. *Chem. Eur. J.* **8,** 2454–2463.

Geyer, A., Gege, C., and Schmidt, R. R. (2000). Calcium-dependent carbohydrate-carbohydrate recognition between Lewis^x blood group antigens. *Angew. Chem. Int. Ed.* **39,** 3245–3249.

Gourier, C., Pincet, F., Perez, E., Zhang, Y. M., Mallet, J.-M., and Sinaÿ, P. (2004). Specific and non specific interactions involving Le^x determinant quantified by lipid vesicle micromanipulation. *Glycoconjugate J.* **21,** 165–174.

Gutiérrez-Wing, C., Ascencio, J. A., Pérez-Alvarez, M., Marin-Almazo, M., and José-Yacamán, M. (1998). On the structure and formation of self-assembled lattices of gold nanoparticles. *J. Cluster Sci.* **9,** 529–545.

Hakomori, S. (2003). Structure, organization, and function of glycosphingolipids in membranes. *Curr. Opin. Hematol.* **10,** 16–24.

Hakomori, S. (2004a). Carbohydrate-to-carbohydrate interaction in basic cell biology: A brief overview. *Arch. Biochem. Biophys.* **426,** 173–181.

Hakomori, S. (2004b). Carbohydrate-to-carbohydrate interaction, through glycosynapse, as a basis of cell recognition and membrane organization. *Glycoconjugate J.* **21,** 125–137.

Halkes, K. M., Carvalho de Souza, A., Maljaars, C. E. P., Gerwig, G. J., and Kamerling, J. P. (2005). A facile method for the preparation of gold glyconanoparticles from free oligosaccharides and their applicability in carbohydrate-protein interaction studies. *Eur. J. Org. Chem.* 3650–3659.

Haseley, S. R., Vermeer, H. J., Kamerling, J. P., and Vliegenthart, J. F. G. (2001). Carbohydrate self-recognition mediates marine sponge cellular adhesion. *Proc. Natl. Acad. Sci. USA* **98**, 9419–9424.

Hernáiz, M. J., de la Fuente, J. M., Barrientos, A. G., and Penadés, S. (2002). A model system mimicking glycosphingolipid clusters to quantify carbohydrate self-interactions by surface plasmon resonance. *Angew. Chem. Int. Ed.* **41**, 1554–1557.

Hostetler, M. J., Wingate, J. E., Zhong, C.-J., Harris, J. E., Vachet, R. W., Clark, M. R., Londono, J. D., Green, S. J., Stokes, J. J., Wignall, G. D., Glish, G. L., Porter, M. D., Evans, N. D., and Murray, R. W. (1998). Alkanethiolate gold cluster molecules with core diameters from 1.5 to 5.2 nm: Core and monolayer properties as a function of core size. *Langmuir* **14**, 17–30.

Jarchow, J., Fritz, J., Anselmetti, D., Calabro, A., Hascall, V. C., Gerosa, D., Burger, M. M., and Fernàndez-Busquets, X. (2000). Supramolecular structure of a new family of circular proteoglycans mediating cell adhesion in sponges. *J. Struct. Biol.* **132**, 95–105.

Kamerling, J. P., and Vliegenthart, J. F. G. (1989). Carbohydrates. *In* "Clinical Biochemistry—Principles, Methods, Applications". (A. M. Lawson, ed.), Vol. 1, pp. 175–263. Walter de Gruyter. Berlin, New York.

Kansas, G. S. (1996). Selectins and their ligands: Current concepts and controversies. *Blood* **88**, 3259–3287.

Kleene, R., and Berger, E. G. (1993). The molecular and cell biology of glycosyltransferases. *Biochim. Biophys. Acta* **1154**, 283–325.

Kojima, N., Fenderson, B. A., Stroud, M. R., Goldberg, R. I., Habermann, R., Toyokuni, T., and Hakomori, S. (1994). Further studies on cell adhesion based on Le^x-Le^x interaction, with new approaches: Embryoglycan aggregation of F9 teratocarcinoma cells, and adhesion of various tumor cells based on Le^x expression. *Glycoconjugate J.* **11**, 238–248.

Kojima, N., and Hakomori, S. (1989). Specific interaction between gangliotriaosylceramide (Gg3) and sialosyllactosylceramide (GM3) as a basis for specific cellular recognition between lymphoma and melanoma cells. *J. Biol. Chem.* **264**, 20159–20162.

Kojima, N., and Hakomori, S. (1991). Cell adhesion, spreading, and motility of GM3-expressing cells based on glycolipid-glycolipid interaction. *J. Biol. Chem.* **266**, 17552–17558.

Koshy, K. M., and Boggs, J. M. (1996). Investigation of the calcium-mediated association between the carbohydrate head groups of galactosylceramide and galactosylceramide I^3 sulfate by electrospray ionization mass spectrometry. *J. Biol. Chem.* **271**, 3496–3499.

Leff, D. V., Ohara, P. C., Heath, J. R., and Gelbart, W. M. (1995). Thermodynamic control of gold nanocrystal size: Experiment and theory. *J. Phys. Chem.* **99**, 7036–7041.

Lin, C.-C., Yeh, Y.-C., Yang, C.-Y., Chen, C.-L., Chen, G.-F., Chen, C.-C., and Wu, Y.-C. (2002). Selective binding of mannose-encapsulated gold nanoparticles to type 1 pili in *Escherichia coli*. *J. Am. Chem. Soc.* **124**, 3508–3509.

Mammen, M., Dahmann, G., and Whitesides, G. M. (1995). Effective inhibitors of hemagglutination by influenza virus synthesized from polymers having active ester groups. Insight into mechanism of inhibition. *J. Med. Chem.* **38**, 4179–4190.

Marchand, P., and Marmet, L. (1983). Binomial smoothing filter: A way to avoid some pitfalls of least-squares polynomial smoothing. *Rev. Sci. Instrum.* **54**, 1034–1041.

Matsuura, K., Kitakouji, H., Sawada, N., Ishida, H., Kiso, M., Kitajima, K., and Kobayashi, K. (2000). A quantitative estimation of carbohydrate-carbohydrate interaction using clustered oligosaccharides of glycolipid monolayers and of artificial glycoconjugate polymers by surface plasmon resonance. *J. Am. Chem. Soc.* **122**, 7406–7407.

Matsuura, K., and Kobayashi, K. (2004). Analysis of GM3-Gg3 interaction using clustered glycoconjugate models constructed from glycolipid monolayers and artificial glycoconjugate polymers. *Glycoconjugate J.* **21**, 139–148.

Merkel, R., Nassoy, P., Leung, A., Ritchie, K., and Evans, E. (1999). Energy landscapes of receptor-ligand bonds explored with dynamic force spectroscopy. *Nature* **397**, 50–53.

Misevic, G. N., and Burger, M. M. (1990). The species-specific cell-binding site of the aggregation factor from the sponge *Microciona prolifera* is a highly repetitive novel glycan containing glucuronic acid, fucose, and mannose. *J. Biol. Chem.* **265**, 20577–20584.

Misevic, G. N., and Burger, M. M. (1993). Carbohydrate-carbohydrate interactions of a novel acidic glycan can mediate sponge cell adhesion. *J. Biol. Chem.* **268**, 4922–4929.

Misevic, G. N., Finne, J., and Burger, M. M. (1987). Involvement of carbohydrates as multiple low affinity interaction sites in the self-association of the aggregation factor from the marine sponge *Microciona prolifera*. *J. Biol. Chem.* **262**, 5870–5877.

Nolting, B., Yu, J.-J., Liu, G.-Y., Cho, S.-J., Kauzlarich, S., and Gervay-Hague, J. (2003). Synthesis of gold glyconanoparticles and biological evaluation of recombinant Gp120 interactions. *Langmuir* **19**, 6465–6473.

Papandreou, M.-J., Sergi, I., Benkirane, M., and Ronin, C. (1990). Carbohydrate-dependent epitope mapping of human thyrotropin. *Mol. Cell. Endocrinol.* **73**, 15–26.

Popescu, O., and Misevic, G. N. (1997). Self-recognition by proteoglycans. *Nature* **386**, 231–232.

Popescu, O., Checiu, I., Gherghel, P., Simon, Z., and Misevic, G. N. (2003). Quantitative and qualitative approach of glycan-glycan interactions in marine sponges. *Biochimie* **85**, 181–188.

Rabinovich, G. A., Rubinstein, N., and Toscano, M. A. (2002). Role of galectins in inflammatory and immunomodulatory processes. *Biochim. Biophys. Acta* **1572**, 274–284.

Rojo, J., Díaz, V., de la Fuente, J. M., Segura, I., Barrientos, A. G., Riese, H. H., Bernad, A., and Penadés, S. (2004). Gold glyconanoparticles as new tools in antiadhesive therapy. *ChemBioChem* **5**, 291–297.

Sacchettini, J. C., Baum, L. G., and Brewer, C. F. (2001). Multivalent protein-carbohydrate interactions. A new paradigm for supermolecular assembly and signal transduction. *Biochemistry* **40**, 3009–3015.

Santacroce, P. V., and Basu, A. (2003). Probing specificity in carbohydrate-carbohydrate interactions with micelles and Langmuir monolayers. *Angew. Chem. Int. Ed.* **42**, 95–98.

Santacroce, P. V., and Basu, A. (2004). Studies of the carbohydrate-carbohydrate interaction between lactose and GM3 using Langmuir monolayers and glycolipid micelles. *Glycoconjugate J.* **21**, 89–95.

Spillmann, D., Hård, K., Thomas-Oates, J., Vliegenthart, J. F. G., Misevic, G., Burger, M. M., and Finne, J. (1993). Characterization of a novel pyruvylated carbohydrate unit implicated in the cell aggregation of the marine sponge *Microciona prolifera*. *J. Biol. Chem.* **268**, 13378–13387.

Spillmann, D., Thomas-Oates, J. E., van Kuik, J. A., Vliegenthart, J. F. G., Misevic, G., Burger, M. M., and Finne, J. (1995). Characterization of a novel sulfated carbohydrate unit implicated in the carbohydrate-carbohydrate-mediated cell aggregation of the marine sponge *Microciona prolifera*. *J. Biol. Chem.* **270**, 5089–5097.

St. Hilaire, P. M., Boyd, M. K., and Toone, E. J. (1994). Interaction of the Shiga-like toxin type 1 B-subunit with its carbohydrate receptor. *Biochemistry* **33**, 14452–14463.

Sung, M.-A., Chen, H. A., and Matthews, S. (2001). Sequential assignment and secondary structure of the triple-labelled carbohydrate-binding domain of papG from uropathogenic. *E. coli. J. Biomol. NMR* **19**, 197–198.

Svarovsky, S. A., Szekely, Z., and Barchi, J. J. (2005). Synthesis of gold nanoparticles bearing the Thomsen-Friedenreich disaccharide: A new multivalent presentation of an important tumor antigen. *Tetrahedron: Asymmetry* **16**, 587–598.

Templeton, A. C., Chen, S., Gross, S. M., and Murray, R. W. (1999). Water-soluble, isolable gold clusters protected by tiopronin and coenzyme A monolayers. *Langmuir* **15**, 66–76.

Templeton, A. C., Wuelfing, W. P., and Murray, R. W. (2000). Monolayer-protected cluster molecules. *Acc. Chem. Res.* **33**, 27–36.

Terrill, R. H., Postlethwaite, T. A., Chen, C.-H., Poon, C.-D., Terzis, A., Chen, A., Hutchison, J. E., Clark, M. R., Wignall, G., Londono, J. D., Superfine, R., Falvo, M., Johnson, C. S., Jr., Samulski, E. T., and Murray, R. W. (1995). Monolayers in three dimensions: NMR, SAXS, thermal, and electron hopping studies of alkanethiol stabilized gold clusters. *J. Am. Chem. Soc.* **117**, 12537–12548.

Tiemeyer, M., and Goodman, C. S. (1996). Gliolectin is a novel carbohydrate-binding protein expressed by a subset of glia in the embryonic *Drosophila* nervous system. *Development* **122**, 925–936.

Tromas, C., Rojo, J., de la Fuente, J. M., Barrientos, A. G., García, R., and Penadés, S. (2001). Adhesion forces between Lewis[x] determinant antigens as measured by atomic force microscopy. *Angew. Chem. Int. Ed.* **40**, 3052–3055.

Turkevitch, J., Stevenson, P. C., and Hilier, J. (1951). Nucleation and growth process in the synthesis of colloidal gold. *Discuss. Faraday Soc.* **11**, 55–75.

Ulman, A. (1996). Formation and structure of self-assembled monolayers. *Chem. Rev.* **96**, 1533–1554.

Vacquier, V. D., and Moy, G. W. (1997). The fucose sulfate polymer of egg jelly binds to sperm REJ and is the inducer of the sea urchin sperm acrosome reaction. *Dev. Biol.* **192**, 125–135.

Whetten, R. L., Khoury, J. T., Alvarez, M. M., Murthy, S., Vezmar, I., Wang, Z. L., Stephens, P. W., Cleveland, C. L., Luedtke, W. D., and Landman, U. (1996). Nanocrystal gold molecules. *Adv. Mater.* **8**, 428–433.

Willemsen, O. H., Snel, M. M. E., Cambi, A., Greve, J., De Grooth, B. G., and Figdor, C. G. (2000). Biomolecular interactions measured by atomic force microscopy. *Biophys. J.* **79**, 3267–3281.

Wolfhagen, M. J. H. M., Torensma, R., Fluit, A. C., Aarsman, C. J. M., Jansze, M., and Verhoef, J. (1994). Multivalent binding of toxin A from *Clostridium difficile* to carbohydrate receptors. *Toxicon* **32**, 129–132.

Yonezawa, T., and Kunitake, T. (1999). Practical preparation of anionic mercapto ligand-stabilized gold nanoparticles and their immobilization. *Colloids Surf. A: Physicochem. Eng. Asp.* **149**, 193–199.

Yu, S., Kojima, N., Hakomori, S., Kudo, S., Inoue, S., and Inoue, Y. (2002). Binding of rainbow trout sperm to egg is mediated by strong carbohydrate-to-carbohydrate interaction between (KDN)GM3 (deaminated neuraminyl ganglioside) and Gg3-like epitope. *Proc. Natl. Acad. Sci. USA* **99**, 2854–2859.

Section VI

Galectin Function

[17]   Galectin Interactions with Extracellular Matrix and Effects on Cellular Function

By Jiale He and Linda G. Baum

Abstract

In vivo, cells exist within a complex mixture of glycoproteins and proteoglycans termed the extracellular matrix (ECM). The components of the ECM are secreted by the cells in that site, and the ECM provides not only a physical support but also outside-in signals that regulate many cellular functions, including cell proliferation, differentiation, migration, and survival. Altering the composition of the ECM can thus significantly alter cell behavior. Many types of cells, including normal and malignant epithelial and mesenchymal cells, secrete galectin-1, a member of the galectin family of lectins, into the ECM surrounding the cells. Galectin-1 is known to regulate many of these same cellular functions (i.e., proliferation, differentiation, migration, and death), so that the presence of galectin-1 in ECM will modify the effects of ECM on cells. In this chapter, we present three types of assays that allow interrogation of the effects of galectin-1 on cell adhesion to ECM, cell migration through ECM, and cell death in ECM, using T-cell lines as a model cell type.

Overview

The extracellular matrix (ECM) is a complex mixture of soluble glyco-proteins secreted by various adherent cell types to create a microenvironment around the cells *in vivo*. The ECM provides not only structural support for cells but modulates a variety of functions such as cell proliferation, adhesion, differentiation, and morphogenesis by binding and concentrating cytokines and chemokines, by promoting cell anchorage by integrins, and by directly providing outside-in signals to cells that regulate gene expression (Davis and Senger, 2005; DeClerck *et al.*, 2004; Jenniskens *et al.*, 2006; Radisky *et al.*, 2002; Rot and von Andrian, 2004). The ECM also controls normal cell migration during development and aberrant cell migration during tumor cell metastasis. In cancer, ECM surrounding the tumor cells is a mixture of matrix secreted by the nonmalignant supporting cells, such as fibroblasts, and the matrix secreted by the tumor cells. Tumor-associated ECM may be further modified by the tumor (e.g., degraded by matrix metalloproteinases secreted by tumor cells) and thus provide a micro-environment that is permissive for tumor cell metastasis and escape from

METHODS IN ENZYMOLOGY, VOL. 417
0076-6879/06 $35.00
DOI: 10.1016/S0076-6879(06)17017-2

immune detection (Bellail *et al.*, 2004; Loberg *et al.*, 2005; Parks *et al.*, 2004; Reddig and Juliano, 2005).

A common modification of ECM by tumor cells is secretion of galectin-1 into the extracellular milieu, where galectin-1 can bind to glycans on ECM glycoproteins (van den Brule *et al.*, 1995, 2001, 2003; Horiguchi *et al.*, 2003). Galectin-1 is a member of the galectin family of soluble lectins and is synthesized as a 14-kDa monomer that has no posttranslational modifications. Galectin-1 spontaneously forms noncovalent homodimers that can cross-link glycan ligands. Galectin-1 is made in the cytosol and can remain intracellular or be secreted through a nonclassical secretion pathway. Secreted galectin-1 typically concentrates in the vicinity of the cells that produced it by binding back to glycan ligands on the cell surface or to glycan ligands in the surrounding ECM (He and Baum, 2004). A variety of ECM proteins bind to galectin-1, including laminin and fibronectin (Cooper *et al.*, 1991; Moiseeva *et al.*, 1999, 2003a; Ozeki *et al.*, 1995; Zhou and Cummings, 1993). Galectin-1 does not just passively bind to ECM glycoproteins but also regulates the relative abundance of various glycoproteins in the ECM (Gu *et al.*, 1994; Moiseeva *et al.*, 2003b); for example, the presence of galectin-1 reduced ECM incorporation of vitronectin and chondroitin sulfate compared with ECM depleted of galectin-1. The presence of galectin-1 in ECM affects many interactions of cells with ECM, such as cell adhesion, cell migration, and cell survival. In patients with ovarian and prostate cancer, a correlation has been established between galectin-1 expression in the tumor-associated matrix and aggressiveness of the tumor (van den Brule *et al.*, 2001, 2003). Our group has shown that the presence of galectin-1 in ECM reduced T cell migration through the ECM and induced cell death of susceptible T cells (He and Baum, 2004, 2006).

This chapter focuses on experimental procedures for *in vitro* cell biology studies of the effects of galectin-1 on ECM interactions with cells. These protocols use two murine T lymphocyte cell lines, BW5147 and Pha[R]2.1, because our group has previously characterized these two T cell lines as resistant and susceptible, respectively, to galectin-1 cell death (Galvan *et al.*, 2000). However, these protocols could be easily adapted to study other cell types. In these protocols, the commercial ECM preparation Matrigel is used to assess the effects of galectin-1 on cell adhesion to ECM, cell migration through ECM, and cell death after binding to ECM.

Chemicals and Reagents

Recombinant human galectin-1 is prepared as previously described (Pace *et al.*, 2003). Because recombinant galectin-1 is stored in 0.1 M lactose, the stock solution of galectin-1 (approx. 6 mg/ml) is dialyzed in 8 mM

dithiothreitol (DTT) in PBS (10 mM NaPO$_4$, 140 mM NaCl, pH 7.4) before use. The following antibodies are used: biotinylated rabbit anti-human galectin-1 Ig prepared in our laboratory as previously described (He and Baum, 2004), goat anti-rabbit Ig-horseradish peroxidase (HRP), and goat anti-rabbit Ig-FITC from Jackson ImmunoResearch Laboratories Inc. (West Grove, PA). Reagents are purchased from the indicated suppliers: annexin V/propidium iodide (P.I.) from R&D Systems (Minneapolis, MN); 3-amino, 9-ethylcarbozole (AEC) from Biomeda Corp. (Foster City, CA); DTT from Fisher Scientific (Fairlawn, NJ); PBS from Sigma (St Louis, MO); 5, 6-carboxyfluorescein diacetate succinimidyl ester (CFSE) and Prolong anti-fade mounting medium from Molecular Probes (Eugene, OR); Matrigel and BioCoat Matrigel Invasion chambers (8 μm) from BD Biosciences (San Jose, CA); cell culture inserts (0.4 μm) from Falcon/Becton Dickinson (Franklin Lakes, NJ); Diff-Quick staining solution from Dade Behring (Newark, DE).

Preparation of ECM

Preparation of ECM from Cultured Cells

ECM can be prepared from a variety of cells that secrete matrix glycoproteins, such as endothelial cells, epithelial cells, fibroblasts, and tumor cells. An experimental protocol is detailed in Gospodarowicz et al. (1983). Although the original description of this method used bovine corneal epithelial cells, we have successfully used this method to harvest ECM from CHO cells. In this method, cells are cultured with 4% (w/v) dextran T40, to increase the viscosity of the growth medium. The medium is changed every other day until the cells are confluent, and the confluent cells are cultured an additional 5–8 days without medium change. The culture medium is aspirated from the plate, and cells are lysed by the addition of Triton/NH$_4$OH cell lysis solution (0.5% (v/v) Triton X-100, 20 mM NH$_4$OH in 1$\times$ PBS) for 3–5 min with gentle shaking at room temperature. Cells are examined by phase-contrast microscopy, and if cell nuclei remain bound to the plate, incubate the plate for an additional 5–10 min at 37° to completely solubilize the cell layer. Cell debris is removed by washing four times with PBS, and ECM material remains adherent to the culture plate. Cover ECM-coated plates with 2 ml PBS supplemented with 50 μg/ml gentamicin and 0.25 μg/ml Fungizone. Store plates at 4°. The ECM layer can be scratched with a needle to examine the thickness and homogeneity of the layer. To remove the ECM layer from the plate, scratch the plate with a needle, which will detach and lift the ECM membrane along the edge of the scratch. Detached ECM can be used for cell binding and migration assays or for biochemical analysis.

Preparation of Matrigel

Matrigel is a commercial ECM mixture obtained from cultures of Engelbreth-Holm-Swarm fibrosarcoma cells. Matrigel contains primarily laminin, collagen IV, heparan sulfate proteoglycans, and entactin. Matrigel is liquid at 4°, while at room temperature and higher (22–35°) Matrigel polymerizes into a gel. We have analyzed Matrigel preparations by immunoblotting and have not detected galectin-1 in Matrigel. However, Matrigel preparations can vary from lot to lot, and each lot should be examined for characteristics of interest.

To make a thin gel layer of Matrigel (0.5 mm), use a cooled pipette to mix the Matrigel (10 mg/ml) to homogeneity. Keeping culture plates on ice, add 50 μl Matrigel per square centimeter of growth surface. Place plates at room temperature to solidify the gel. To make a Matrigel layer on glass coverslips, add 100 μl liquid Matrigel (10 mg/ml), using a cooled pipette, onto coverslips (Ø 1.8 cm) placed into the wells of 6-well culture plates that are on ice. Place plates at room temperature for 1 h. Add 100 μl of media onto the solidified gel to keep the gel hydrated (He and Baum, 2004). Alternately, Matrigel layers can be made on invasion chamber inserts that have polyethylene terephthalate membranes with pores of defined size for cell migration assays, or commercially available precoated Matrigel invasion chambers can be used.

Coating Matrigel with Galectin-1

To add galectin-1 directly to Matrigel, recombinant human galectin-1 is coated directly on the surface of the solidified gel (He and Baum, 2004); 100 μl galectin-1 in PBS (0.3–20 μM solution) is layered on Matrigel; if the tissue culture transwell inserts or invasion chambers are used, the final concentration of galectin-1 is 0.1–7.1 μg/cm^2 on a 4.2 cm^2 insert. Galectin-1 is allowed to bind to the gel for 20–30 min at room temperature and the inserts or coverslips washed once with PBS before the coated Matrigel layers are used for assays.

To deposit galectin-1 secreted from cells onto Matrigel, seed galectin-1 secreting cells, such as CHO cells, directly on solidified Matrigel for 72 h at 37°; 2×10^5 CHO cells per tissue culture well with Matrigel-coated coverslips were used in our assays to detect secreted galectin-1 on Matrigel (He and Baum, 2004). Alternately, galectin-1 secreting cells can be placed in upper transwell cell culture inserts with 0.4-μm pores (He and Baum, 2004). The inserts are placed over lower wells containing culture medium over Matrigel layers solidified on cover slips. This approach allows galectin-1 to diffuse through the 0.4-μm pores in the membrane to the solidified Matrigel on the coverslips with no direct cell-Matrigel contact.

Matrigel-coated coverslips can be left in culture plates for further analysis or removed with tweezers. To detect galectin-1 bound to Matrigel, Matrigel can be analyzed by immunochemical staining. Purified anti-galectin-1 IgG (rabbit) (1:100) is added to tissue culture wells containing Matrigel-coated coverslips with or without galectin-1. The antibody is incubated with the Matrigel for 1 h at room temperature. After washing the coverslips four times (5 min each wash) with PBS, bound antibody is detected with mouse anti rabbit-HRP (1:200) for 1 h at 37°. Bound secondary reagent is detected colorimetrically using AEC (He and Baum, 2004). After color development, cover slips are rinsed three times with distilled water, removed from the culture wells with tweezers, inverted, mounted on glass slides with 50 μl Faramount aqueous mounting medium (DAKO Inc, Carpinteria, CA), and examined by light microscopy.

T Cell–ECM Adhesion Assays

Labeling T Cells with Fluorescent Probe

To avoid confounding effects of galectin-1 and cell death, we examine T cell adhesion using BW5147 murine T cells that are resistant to galectin-1 death (Galvan *et al.*, 2000). BW5147 cells are maintained in complete DMEM supplemented with 10% heat inactivated FBS, 2 mM L-glutamine, 1 mM MEM sodium pyruvate solution. Before labeling, wash cells twice in serum-free media. Cells are labeled with carboxyfluorescein diacetate succinimidyl ester (CFSE), a nonpolar dye that penetrates cell membranes; once inside the cell, anionic CFSE is formed by intracellular esterases. The ester form irreversibly binds to intracellular proteins, stably labeling the cells. Resuspend 10^6 cells in 2 ml prewarmed (37°) PBS containing 10 μM CFSE and incubate for 15 min in a foil-covered centrifuge tube in at 37°. Mix the cells gently twice during this period to ensure adequate labeling consistency. Stop the reaction by adding complete medium with 10% FBS to dilute out the dye. Pellet cells by centrifugation at 1000 rpm for 5 min and resuspend in fresh prewarmed medium. Incubate cells for another 30 min at 37° to ensure complete intracellular modification of CFSE.

CFSE-labeled Cell Adhesion Assay

To investigate the ability of cells to bind to Matrigel in the presence or absence of galectin-1, add 2 × 10^5 CFSE-labeled T cells onto Matrigel-coated coverslips in 6-well tissue culture plates. Allow cells to adhere to Matrigel for 1 h at 37°. Wash the Matrigel twice to remove nonadherent cells. Fix the samples with 2% paraformaldehyde for 30 min at 4°, and quench the fixation by washing three times with 0.2 M glycine at 4° (10 min

FIG. 1. BW5147 cells are labeled with the green CFSE dye, and labeled BW5147 T cells adhere to Matrigel-coated coverslips, as observed by fluorescent microscopy (100×).

each wash) in PBS. Lift the coverslips with tweezers and mount on glass slides as described previously. Examine the CFSE-labeled cells by fluorescence microscopy (excitation/emission wavelengths 488/535), which will detect the green CFSE fluorescence (Fig. 1). Determine the number of adherent cells by counting five nonoverlapping random fields on each coverslip at 100× magnification on four to five coverslips (400–500 cells total) for each experimental point. The absolute number of fluorescent cells on the Matrigel layer can be reported or the data normalized to number of cells/high power field. An alternative way of quantifying the number of adherent cells is to use a fluorescence plate reader after determining a standard curve of fluorescence vs. cell number.

Cell Migration through ECM

The T cell–ECM migration assay is performed using a modification of the procedure of Li *et al.* (2001). Rehydrate 6-well Matrigel transwell (8 μm) invasion chambers for 2 h at 37°, 5% CO_2. Coat the Matrigel with recombinant galectin-1 or leave uncoated as described previously. Add BW5147 T cells (1 × 10^6, precultured for 24 h in serum-free media) to each upper insert; place the inserts over the lower wells that contain media with 10% FBS as a chemoattractant. Allow the cells to migrate through the Matrigel layer to the underside of the chamber insert for 24 h at 37°. After 24 h, nonmigratory cells are removed from the upper surface of the membrane by scrubbing with a cotton swab. Cells that have

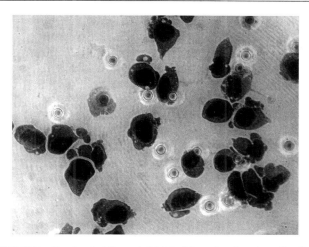

FIG. 2. BW5147 T cells migrated through Matrigel Invasion chambers. T cells are placed in the upper well of the insert, above a Matrigel-coated membrane that has 8-μm pores (open circles in the membrane). Media with 10% FBS as an attractant is placed in the lower well, and cells migrate through the Matrigel-covered membrane into the lower well for 24 h. Inserts are removed, and adherent cells (purple) on the lower surface are stained and observed by light microscopy (400×). (See color insert.)

migrated to the underside of the membrane are fixed and stained with Diff-Quick staining solution and the membranes examined by light micros-copy (Fig. 2). Cell counts are performed on five nonoverlapping random fields for each well, and at least four wells should be counted for each experimental point. The percent migrating cells is normalized to corresponding controls (i.e., migration with and without Matrigel or migra-tion through Matrigel with and without galectin-1) to describe the percent migrating cells (e.g., number of cells migrated through galectin-1 coated Matrigel membrane/number of cells migrated through control Matrigel membrane).

Assessing Cell Death on Matrigel

For this assay, our laboratory uses the Pha^{R}2.1 murine T cell line that is susceptible to galectin-1 death (Galvan *et al.*, 2000). Pha^{R}2.1 cells are maintained in complete DMEM supplemented with 10% heat inactivated FBS, 2 mM L-glutamine, 1 mM MEM sodium pyruvate solution.

To assess T cell death by contact with galectin-1–coated Matrigel, add 10^5 Pha^R 2.1 T cells to galectin-1–coated Matrigel solidified on coverslips placed in 6-well tissue culture plates. Allow the T cells to bind to the

Matrigel for 1–6 h at 37°. Wash once with PBS to remove unbound cells. Add 100 μl annexin V-FITC (1 μg/ml) and propidium iodide (1 μg/ml) in binding buffer (10 mM HEPES, pH 7.4, 150 mM NaCl, 2.5 mM CaCl$_2$, 1 mM MgCl$_2$) to the wells for 10 min at room temperature in the dark. Wash the coverslips with PBS, then fix with 2% paraformaldehyde, and quench with 0.2 M glycine, as described previously. Remove the cover slips from the wells, invert, and mount the coverslips on glass microscope slides with 25 μl Prolong Anti-fade mounting medium. For confocal microscopy, samples are excited at 488 and 568 nm with argon and krypton lasers to detect FITC-labeled annexin V and P.I. respectively; light emitted between 525 and 540 nm is recorded for FITC and greater than 630 nm for P.I. dual emission fluorescent images are collected in separate channels. To compensate the microscope, single stains with corresponding negative controls are performed. T-cell death is quantified by counting 100–200 cells randomly selected in five to eight microscopic fields for each experiment. The percent annexin V–positive T cells is calculated as the number of annexin V–positive T cells/total number of adherent T cells.

Note: Annexin V binding depends on the presence of calcium ions. It is important to include the proper concentration of calcium in the binding buffer and to avoid chelators such as EDTA that will prevent annexin V binding. We typically observe robust annexin V labeling of T cells after 1 h on galectin-1–coated Matrigel, but longer incubation times are required for significant P.I. uptake by the cells.

Conclusions

It is becoming increasingly clear that it is critical to examine cellular interactions in conditions that recapitulate *in vivo* biology to understand the complex factors that regulate cell proliferation, differentiation, morphogenesis, and migration. The extracellular matrix plays an essential role in regulating these functions; thus, developing experimental methods to investigate cell functions and cellular interactions within ECM will provide important mechanistic information. Defined ECM, such as Matrigel, offers an experimentally tractable system that can be modified by addition of specific signaling molecules, such as cytokines, integrins, and galectins, to interrogate the contributions of these molecules to cell–matrix interactions. In addition to the assays described previously, systems that examine interactions of more than one cell type within ECM (e.g., tumor cells and cytotoxic T cells) can be developed, and microenvironments of cells and ECM can be implanted in animal models to investigate the contribution of ECM to cell migration and metastasis. Understanding the molecular cross-talk between cells and ECM will provide novel approaches to regulating cell function in development and disease.

Acknowledgments

This work was supported by NIH grant R01GM63281, Cancer Research Institute grant 01103591, and DOD grant DAMD 17-02-1-0022 to L. G. B.

References

Bellail, A. C., Hunter, S. B., Brat, D. J., Tan, C, and Van and Meir, E. G. (2004). Microregional extracellular matrix heterogeneity in brain modulates glioma cell invasion. *Int. J. Biochem. Cell Biol.* **36,** 1046–1069.

Cooper, D. N., Massa, S. M., and Barondes, S. H. (1991). Endogenous muscle lectin inhibits myoblast adhesion to laminin. *J. Cell. Biol.* **115,** 1437–1448.

Davis, G. E., and Senger, D. R. (2005). Endothelial extracellular matrix: Biosynthesis, remodeling, and functions during vascular morphogenesis and neovessel stabilization. *Circ. Res.* **97,** 1093–1107.

DeClerck, Y. A., Mercurio, A. M., Stack, M. S., Chapman, H. A., Zutter, M. M., Muschel, R. J., Raz, A., Matrisian, L. M., Sloane, B. F., Noel, A., Hendrix, M. J., Coussens, L., and Padarathsingh, M. (2004). Proteases, extracellular matrix, and cancer: A workshop of the path B study section. *Am. J. Pathol.* **164,** 1131–1139.

Galvan, M., Tsuboi, S., Fukuda, M., and Baum, L. G. (2000). Expression of a specific glycosyltransferase enzyme regulates T cell death mediated by galectin-1. *J. Biol. Chem.* **275,** 16730–16737.

Gospodarowicz, D., Gonzalez, R., and Fujii, D. K. (1983). Are factors originating from serum, plasma, or cultured cells involved in the growth-promoting effect of the extracellular matrix produced by cultured bovine corneal endothelial cells? *J. Cell. Physiol.* **114,** 191–202.

Gu, M., Wang, W., Song, W. K., Cooper, D. N., and Kaufman, S. J. (1994). Selective modulation of the interaction of alpha 7 beta 1 integrin with fibronectin and laminin by L-14 lectin during skeletal muscle differentiation. *J. Cell Sci.* **107,** 175–181.

He, J., and Baum, L. G. (2004). Presentation of galectin-1 by extracellular matrix triggers T cell death. *J. Biol. Chem.* **279,** 4705–4712.

He, J., and Baum, L. G. (2006). Endothelial cell expression of galectin-1 induced by prostate cancer cells inhibits T cell transendothelial migration. *Lab Invest.* **86,** 578–590.

Horiguchi, N., Arimoto, K., Mizutani, A., Endo-Ichikawa, Y., Nakada, H., and Taketani, S. (2003). Galectin-1 induces cell adhesion to the extracellular matrix and apoptosis of non-adherent human colon cancer Colo201 cells. *J. Biochem.* (Tokyo). **134,** 869–874.

Jenniskens, G. J., Veerkamp, J. H., and van Kuppevelt, T. H. (2006). Heparan sulfates in skeletal muscle development and physiology. *J. Cell. Physiol.* **206,** 283–294.

Li, Y., Tondravi, M., Liu, J., Smith, E., Haudenschild, C. C., Kaczmarek, M., and Zhan, X. (2001). Cortactin potentiates bone metastasis of breast cancer cells. *Cancer Res.* **61,** 6906–6911.

Loberg, R. D., Gayed, B. A., Olson, K. B., and Pienta, K. J. (2005). A paradigm for the treatment of prostate cancer bone metastases based on an understanding of tumor cell-microenvironment interactions. *J. Cell. Biochem.* **96,** 439–446.

Moiseeva, E. P., Spring, E. L., Baron, J. H., and de Bono, D. P. (1999). Galectin 1 modulates attachment, spreading and migration of cultured vascular smooth muscle cells via interactions with cellular receptors and components of extracellular matrix. *J. Vasc. Res.* **36,** 47–58.

Moiseeva, E. P., Williams, B., Goodall, A. H., and Samani, N. J. (2003a). Galectin-1 interacts with beta-1 subunit of integrin. *Biochem. Biophys. Res. Commun.* **310,** 1010–1016.

Moiseeva, E. P., Williams, B., and Samani, N. J. (2003b). Galectin 1 inhibits incorporation of vitronectin and chondroitin sulfate B into the extracellular matrix of human vascular smooth muscle cells. *Biochim. Biophys. Acta* **1619,** 125–132.

Ozeki, Y., Matsui, T., Yamamoto, Y., Funahashi, M., Hamako, J., and Titani, K. (1995). Tissue fibronectin is an endogenous ligand for galectin-1. *Glycobiology* **5,** 255–261.

Pace, K. E., Hahn, H. P., and Baum, L. G. (2003). Preparation of recombinant human galectin-1 and use in T-cell death assays. *Methods Enzymol.* **363,** 499–518.

Parks, W. C., Wilson, C. L., and Lopez-Boado, Y. S. (2004). Matrix metalloproteinases as modulators of inflammation and innate immunity. *Nat. Rev. Immunol.* **4,** 617–629.

Radisky, D., Muschler, J., and Bissell, M. J. (2002). Order and disorder: The role of extracellular matrix in epithelial cancer. *Cancer Invest.* **20,** 139–153.

Reddig, P. J., and Juliano, R. L. (2005). Clinging to life: Cell to matrix adhesion and cell survival. *Cancer Metastasis Rev.* **24,** 425–439.

Rot, A., and von Andrian, U. H. (2004). Chemokines in innate and adaptive host defense: Basic chemokinese grammar for immune cells. *Annu. Rev. Immunol.* **22,** 891–928.

van den Brule, F. A, Buicu, C., Baldet, M., Sobel, M. E., Cooper, D. N., Marschal, P., and Castronovo, V. (1995). Galectin-1 modulates human melanoma cell adhesion to laminin. *Biochem. Biophys. Res. Commun.* **209,** 760–767.

van den Brule, F., Califice, S., Garnier, F., Fernandez, P. L., Berchuck, A., and Castronovo, V. (2003). Galectin-1 accumulation in the ovary carcinoma peritumoral stroma is induced by ovary carcinoma cells and affects both cancer cell proliferation and adhesion to laminin-1 and fibronectin. *Lab. Invest.* **83,** 377–386.

van den Brule, F. A., Waltregny, D., and Castronovo, V. (2001). Increased expression of galectin-1 in carcinoma-associated stroma predicts poor outcome in prostate carcinoma patients. *J. Pathol.* **193,** 80–87.

Zhou, Q., and Cummings, R. D. (1993). L-14 lectin recognition of laminin and its promotion of *in vitro* cell adhesion. *Arch. Biochem. Biophys.* **300,** 6–17.

[18] Galectins in Apoptosis

By Daniel K. Hsu, Ri-Yao Yang, and Fu-Tong Liu

Abstract

Galectins are a family of animal lectins with affinity for β-galactosides. By using recombinant proteins, a number of galectins have been shown to interact with cell-surface and extracellular matrix glycoconjugates through lectin–carbohydrate interactions. Through this action, they can affect a variety of cellular processes, and the most extensively documented function is induction of apoptosis. By using gene transfection approaches, galectins have been shown to regulate various cellular processes, including apoptosis. Evidence has been provided that some of these functions involve binding to cytoplasmic and nuclear proteins, through protein–protein interactions, and

METHODS IN ENZYMOLOGY, VOL. 417
0076-6879/06 $35.00
DOI: 10.1016/S0076-6879(06)17018-4

modulation of intracellular signaling pathways. Thus, some galectins are pro-apoptotic, whereas others are anti-apoptotic; some galectins induce apoptosis by binding to cell surface glycoproteins, whereas others regulate apoptosis through interactions with intracellular proteins. This review describes involvement of galectin-1, -2, -3, -7, -8, -9, and -12 in apoptosis.

Overview

Galectins are a family of animal lectins that recognize β-galactose and contain consensus amino-acid sequences. In mammals, 15 galectins have been studied and reported to date (reviewed in Cooper (2002), all containing conserved carbohydrate-recognition domains (CRDs) of approximately 130 amino acids responsible for carbohydrate binding. Some galectins have one CRD (galectin-1, -2, -3, -5, -7, -10, -11, -13, -14, and -15), whereas others are composed of two homologous CRDs in a single polypeptide chain, separated by a linker of up to 70 amino acids (galectin-4, -6, -8, -9, and -12). Among the single-CRD type, galectin-3 is unique in that it contains a long N-terminal region (approximately 120 amino acids) connected to the CRD. Different galectins are specific for different oligosaccharides (Hirabayashi *et al.*, 2002; Rini and Lobsanov, 1999).

Galectins do not have a signal sequence and have other features characteristic of intracellular proteins; consistent with these properties they reside intracellularly (reviewed in Liu *et al.* (2002). However, they can be secreted by cells (Hughes, 1999). Recombinant galectins have been shown to bind to extracellular glycoconjugates containing suitable galactose-bearing oligosaccharides. Because many galectins are either bivalent or multivalent with regard to their carbohydrate-binding activities (Ahmad *et al.*, 2004), they can form ordered arrays of complexes on binding to multivalent glycoconjugates (Brewer *et al.*, 2002), thus triggering a cascade of transmembrane signaling events. Intracellularly, galectins are detected in both the cytoplasm and the nucleus and observed to shuttle between these two compartments (Davidson *et al.*, 2002). A number of intracellular functions have been demonstrated for some galectins (Liu *et al.*, 2002). They seem to exert these functions by interacting with intracellular proteins (Liu *et al.*, 2002; Wang *et al.*, 2004).

Regulation of Apoptosis by Extracellular Galectins

Galectin-1

By using recombinant protein, a large number of studies have demonstrated that galectin-1 induces apoptosis in activated human T cells and T leukemia cell lines (Perillo *et al.*, 1995; Rabinovich *et al.*, 2002). More recently, it has been shown that galectin-1 can sensitize human resting T cells to Fas-mediated apoptosis (Matarrese *et al.*, 2005). It has been shown that galectin-1 presented by the extracellular matrix is significantly more effective in inducing apoptosis in T cell lines (He and Baum, 2004). In addition, by using an antisense strategy, it has been shown that galectin-1 secreted by murine melanoma cells can induce apoptosis of activated T cells specific for the tumor cells (Rubinstein *et al.*, 2004).

The effects of galectin-1 are dependent on expression of glycosyltransferase in the targeted cells (Galvan *et al.*, 2000) and presence of N-glycans on the cell surfaces (Amano *et al.*, 2003; Lanteri *et al.*, 2003). Galectin-1 binds to a restricted set of T-cell surface glycoproteins, including CD45, CD43, and CD7 (Pace *et al.*, 1999). It causes redistribution of some of these proteins into segregated microdomains, and some of them aggregate with the apoptotic blebs.

The signal transduction events induced by galectin-1 leading to apoptosis have been investigated, and the studies revealed the involvement of various known apoptotic pathways. These include activation of extracellular signal–regulated kinase-2 (ERK-2) (Vespa *et al.*, 1999), induction of the transcription factor AP-1 (Rabinovich *et al.*, 2000), a down-regulation of the anti-apoptotic protein Bcl-2, thus a predominance of the proapoptotic protein Bax (Novelli *et al.*, 1999; Rabinovich *et al.*, 2002), and activation of caspases (Rabinovich *et al.*, 2002).

More recently, it was documented that galectin-1 triggers translocation of endonuclease G from the mitochondria to the nucleus in human T-cell lines. However, cytochrome *c* release from the mitochondria and caspase activation, commonly associated with apoptosis, was not detected (Hahn *et al.*, 2004). Another study found that galectin-1 induced the exposure of phosphatidylserine on the cell surface of a T-cell line, MOLT-4, which is an early event of apoptosis that causes the cell to become recognizable by annexin-V. However, there was no detectable DNA fragmentation, which is also commonly associated with apoptosis (Dias-Baruffi *et al.*, 2003). Therefore, it seems that dependent on the cell types, galectin-1 may affect different end points in the apoptotic pathway and may or may not cause apoptosis.

Induction of apoptosis by galectins in other immune cells has also been studied. Galectin-1 was found to induce apoptosis in a B lymphoma cell line; it binds to CD45 on this cell line and triggers the latter's phosphatase activity (Fouillit *et al.*, 2000). Galectin-1 was also found to induce phosphatidylserine exposure on the human promyelocytic HL-60 cells and fMet-Leu-Phe-activated, but not resting, human neutrophils. However, as in the case of MOLT-4 cells described previously, DNA fragmentation was not detected in these cells (Dias-Baruffi *et al.*, 2003), but the treatment significantly promotes phagocytosis of these cells by activated mouse macrophages, apparently because of the recognition of phosphatidylserine on the cell surface by phagocytes.

Other Galectins

Galectin-2 can also induce T-cell apoptosis (Sturm *et al.*, 2004). Galectin-2's effect involves activation of caspases-3 and -9, cytochrome *c* release, disruption of the mitochondrial membrane potential, as well as DNA fragmentation. There was also an increase in the ratio of the pro-apoptotic and anti-apoptotic proteins, Bax and Bcl-2 (Sturm *et al.*, 2004).

Galectin-3 induces apoptosis of human T leukemia cell lines, human peripheral blood mononuclear cells, and activated mouse T cells (Fukumori *et al.*, 2003). Galectin-3 was shown to bind to CD7 and CD29 (β1 integrin) on the cell surface, and neutralizing antibodies to these antigens inhibited galectin-3-induced apoptosis (Fukumori *et al.*, 2003), suggesting that they mediate apoptosis induced by this lectin. Galectin-3 causes activation of mitochondrial apoptosis events, including cytochrome *c* release and caspase-3, but not caspase-8, activation (Fukumori *et al.*, 2003). Galectin-3 has also been shown to induce apoptosis in neutrophils (Fernandez *et al.*, 2005).

Distinction between apoptosis induced by galectin-1 and galectin-3 in T cells was recently described, in which each galectin was found not to be additive or synergistic to the effects of the other, thus suggesting independent cell surface pathways (Stillman *et al.*, 2006). Glycoproteins CD45 and CD71 were required for induction of T-cell apoptosis, but not CD7, CD29, or CD43, which are involved in cell death induced by galectin-1. The common receptor, CD45, was uniformly distributed on the cell surface after galectin-3 binding, as opposed to patching observed after galectin-1 binding (Pace *et al.*, 1999). In addition, galectin-1 was found to effectively induce apoptosis in CD4$^-$ CD8$^-$ and CD4$^+$ CD8$^+$ thymocytes, but galectin-3 preferentially acted on CD4$^-$ CD8$^-$ cells.

Galectin-8 has been shown to induce apoptosis in a human carcinoma cell line, 1299, in a fashion that is dependent on its inhibition of cell adhesion (Hadari *et al.*, 2000). A more recently study showed that galectin-8 can induce either cell growth arrest or apoptosis in a fashion that is dependent on the level of cyclin-dependent kinase inhibitor p21 (Arbel-Goren *et al.*, 2005).

Galectin-9 can also induce T-cell apoptosis (Kashio *et al.*, 2003; Wada *et al.*, 1997). Galectin-9's apoptotic induction effect was demonstrated with Jurkat cells, as well as with human peripheral blood T cells. Both CD4$^+$ and CD8$^+$ cells are susceptible to apoptosis induced by this lectin (Kashio *et al.*, 2003). Apoptosis in these cells is more pronounced if they are also activated with anti-CD3 antibody. Galectin-9 induces apoptosis by means of the caspase-1 pathway, and caspase-8, -9, and -10 are not involved (Kashio *et al.*, 2003). Galectin-9 also induces apoptosis in a human B-cell line (BALL-1), a monocytic cell line (THP-1), and a promyelocytic cell line (HL-60) (Kashio *et al.*, 2003). Finally, galectin-9 induces apoptosis in human melanoma cell lines (Kageshita *et al.*, 2002).

Galectin-9 was recently identified as a ligand for Tim-3, a T helper 1 (Th1)–specific cell surface molecule, and acts by way of this receptor to selectively induce apoptosis in these cells (Zhu *et al.*, 2005). Cells treated with galectin-9 demonstrated activation through intracellular Ca^{2+} release, resulting in cell aggregation and Th1 cell death by both apoptosis and necrosis. The ability of galectin-9 to kill Th1 cells *in vivo* was demonstrated by treatment of mice with recombinant galectin-9 in a mouse model of experimental autoimmune encephalitis (EAE). Substantial elimination of antigen-specific IFN-γ–producing $CD4^+$ cells was observed.

Galectin-9 has been shown to significantly suppress apoptosis in eosinophils from eosinophilic patients but enhance apoptosis in those from healthy subjects (Saita *et al.*, 2002). Furthermore, it suppresses dexamethasone-induced apoptosis of eosinophils from healthy subjects but enhances apoptosis induced by anti-Fas antibody in eosinophils from both eosinophilic patients and healthy subjects (Saita *et al.*, 2002).

Regulation of Apoptosis Demonstrated by Use of the Gene Transfection Approach

Galectin-3

Galectins have also been found to affect apoptosis by using the gene transfection approach, in which cells were transfected with galectin cDNA or anti-sense DNA or oligonucleotides. Additional studies provided substantial evidence that galectins can control apoptosis by functioning intracellularly, without encountering cell surface glycoconjugates. The most extensively documented example is the anti-apoptotic activity of galectin-3. This has been demonstrated with a range of cell types exposed to diverse apoptotic stimuli (reviewed in Liu *et al.* [2002]). The human T lymphoma Jurkat cell line transfected with galectin-3 cDNA is more resistant to apoptosis induced by anti-Fas antibody and staurosporine (Yang *et al.*, 1996). Similarly, another human T-cell line CEM transfected with galectin-3 cDNA is more resistant to apoptosis induced by galectin-1 (Hahn *et al.*, 2004). Furthermore, galectin-3–transfected CEM cells were found to be more resistant to C(2)-ceramide–induced apoptosis (Fukumori *et al.*, 2003).

Human B lymphoma cells transfected with galectin-3 cDNA are also more resistant to apoptosis induced by anti-Fas antibody (Hoyer *et al.*, 2004). Conversely, a galectin-3–positive B cell line transfected with a plasmid containing cDNA coding for amino-terminal truncated galectin-3 (that functions in a dominant negative fashion) exhibited increased sensitivity to anti-Fas–induced apoptosis (Hoyer *et al.*, 2004). Overexpression of galectin-3 in breast cancer cell lines rendered the cells more resistance to apoptosis induced by a variety of apoptotic stimuli (Akahani *et al.*, 1997; Choi *et al.*, 2004; Matarrese *et al.*, 2000a; Moon *et al.*, 2001; Takenaka *et al.*, 2004). In addition, overexpression of galectin-3 in a bladder carcinoma cell line conferred resistance to apoptosis induced by engagement of TRAIL (TNF-related apoptosis-inducing ligand) (Oka *et al.*, 2005). Finally, the anti-apoptotic activity of galectin-3 has been confirmed by comparing macrophages from galectin-3–deficient mice and wild-type mice (Hsu *et al.*, 2000).

The mechanism of the anti-apoptotic activity by galectin-3 is not fully understood but probably involves its engagement in the intracellular apoptosis-regulation pathways. Current evidence suggests that the activity involves translocation of galectin-3 either from the cytosol or the nucleus to the mitochondria after exposure to apoptotic stimuli (Matarrese *et al.*, 2000b; Yu *et al.*, 2002). In addition, galectin-3 expression inhibited the loss of mitochondrial membrane potential, which is commonly associated with apoptosis (Matarrese *et al.*, 2000). Therefore, galectin-3 may exert its anti-apoptotic activity by interacting with other apoptosis regulators that function in the mitochondria. The location of galectin-3 is critical for its activity in apoptosis, because it has been shown that galectin-3 localized in the cytosol protects the cell from apoptosis but has an opposite effect when localized in the nucleus (Califice *et al.*, 2004).

Some structural features of galectin-3 are essential for its apoptosis regulating activity. First, phosphorylation of galectin-3 on the serine residue at the sixth amino acid position is critical for the protein's anti-apoptotic activity (Yoshii *et al.*, 2002). This phosphorylation is necessary for export of galectin-3 from the nucleus when the cells are exposed to apoptotic stimuli (Takenaka *et al.*, 2004). Second, an Asn-Trp-Gly-Arg (NWGR) motif in the C-terminal part of the molecule, required for carbohydrate-binding activity, is essential for its anti-apoptotic activity. Substitution of glycine to alanine in this motif abrogates galectin-3's anti-apoptotic activity (Akahani *et al.*, 1997).

Galectin-3 has been shown to regulate the expression of some molecules in the apoptosis-regulating pathway. First, phosphorylated galectin-3 has been shown to upregulate the mitogen-activated protein kinase (MAPK) pathway, which is known to be involved in regulation of apoptosis

(Takenaka *et al.*, 2004). Second, galectin-3 has been shown to upregulate levels of AKT (Oka *et al.*, 2005).

Galectin-3 has been shown to interact with a number of molecular components of apoptosis-regulating pathways. One study found that galectin-3 is complexed with Fas receptor (CD95), the engagement of which is known to induce apoptosis (Fukumori *et al.*, 2004). Galectin-3 was shown to bind to Bcl-2, a well-characterized anti-apoptotic protein (Yang *et al.*, 1996). However, direct interaction between these two proteins and their intracellular interaction have not been established. Another molecule that is possibly involved in galectin-3's anti-apoptotic activity is synexin, which is a Ca^{2+}- and phospholipid-binding protein. This protein is necessary for translocation of galectin-3 to the perinuclear membranes in cells treated with apoptotic stimuli (Yu *et al.*, 2002).

A powerful technique used successfully to isolate interacting partners is the yeast two-hybrid system, which uses a protein of interest (bait) in fusion with a DNA-binding protein (Brent and Ptashne, 1984; Fields and Song, 1989; Ma and Ptashne, 1987). The bait-DNA binding protein is then used to identify binding partners from an expression library constructed to contain an attached transcriptional activation motif by the induction of transcriptional activation of a reporter gene. Thus, proteins interacting with the bait can be identified by subsequent subcloning, and this method has been used successively to identify galectin interaction partners (Menon *et al.*, 2000; Park *et al.*, 2001; Yu *et al.*, 2002). We have performed yeast two-hybrid screening and identified another protein involved in apoptosis. This protein was independently identified as Alg-2 interacting protein X (ALIX) (Missotten *et al.*, 1999), AIP1 (ALG-2 interacting protein 1) (Vito *et al.*, 1999), and Hp95 (Wu *et al.*, 2001). Alg-2 (apoptosis-linked gene 2) was described to mediate apoptosis through Ca^{2+} regulated signals (Vito *et al.*, 1999). Our results suggest that galectin-3 is also involved in apoptosis through a pathway mediated by Alg-2.

Other Galectins

Other galectins can also regulate apoptosis as demonstrated by the use of gene transfection approaches, but much less is known about possible mechanisms. We and others have shown that transfected tumor cells over-expressing galectin-7 are more prone to undergo apoptosis induced by a number of different apoptotic stimuli that trigger different signaling pathways (Bernerd *et al.*, 1999; Kuwabara *et al.*, 2002). Galectin-12 has been shown to promote apoptosis when overexpressed in a fibroblast cell line (Hotta *et al.*, 2001). Both galectins are believed to regulate apoptosis by functioning intracellularly; however, definitive evidence for this is lacking.

Many studies demonstrating pro-apoptotic cell death require highly purified protein, substantially free of endotoxin. Conveniently, lectin activities of many galectins can be used advantageously in purification of protein from expression systems, and an example of efficient bacterial expression and detoxification is described here. In addition, many studies performed to demonstrate activities of galectins *in vitro* use transfection to introduce lectin into cells deficient in these proteins, and an example of an eukaryotic expression system is described.

Preparation of Recombinant Galectin-3

Construction of a Bacterial Galectin-3 Expression Vector

A number of vectors are available for expression in *Escherichia coli*, and a versatile system is the pET system of vectors (Studier and Moffatt, 1986) (e.g., Novagen-EMD Biosciences, Inc.) that provide a number of options, including high-stringency basal expression, several fusion protein partners, and periplasmic targeting. In the given example, human galectin-3 cDNA (Robertson *et al.*, 1990) was amplified by PCR to create an *Nco*I site corresponding to the start codon (primer AGCGGACCATGGCAGAC AATTTTTCG) and a downstream *Hind*III site (primer CCCCAAGCTTT CAGATTATATCATGG), followed by insertion into the corresponding sites in the pET-25b vector by standard cloning techniques (Sambrook *et al.*, 1989). Successful constructs were sequenced to verify integrity and in-frame location of the cDNA, and the low-protease BL21(DE3) strain (Grodberg and Dunn, 1988) was transformed. Transformants were evaluated for expression of galectin-3 by SDS-PAGE in mini-cultures on induction with isopropylthiodigalactoside (IPTG).

Purification of Endotoxin-free Galectin-3

Each expression construct must be optimized under conditions determined by empirical methods to achieve maximal yields. A set of conditions was established for this construct as follows. A 2 ml stock in 2× YT medium is cultured overnight at 37° under ampicillin selection to produce a starting culture. This culture is expanded into 200 ml of the same culture broth the next morning and maintained under vigorous agitation at 37° for 8–10 h ($A_{600} = 0.8$–1.0). A 1-l culture is now prepared in the same culture broth at 25° and induced overnight with 1 mM IPTG. This concentration of IPTG induces high-level expression of the T7lac promoter in pET-25b. The following procedure describes simultaneous affinity purification and endotoxin removal.

Materials for Purification

Lactosyl-Sepharose prepared by described procedures (Levi and Teichberg, 1981)

DEAE-Trisacryl M (IBF Biotechnics)

Cellulose acetate dialysis membrane, M_r 12,000 cutoff (Spectrapore)

Polyethylene glycol (PEG), average M_r15,000–20,000 (Sigma-Aldrich)

Phosphate-buffered saline (PBS), endotoxin free

Endotoxin-free lactose (Sigma Ultra L8783)

Limulus amebocyte lysate endotoxin test kit (e.g., Cambrex-Bio Whittaker)

Endotoxin-free water

Low-pressure columns with adapters (e.g., BioRad Econo-Column)

Bradford protein assay reagent (e.g., Pierce Chemical)

Solutions

Lysis buffer: 20 mM Tris HCl, pH 7.5, 5 mM EDTA, 10 mM sucrose, 20 mM 2-mercaptoethanol

Wash buffer A: PBS, 5 mM sucrose, 10 mM 2-mercaptoethanol

Wash buffer B: PBS, 10 mM 2-mercaptoethanol

Wash buffer C: 10 mM Na phosphate, pH 7.4

Protease inhibitors: pepstatin (1 μM), leupeptin (10 μM), phenyl-methylsulfonyl fluoride (0.2 μM prepared fresh)

0.5 M lactose in wash buffer C

PBS containing 10 % (v/v) glycerol (PBS/glycerol)

Affinity Chromatography of Galectin-3

This affinity chromatography uses the lectin activity of galectin-3 for purification. The ability of galectin-3 to oligomerize (Ahmad *et al.*, 2004; Hsu *et al.*, 1992; Massa *et al.*, 1993) and its ability to avidly bind bacterial lipopolysaccharide (LPS) (Mey *et al.*, 1996) will result in large amounts of endotoxin in the preparation if this is not removed. *E. coli* LPS is anionic and can be efficiently removed by interaction with the cationic matrix DEAE-Trisacryl M at low ionic strength (Petsch and Anspach, 2000).

1. Centrifuge the cooled bacterial culture at 1500*g* at 4° for 30 min and resuspend in 60 ml lysis buffer with protease inhibitors. Sonicate with a 1-cm probe at 120–200 W for 1 min at 1 sec on–1 sec off in a glass beaker immersed in ice slush. Rest 2 min and repeat the sonication cycle three more times. Centrifuge the bacterial lysate at 48,000*g* at 4° for 45 min and the decant supernatant into a cold container.

2. Load the clarified lysate on a lactosyl-Sepharose column (15–20 ml bed volume) equilibrated with wash buffer A. Wash with 100 ml wash buffer A, followed by 100 ml wash buffer B. Replace the buffer container with one holding 100 ml wash buffer C and continue washing the column, making sure that wash buffer B has been almost completely depleted before adding buffer C (to prevent excessively high ionic strength in the subsequent column).

3. Attach the effluent line from the lactosyl-Sepharose column to the DEAE-Trisacryl column (15–20 ml bed volume), pre-equilibrated with wash buffer C. Proceed to elute galectin-3 from the lactosyl-Sepharose column with 100 ml 0.5 M lactose. The first 10 ml effluent may be discarded and 1.5 ml fractions collected thereafter.

4. Evaluate protein content with 10 μl in alternating fractions and 30 μl of Bradford reagent, and visually identify fractions containing galectin-3 against a white background. Test alternating fractions for the presence of endotoxin by using small aliquots of each fraction directly, or diluted in endotoxin free water, using an endotoxin assay kit. Pool fractions that are endotoxin free. Typically, all fractions should be endotoxin free.

5. Place endotoxin-free fractions containing galectin-3 into dialysis tubing for concentration with PEG. The dialysis bag is surrounded with PEG granules and placed in a 4° refrigerator until a total volume of 1μ2 ml is attained. During concentration, no part of the dialysis tubing should be completely dry, because this would cause protein to precipitate.

6. After suitable concentration, adherent PEG is gently scraped off, and the dialysis bag is placed in cold PBS/glycerol under constant gentle stirring at 4°. Buffer is replaced at 4-h intervals (one or two intervals can occur overnight), for a total of three additional buffer changes with PBS/glycerol. After this period, the protein concentration is determined, and endotoxin levels are measured again in the galectin-3 concentrate and in serially diluted samples. The levels of endotoxin should be less than 0.05 endotoxin units, the limits of sensitivity of the *Limulus* amebocyte lysate procedure. Additional analytical tests may be performed to determine purity and activity by SDS-PAGE and hemagglutination (Frigeri *et al.*, 1990), respectively.

Overexpression of Galectin-3 in Eukaryotic Cells

Transfection is defined as the introduction of nucleic acids into cells by nonviral methods (Chisholm, 1995). The ability to express an exogenous gene in cells by transfection has greatly facilitated the functional studies of many galectins. Transfection is "transient" or "stable" on the basis of the period of expression achieved. Transient transfection is usually used for brief assays (e.g., promoter activities) that can be completed in a few days. Stable transfection is often required for the study of the function of genes that may require a relatively long time to manifest. This is usually achieved by the inclusion of a drug resistance gene in the construct and selection for transfectants by virtue of their resistance to the drug after culturing in selection medium (Southern and Berg, 1982).

Construction of the Expression Vectors pEF1 and pEF1-Neo

The expression vector pEF1-neo (Fig. 1) is a bi-cistronic vector designed to coexpress the target gene and a neomycin resistance (Neo^r) gene in one transcript under the control of the constitutive elongation factor-1α (EF-1?α) promoter (Uetsuki *et al.*, 1989; Wakabayashi-Ito and Nagata, 1994) was constructed in two steps:

1. The human EF1α promoter was excised from the vector pEF1-BOS with *Hind*III and *Eco*RI, and cloned into pBK-CMV (Stratagene) at the same sites.

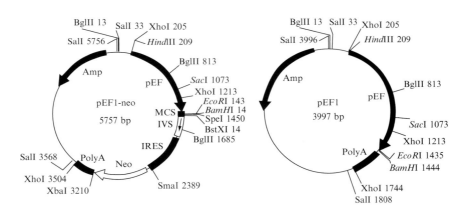

FIG. 1. Mammalian expression vectors pEF1-neo and pEF1. A gene of interest (usually a cDNA with a complete open-reading frame) can be cloned into the multiple cloning site (MCS) of pEF1-neo (left panel). An ampicillin-resistance gene (Amp) permits selective propagation in *E. coli*. After transfection of the construct into mammalian cells, the constitutive human elongation factor-1α promoter will drive the transcription of the bi-cistronic cassette of the gene of interest and the neo gene, whereas the bovine growth hormone polyadenylation sequence (PolyA) provides the termination signal for transcription. The internal ribosome entry site (IRES) permits the simultaneous translation of the gene of interest and the neo gene from a single transcript, and selection for stable transfectants can be achieved with G418. The vector pEF1 (right panel) is useful for transient transfection when high expression for a few days is sufficient. The gene of interest can be cloned into the unique *Eco*RI and *Bam*HI sites.

2. The resulting plasmid was digested with *Sca*I and *Eco*RI, and the shorter fragment containing the EF1α promoter was cloned into the bi-cistronic vector pIRES1-neo (Clontech) from which the CMV promoter had been removed by digestion with *Nru*I and *Eco*RI. The expression vector pEF1 (Fig. 1) was produced by removing the SpeI-XbaI fragment containing the IRES and Neo sequences from pEF1-neo. This vector does not carry a selection marker but may be capable of directing higher expression of the target gene and, therefore, be suitable for transient transfection if higher expression is desired.

The pEF1-neo vector contains the internal ribosome entry site (IRES) of the encephalomyocarditis virus, which permits the translation of two open reading frames from one messenger RNA. The constitutive EF1 promoter directs strong expression in a variety of mammalian cell types (Mizushima and Nagata, 1990). We tested expression of a green fluorescence protein (GFP) gene with this vector in several cell lines and found that more than 85% of G418-resistant cells express GFP. These two vectors have been used successfully for transfection of galectin-3 and galectin-12 in all cell types tested, including HeLa, MCF7, COS-1, HEK293, NIH3T3, Jurkat, U937, mouse embryonic fibroblasts, and mouse bone marrow–derived mast cells. Overexpression of some genes may not be compatible with cell viability or cell proliferation, and stable transfectants of these genes will be difficult to establish. In such a situation, an inducible expression system may be considered (e.g., tetracycline inducible (Jones *et al.*, 2005), Gateway System, Invitrogen).

Preparation of Plasmid DNA for Transfection

It is essential that DNA is highly purified before its use in transfection to remove impurities toxic to cells, because this will greatly increase transfection efficiency. We found that PEG precipitation method described by Wang (Wang *et al.*, 1994) produces high-quality DNA suitable for transfection to be convenient and reproducible, eliminating the need for CsCl density gradient preparations.

Transfection of Mammalian Cells

Perturbation of cell membrane by an electrical pulse creates pores that allow the entry of DNA into the cell (Neumann *et al.*, 1982). Traditionally, this techniques required extensive optimization for each cell type, with regard to the duration and strength of the electrical pulse. Baum reported a method (Baum *et al.*, 1994) that was applicable to a variety of cell types, and this method has been routinely used with satisfactory results. Transfection with lipid reagents can also be an effective method, and a variety of high-efficiency reagents are commercially available (e.g., LipofectAMINE—Invitrogen, SuperFect—Qiagen, FuGENE—Roche).

Apoptosis Assays

Apoptosis, or programmed cell death, is an integral part of the development of multicellular organisms (Jacobson *et al.*, 1997; Pellegrini and Strasser, 1999). Apoptosis can be induced by two main pathways: the extrinsic pathway mediated by cell surface receptors such as those death receptors of the TNF receptor family and the intrinsic pathway acting on the mitochondria (Green and Kroemer, 2005). Morphologically, apoptosis can be readily distinguished from the random death of necrosis by features such as membrane blebbing, cell shrinkage, and chromatin condensation, which can be observed microscopically. Biochemical features of apoptosis include DNA fragmentation, specific cleavage of certain proteins, increased mitochondrial membrane permeability, and exposure of phosphatidylserine (PS) on the cell surface, and a variety of assays have been designed to quantitate apoptosis on the basis of these features (Gavrieli *et al.*, 1992; Squier and Cohen, 2001; Telford *et al.*, 1994).

Acknowledgments

These studies were supported by grants from the National Institute of Allergy and Infectious Diseases, National Institutes of Health.

References

Ahmad, N., Gabius, H. J., Andre, S., Kaltner, H., Sabesan, S., Roy, R., Liu, B., Macaluso, F., and Brewer, C. F. (2004). Galectin-3 precipitates as a pentamer. with synthetic multivalent carbohydrates and forms heterogeneous cross-linked complexes. *J. Biol. Chem.* **279**, 10841–10847.

Akahani, S., Nangia-Makker, P., Inohara, H., Kim, H. R. C., and Raz, A. (1997). Galectin-3: A novel antiapoptotic molecule with a functional BH1 (NWGR) domain of Bcl-2 family. *Cancer Res.* **57**, 5272–5276.

Amano, M., Galvan, M., He, J., and Baum, L. G. (2003). The ST6Gal I sialyltransferase selectively modifies N-glycans on CD45 to negatively regulate galectin-1-induced CD45 clustering, phosphatase modulation, and T cell death. *J. Biol. Chem.* **278**, 7469–7475.

Arbel-Goren, R., Levy, Y., Ronen, D., and Zick, Y. (2005). Cyclin-dependent kinase inhibitors and JNK act as molecular switches, regulating the choice between growth arrest and apoptosis induced by galectin-8. *J. Biol. Chem.* **280**, 19105–19114.

Baum, C., Forster, P., Hegewisch-Becker, S., and Harbers, K. (1994). An optimized electroporation protocol applicable to a wide range of cell lines. *Biotechniques* **17**, 1058–1062.

Bernerd, F., Sarasin, A., and Magnaldo, T. (1999). Galectin-7 overexpression is associated with the apoptotic process in UVB-induced sunburn keratinocytes. *Proc. Natl. Acad. Sci. USA* **96**, 11329–11334.

Brent, R., and Ptashne, M. (1984). A bacterial repressor protein or a yeast transcriptional terminator can block upstream activation of a yeast gene. *Nature* **312**, 612–615.

Brewer, C. F., Miceli, M. C., and Baum, L. G. (2002). Clusters, bundles, arrays and lattices: novel mechanisms for lectin-saccharide-mediated cellular interactions. *Curr. Opin. Struct. Biol.* **12,** 616–623.

Califice, S., Castronovo, V., Bracke, M., and Van Den Brule, F. (2004). Dual activities of galectin-3 in human prostate cancer: Tumor suppression of nuclear galectin-3 vs tumor promotion of cytoplasmic galectin-3. *Oncogene* **45,** 7527–7536.

Chisholm, V. (1995). "High Efficiency Gene Transfer into Mammalian Cells." *In* DNA Cloning 4: Mammalian Systems: A Practical Approach, Vol. 1, pp. 1–39. Oxford University Press, Oxford.

Choi, J. H., Chun, K. H., Raz, A., and Lotan, R. (2004). Inhibition of N-(4-hydroxyphenyl) retinamide-induced apoptosis in breast cancer cells by galectin-3. *Cancer Biol. Ther.* **3,** 447–452.

Cooper, D. N. (2002). Galectinomics: Finding themes in complexity. *Biochim. Biophys. Acta* **1572,** 209–231.

Davidson, P. J., Davis, M. J., Patterson, R. J., Ripoche, M. A., Poirier, F., and Wang, J. L. (2002). Shuttling of galectin-3 between the nucleus and cytoplasm. *Glycobiology* **12,** 329–337.

Dias-Baruffi, M., Zhu, H., Cho, M., Karmakar, S., McEver, R. P., and Cummings, R. D. (2003). Dimeric galectin-1 induces surface exposure of phosphatidylserine and phagocytic recognition of leukocytes without inducing apoptosis. *J. Biol. Chem.* **278,** 41282–41293.

Fernandez, G. C., Ilarregui, J. M., Rubel, C. J., Toscano, M. A., Gomez, S. A., Beigier Bompadre, M., Isturiz, M. A., Rabinovich, G. A., and Palermo, M. S. (2005). Galectin-3 and soluble fibrinogen act in concert to modulate neutrophil activation and survival: Involvement of alternative MAPK pathways. *Glycobiology* **15,** 519–527.

Fields, S., and Song, O. (1989). A novel genetic system to detect protein-protein interactions. *Nature* **340,** 245–246.

Fouillit, M., Joubert-Caron, R., Poirier, F., Bourin, P., Monostori, E., Levi-Strauss, M., Raphael, M., Bladier, D., and Caron, M. (2000). Regulation of CD45-induced signaling by galectin-1 in Burkitt lymphoma B cells. *Glycobiology* **10,** 413–419.

Frigeri, L. G., Robertson, M. W., and Liu, F.-T. (1990). Expression of biologically active recombinant rat IgE-binding protein in *Escherichia coli. J. Biol. Chem.* **265,** 20763–20769.

Fukumori, T., Takenaka, Y., Yoshii, T., Kim, H. R., Hogan, V., Inohara, H., Kagawa, S., and Raz, A. (2003). CD29 and CD7 mediate galectin-3-induced type II T-cell apoptosis. *Cancer Res.* **63,** 8302–8311.

Fukumori, T., Takenaka, Y., Oka, N., Yoshii, T., Hogan, V., Inohara, H., Kanayama, H. O., Kim, H. R., and Raz, A. (2004). Endogenous galectin-3 determines the routing of CD95 apoptotic signaling pathways. *Cancer Res.* **64,** 3376–3379.

Galvan, M., Tsuboi, S., Fukuda, M., and Baum, L. G. (2000). Expression of a specific glycosyltransferase enzyme regulates T cell death mediated by galectin-1. *J. Biol. Chem.* **275,** 16730–16737.

Gavrieli, Y., Sherman, Y., and Ben-Sasson, S. A. (1992). Identification of programmed cell death *in situ* via specific labeling of nuclear DNA fragmentation. *J. Cell. Biol.* **119,** 493–501.

Green, D. R., and Kroemer, G. (2005). Pharmacological manipulation of cell death: Clinical applications in sight? *J. Clin. Invest.* **115,** 2610–2617.

Grodberg, J., and Dunn, J. J. (1988). ompT encodes the *Escherichia coli* outer membrane protease that cleaves T7 RNA polymerase during purification. *J. Bacteriol.* **170,** 1245–1253.

Hadari, Y. R., Arbel-Goren, R., Levy, Y., Amsterdam, A., Alon, R., Zakut, R., and Zick, Y. (2000). Galectin-8 binding to integrins inhibits cell adhesion and induces apoptosis. *J. Cell Sci.* **113**(Pt 13), 2385–2397.

Hahn, H. P., Pang, M., He, J., Hernandez, J. D., Yang, R. Y., Li, L. Y., Wang, X., Liu, F. T., and Baum, L. G. (2004). Galectin-1 induces nuclear translocation of endonuclease G in caspase- and cytochrome c-independent T cell death. *Cell Death Differ.* **11**, 1277–1286.

He, J., and Baum, L. G. (2004). Presentation of galectin-1 by extracellular matrix triggers T cell death. *J. Biol. Chem.* **279**, 4705–4712.

Hirabayashi, J., Hashidate, T., Arata, Y., Nishi, N., Nakamura, T., Hirashima, M., Urashima, T., Oka, T., Futai, M., Muller, W. E., Yagi, F., and Kasai, K. (2002). Oligosaccharide specificity of galectins: a search by frontal affinity chromatography. *Biochim. Biophys. Acta* **1572**, 232–254.

Hotta, K., Funahashi, T., Matsukawa, Y., Takahashi, M., Nishizawa, H., Kishida, K., Matsuda, M., Kuriyama, H., Kihara, S., Nakamura, T., Tochino, Y., Bodkin, N. L., Hansen, B. C., and Matsuzawa, Y. (2001). Galectin-12, an adipose-expressed galectin-like molecule possessing apoptosis-inducing activity. *J. Biol. Chem.* **276**, 34089–34097.

Hoyer, K. K., Pang, M., Gui, D., Shintaku, I. P., Kuwabara, I., Liu, F. T., Said, J. W., Baum, L. G., and Teitell, M. A. (2004). An anti-apoptotic role for galectin-3 in diffuse large B-cell lymphomas. *Amer. J. Pathol.* **164**, 893–902.

Hsu, D. K., Zuberi, R., and Liu, F.-T. (1992). Biochemical and biophysical characterization of human recombinant IgE-binding protein, an S-type animal lectin. *J. Biol. Chem.* **267**, 14167–14174.

Hsu, D. K., Yang, R.-Y., Yu, L., Pan, Z., Salomon, D. R., Fung-Leung, W.-P., and F.-T., Liu (2000). Targeted disruption of the galectin-3 gene results in attenuated peritoneal inflammatory responses. *Amer. J. Pathol.* **156**, 1073–1083.

Hughes, R. C. (1999). Secretion of the galectin family of mammalian carbohydrate-binding proteins. *Biochim. Biophys. Acta* **1473**, 172–185.

Jacobson, M. D., Weil, M., and Raff, M. C. (1997). Programmed cell death in animal development. *Cell* **88**, 347–354.

Jones, J., Nivitchanyong, T., Giblin, C., Ciccarone, V., Judd, D., Gorfien, S., Krag, S. S., and Betenbaugh, M. J. (2005). Optimization of tetracycline-responsive recombinant protein production and effect on cell growth and ER stress in mammalian cells. *Biotechnol. Bioeng.* **91**, 722–732.

Kageshita, T., Kashio, Y., Yamauchi, A., Seki, M., Abedin, M. J., Nishi, N., Shoji, H., Nakamura, T., Ono, T., and Hirashima, M. (2002). Possible role of galectin-9 in cell aggregation and apoptosis of human melanoma cell lines and its clinical significance. *Int. J. Cancer* **99**, 809–816.

Kashio, Y., Nakamura, K., Abedin, M. J., Seki, M., Nishi, N., Yoshida, N., Nakamura, T., and Hirashima, M. (2003). Galectin-9 induces apoptosis through the calcium-calpain-caspase-1 pathway. *J. Immunol.* **170**, 3631–3636.

Kuwabara, I., Kuwabara, Y., Yang, R. Y., Schuler, M., Green, D. R., Zuraw, B. L., Hsu, D. K., and Liu, F. T. (2002). Galectin-7 (PIG1) exhibits pro-apoptotic function through JNK activation and mitochondrial cytochrome c release. *J. Biol. Chem.* **277**, 3487–3497.

Lanteri, M., Giordanengo, V., Hiraoka, N., Fuzibet, J. G., Auberger, P., Fukuda, M., Baum, L. G., and Lefebvre, J. C. (2003). Altered T cell surface glycosylation in HIV-1 infection results in increased susceptibility to galectin-1-induced cell death. *Glycobiology* **13**, 909–918.

Levi, G., and Teichberg, V. I. (1981). Isolation and physicochemical characterization of electrolectin, a β-D-galactoside binding lectin from the electric organ of *Electrophorus Electricus*. *J. Biol. Chem.* **256**, 5735–5740.

Liu, F. T., Patterson, R. J., and Wang, J. L. (2002). Intracellular functions of galectins. *Biochim. Biophys. Acta* **1572**, 263–273.

Ma, J., and Ptashne, M. (1987). A new class of yeast transcriptional activators. *Cell* **51**, 113–119.

Massa, S. M., Cooper, D. N. W., Leffler, H., and Barondes, S. H. (1993). L-29, an endogenous lectin, binds to glycoconjugate ligands with positive cooperativity. *Biochemistry* **32,** 260–267.

Matarrese, P., Fusco, O., Tinari, N., Natoli, C., Liu, F.-T., Semeraro, M. L., Malorni, W., and Iacobelli, S. (2000a). Galectin-3 overexpression protects from apoptosis by improving cell adhesion properties. *Int. J. Cancer.* **85,** 545–554.

Matarrese, P., Tinari, N., Semeraro, M. L., Natoli, C., Iacobelli, S., and Malorni, W. (2000b). Galectin-3 overexpression protects from cell damage and death by influencing mitochondrial homeostasis. *FEBS Lett.* **473,** 311–315.

Matarrese, P., Tinari, A., Mormone, E., Bianco, G. A., Toscano, M. A., Ascione, B., Rabinovich, G. A., and Malorni, W. (2005). Galectin-1 sensitizes resting human T lymphocytes to Fas (CD95)-mediated cell death via mitochondrial hyperpolarization, budding, and fission. *J. Biol. Chem.* **280,** 6969–6985.

Menon, R. P., Strom, M., and Hughes, R. C. (2000). Interaction of a novel cysteine and histidine-rich cytoplasmic protein with galectin-3 in a carbohydrate-independent manner. *FEBS Lett.* **470,** 227–231.

Mey, A., Leffler, H., Hmama, Z., Normier, G., and Revillard, J. P. (1996). The animal lectin galectin-3 interacts with bacterial lipopolysaccharides via two independent sites. *J. Immunol.* **156,** 1572–1577.

Missotten, M., Nichols, A., Rieger, K., and Sadoul, R. (1999). Alix, a novel mouse protein undergoing calcium-dependent interaction with the apoptosis-linked-gene 2 (ALG-2) protein. *Cell Death Diff.* **6,** 124–129.

Mizushima, S., and Nagata, S. (1990). pEF-BOS, a powerful mammalian expression vector. *Nucleic Acids Res.* **18,** 5322.

Moon, B. K., Lee, Y. J., Battle, P., Jessup, J. M., Raz, A., and Kim, H. R. (2001). Galectin-3 protects human breast carcinoma cells against nitric oxide-induced apoptosis: Implication of galectin-3 function during metastasis. *Am. J. Pathol.* **159,** 1055–1060.

Neumann, E., Schaefer-Ridder, M., Wang, Y., and Hofschneider, P. H. (1982). Gene transfer into mouse lyoma cells by electroporation in high electric fields. *EMBO J.* **1,** 841–845.

Novelli, F., Allione, A., Wells, V., Forni, G., and Malluci, L. (1999). Negative cell cycle control of human T cells by beta-galactoside binding protein (beta GBP): Induction of programmed cell death in leukaemic cells. *J. Cell. Physiol.* **178,** 102–108.

Oka, N., Nakahara, S., Takenaka, Y., Fukumori, T., Hogan, V., Kanayama, H. O., Yanagawa, T., and Raz, A. (2005). Galectin-3 inhibits tumor necrosis factor-related apoptosis-inducing ligand-induced apoptosis by activating Akt in human bladder carcinoma cells. *Cancer Res.* **65,** 7546–7553.

Pace, K. E., Lee, C., Stewart, P. L., and Baum, L. G. (1999). Restricted receptor segregation into membrane microdomains occurs on human T cells during apoptosis induced by galectin-1. *J. Immunol.* **163,** 3801–3811.

Park, J. W., Voss, P. G., Grabski, S., Wang, J. L., and Patterson, R. J. (2001). Association of galectin-1 and galectin-3 with Gemin4 in complexes containing the SMN protein. *Nucleic Acids Res.* **29,** 3595–3602.

Pellegrini, M., and Strasser, A. (1999). A portrait of the Bcl-2 protein family: Life, death, and the whole picture. *J. Clin. Immunol.* **19,** 365–377.

Perillo, N. L., Pace, K. E., Seilhamer, J. J., and Baum, L. G. (1995). Apoptosis of T cells mediated by galectin-1. *Nature* **378,** 736–739.

Petsch, D., and Anspach, F. B. (2000). Endotoxin removal from protein solutions. *J. Biotechnol.* **76,** 97–119.

Rabinovich, G. A., Alonso, C. R., Sotomayor, C. E., Durand, S., Bocco, J. L., and Riera, C. M. (2000). Molecular mechanisms implicated in galectin-1-induced apoptosis: activation of the AP-1 transcription factor and downregulation of Bcl-2. *Cell Death Differ.* **7,** 747–753.

Rabinovich, G. A., Ramhorst, R. E., Rubinstein, N., Corigliano, A., Daroqui, M. C., Kier-Joffe, E. B., and Fainboim, L. (2002). Induction of allogenic T-cell hyporesponsiveness by galectin-1-mediated apoptotic and non-apoptotic mechanisms. *Cell Death Differ.* **9,** 661–670.

Rini, J. M., and Lobsanov, Y. D. (1999). New animal lectin structures. *Curr. Opin. Struct. Biol.* **9,** 578–584.

Robertson, M. W., Albrandt, K., Keller, D., and Liu, F.-T. (1990). Human IgE-binding protein: A soluble lectin exhibiting a highly conserved interspecies sequence and differential recognition of IgE glycoforms. *Biochemistry* **29,** 8093–8100.

Rubinstein, N., Alvarez, M., Zwirner, N. W., Toscano, M. A., Ilarregui, J. M., Bravo, A., Mordoh, J., Fainboim, L., Podhajcer, O. L., and Rabinovich, G. A. (2004). Targeted inhibition of galectin-1 gene expression in tumor cells results in heightened T cell-mediated rejection; A potential mechanism of tumor-immune privilege. *Cancer Cell* **5,** 241–251.

Saita, N., Goto, E., Yamamoto, T., Cho, I., Tsumori, K., Kohrogi, H., Maruo, K., Ono, T., Takeya, M., Kashio, Y., Nakamura, K., and Hirashima, M. (2002). Association of galectin-9 with eosinophil apoptosis. *Int. Arch. Allergy Immunol.* **128,** 42–50.

Sambrook, J., Fritsch, E. F., and Maniatis, T. (1989). Molecular Cloning, 2nd Ed. Cold Spring Harbor Laboratory Press, Cold Spring Harbor.

Southern, P. J., and Berg, P. (1982). Transformation of mammalian cells to antibiotic resistance with a bacterial gene under control of the SV40 early region promoter. *J. Mol. Appl. Genet.* **1,** 327–341.

Squier, M. K., and Cohen, J. J. (2001). Standard quantitative assays for apoptosis. *Mol. Biotechnol.* **19,** 305–312.

Stillman, B. N., Hsu, D. K., Pang, M., Brewer, C. F., Johnson, P., Liu, F. T., and Baum, L. G. (2006). Galectin-3 and galectin-1 bind distinct cell surface glycoprotein receptors to induce T cell death. *J. Immunol.* **176,** 778–789.

Studier, F. W., and Moffatt, B. A. (1986). Use of bacteriophage T7 RNA polymerase to direct selective high-level expression of cloned genes. *J. Mol. Biol.* **189,** 113–130.

Sturm, A., Lensch, M., Andre, S., Kaltner, H., Wiedenmann, B., Rosewicz, S., Dignass, A. U., and Gabius, H. J. (2004). Human galectin-2: Novel inducer of T cell apoptosis with distinct profile of caspase activation. *J. Immunol.* **173,** 3825–3837.

Takenaka, Y., Fukumori, T., Yoshii, T., Oka, N., Inohara, H., Kim, H. R., Bresalier, R. S., and Raz, A. (2004). Nuclear export of phosphorylated galectin-3 regulates its antiapoptotic activity in response to chemotherapeutic drugs. *Mol. Cell. Biol.* **24,** 4395–4406.

Telford, W. G., King, L. E., and Fraker, P. J. (1994). Rapid quantitation of apoptosis in pure and heterogeneous cell populations using flow cytometry. *J. Immunol. Methods* **172,** 1–16.

Uetsuki, T., Naito, A., Nagata, S., and Kaziro, Y. (1989). Isolation and characterization of the human chromosomal gene for polypeptide chain elongation factor-1 alpha. *J. Biol. Chem.* **264,** 5791–5798.

Vespa, G. N. R., Lewis, L. A., Kozak, K. R., Moran, M., Nguyen, J. T., Baum, L. G., and Miceli, M. C. (1999). Galectin-1 specifically modulates TCR signals to enhance TCR apoptosis but inhibit IL-2 production and proliferation. *J. Immunol.* **162,** 799–806.

Vito, P., Pellegrini, L., Guiet, C., and D'Adamio, L. (1999). Cloning of AIP1, a novel protein that associates with the apoptosis- linked gene ALG-2 in a Ca2+-dependent reaction. *J. Biol. Chem.* **274,** 1533–1540.

Wada, J., Ota, K., Kumar, A., Wallner, E. I., and Kanwar, Y. S. (1997). Developmental regulation, expression, and apoptotic potential of galectin-9, a β-galactoside binding lectin. *J. Clin. Invest.* **99,** 2452–2461.

Wakabayashi-Ito, N., and Nagata, S. (1994). Characterization of the regulatory elements in the promoter of the human elongation factor-1 alpha gene. *J. Biol. Chem.* **269,** 29831–29837.

Wang, J. L., Gray, R. M., Haudek, K. C., and Patterson, R. J. (2004). Nucleocytoplasmic lectins. *Biochim. Biophys. Acta.* **1673,** 75–93.

Wang, L. F., Voysey, R., and Yu, M. (1994). Simplified large-scale alkaline lysis preparation of plasmid DNA with minimal use of phenol. *Biotechniques* **17,** 26, 28..

Wu, Y., Pan, S., Che, S., He, G., Nelman-Gonzalez, M., Weil, M. M., and Kuang, J. (2001). Overexpression of Hp95 induces G1 phase arrest in confluent HeLa cells. *Differentiation* **67,** 139–153.

Yang, R.-Y., Hsu, D. K., and Liu, F.-T. (1996). Expression of galectin-3 modulates T cell growth and apoptosis. *Proc. Natl. Acad. Sci. USA* **93,** 6737–6742.

Yoshii, T., Fukumori, T., Honjo, Y., Inohara, H., Kim, H. R., and Raz, A. (2002). Galectin-3 phosphorylation is required for its anti-apoptotic function and cell cycle arrest. *J. Biol. Chem.* **277,** 6852–6857.

Yu, F., Finley, R. L., Jr., Raz, A., and Kim, H. R. (2002). Galectin-3 translocates to the perinuclear membranes and inhibits cytochrome c release from the mitochondria. A role for synexin in galectin-3 translocation. *J. Biol. Chem.* **277,** 15819–15827.

Zhu, C., Anderson, A. C., Schubart, A., Xiong, H., Imitola, J., Khoury, S. J., Zheng, X. X., Strom, T. B., and Kuchroo, V. K. (2005). The Tim-3 ligand galectin-9 negatively regulates T helper type 1 immunity. *Nat. Immunol.* **6,** 1245–1252.

[19]   On the Role of Galectins in Signal Transduction

By Susumu Nakahara and Avraham Raz

Abstract

Galectins are a family of proteins that bind to β-galactoside carbohydrate structure through their carbohydrate recognition domains (CRDs). These proteins have been shown to be involved in multiple biological functions such as cell–matrix and cell–cell interactions, cell proliferation, cell differentiation, cellular transformation, or apoptosis mainly through their binding properties to specific ligands. Signal transduction should be dramatically affected and changed in the process of those biological functions; namely, galectins can also be involved in several signal transduction pathways. This chapter discusses the role of galectins in signal transduction, dividing into extracellular and intracellular galectins. In addition, we will indicate the methods to identify the interactions of galectins in signal transduction.

Overview

Galectins are rapidly growing members of animal lectins that bind β-galactoside carbohydrate structures through evolutionary conserved amino acid sequences of the carbohydrate recognition domain (CRD) (Barondes *et al.*, 1994). To date, 15 members of the galectin family have been identified,

METHODS IN ENZYMOLOGY, VOL. 417   0076-6879/06 $35.00
                                                            DOI: 10.1016/S0076-6879(06)17019-6

and all of them can be classified according to the CRD features into three groups on the basis of the content and organization of domains as follows: (1) the prototype group (galectin-1, -2, -5, -7, -10, -11, -13, -14, and -15); (2) the chimera type (galectin-3); (3) the tandem repeat type (galectin-6, -8, -9, and -12) (Cooper and Barondes, 1999; Liu and Rabinovich, 2005). The prototype galectins are composed of a CRD and a short N-terminal sequence, and the tandem repeat type galectins have two nonidentical CRDs with a short linker sequence. Galectin-3 has a unique structure, with one CRD and an extended N-terminal domain containing a short N-terminal end with a serine phosphorylation site and a repeated collagen-like sequence rich in glycine, tyrosine, and proline.

Each galectin protein has distinct sugar binding specificity presumably because of its structurally different properties (i.e., the valence of the galectin molecules). Most prototype galectins can form dimers through their noncovalent interactions to create functionally bivalent molecules, whereas tandem repeat type galectins are constitutively bivalent (Wang et al., 2004). On the other hand, the chimera type galectin-3 can form dimers at high concentrations or even pentamers in the presence of multivalent ligands (Ahmad et al., 2004; Hsu et al., 1992). Thus, the activities of galectins may be exerted because of their dimerized or oligomerized status to function as biological cross-linkers among several glycoproteins. All galectins can efficiently bind to lactose and N-acetyllactosamine, whereas the interaction between a galectin CRD and the monosaccharide ligand, galactose is much weaker (Leffler and Barondes, 1986). Interestingly, galectin-3 showed a remarkably high affinity to almost all oligosaccharide structures compared with other galectins (Hirabayashi et al., 2002). For example, A antigen is a 32-fold better ligand that N-acetyllactosamine for galectin-3, but a threefold worse ligand for galectin-1 (Leffler and Barondes, 1986; Wang et al., 2004). More detailed biochemical properties about the specific structure recognized by each galectin family protein were analyzed (Hirabayashi et al., 2002).

In general, galectins are soluble proteins and have features typical of cytoplasmic proteins. However, their expressions are localized not only in the cytoplasm but also in the nucleus, on cell surfaces, and even in the extracellular space, depending on each galectin protein (Wang et al., 2004). Galectins outside the cell are precisely secreted from cells by nonclassical secretary pathways (Hughes, 1999). These widely distributed proteins have been shown to be involved in diverse biological functions, such as cell growth, cell adhesion, cell differentiation, apoptosis, and pre-mRNA splicing, depending on their specific localization (i.e., extracellular or intracellular compartment) (Liu and Rabinovich, 2005; Liu et al., 2002). Signal transduction should be dramatically affected in the process of those biological functions. In fact, many reports have suggested that galectins can also be involved in several signal transduction changes. Therefore, in this chapter, we will focus on the role of galectins in signal transduction, which is contributed by both the extracellular and intracellular galectins. The roles of extracellular galectins are depicted in Table I, and the roles of intracellular galectins are in Table II. In addition, we will exhibit the experimental procedures to analyze the role of both extracellular and intracellular galectins, in particular, galectin-3.

The Role of Extracellular Galectins in Signal Transduction

Galectins and Integrin-Related Signaling

Integrins are major adhesion- and signaling-receptor proteins that mediate cell–matrix interactions. They also trigger a variety of signal transduction pathways followed by regulation of cytoskeletal organization, specific gene expression, growth control, and apoptosis (Tamura et al., 1999). Generally, integrins are highly glycosylated because of many N-glycosylation sites, and several galectin proteins have been shown to be a binding partner of

TABLE I

THE ROLES OF EXTRACELLULAR GALECTINS AND THEIR LIGANDS IN SIGNAL TRANSDUCTION

Galectins	Ligand(s)	The roles in signaling transduction	Reference(s)
Galectin-8 (immobilized)	$\alpha3\beta1$ integrin $\alpha6\beta1$ integrin	FAK, paxillin ↑ ERK1/2, Akt, p70S6K ↑	Levy et al., 2001, 2003
Galectin-8 (soluble)	$\alpha3\beta1$ integrin $\alpha6\beta1$ integrin	JNK, Akt ↑	Arbel-Goren et al., 2005
Galectin-3	$\alpha3\beta1$ integrin NG2	Up-regulation of integrin signaling?	Fukushi et al., 2004
Galectin-1	$\beta1$ integrin	FAK ↑	Moiseeva et al., 2003
Galectin-1	$\alpha5\beta1$ integrin	MEK1/2, ERK1/2 ↓	Fischer et al., 2005
Galectin-3	EGF-receptor	Nuclear Translocation of ERK1/2 ↑	Partridge et al., 2004
Galectin-3	TGF-β-receptor	Nuclear Translocation of Smad2/3 ↑	Partridge et al., 2004
Galectin-1	?	MEK1/2, ERK1/2 ↑	Maeda et al., 2003
Galectin-3	?	MEK1/2, ERK1/2, PKC, PKA ↑	Maeda et al., 2003
Galectin-1	TCR	ERK2 ↑ TCR-ζ ↑	Vespa et al., 1999 Chung et al., 2000
Galectin-3	TCR	Phosphorylation in T-cells↓	Demetriou et al., 2001

TABLE II
THE ROLES OF INTRACELLULAR GALECTINS AND THEIR LIGANDS IN SIGNAL TRANSDUCTION

Galectins	Ligand(s)	The roles in signal transduction	Reference(s)
Galectin-1	H-Ras	ERK ↑	Paz et al., 2001
		Raf-1 ↑, ERK ↑, PI3K ↓	Elad-Sfadia et al., 2002
Galectin-3	K-Ras	Raf-1 ↑, ERK ↓, PI3K ↑	Elad-Sfadia et al., 2002
		ERK ↑, PI3K ↓, Akt ↓	Shalom-Feuerstein et al., 2005
Galectin-3	?	TRAIL sensitive by Akt ↓	Lee et al., 2003
		TRAIL resistant by Akt ↑	Oka et al., 2005
Galectin-3	CD95	Apoptotic signal change from Type II to Type I	Fukumori et al., 2004
Galectin-3	?	NF-κB signaling ↓ in null mice	Hsu et al., 2001
Galectin-3	β-Catenin	Regulation of Wnt signaling ?	Shimura et al., 2004
	Axin		Shimura et al., 2005
Galectin-12	?	Adipogenic signaling ↑	Yang et al., 2001

the integrins through their CRDs, suggesting that galectin–integrin interactions can affect integrin-mediated signal transduction under certain circumstances.

Galectin-8 can bind to $\alpha3\beta1$, $\alpha6\beta1$ integrins, and physiologically modulate cell adhesion of HeLa or H1299 cells (Hadari et al., 2000). When immobilized, it functions as a matrix protein equivalent to fibronectin in promoting cell adhesion and activates tyrosine phosphorylation of focal adhesion kinase (FAK) and paxillin. In contrast, when present as a soluble form, it negatively regulates cell adhesion through a complex with integrins (Levy et al., 2001). Subsequently, it was revealed that ligation of integrins by galectin-8 can trigger a distinct pattern of cytoskeletal organization compared with fibronectin, after the activation of integrin-mediated signaling cascades such as extracellular signal-regulated kinase (ERK) 1/2, phosphatidylinositol 3-kinase (PI3K), and Akt, p70 S6 kinase (Levy et al., 2003). In a more recent study, it was reported that addition of a high concentration of galectin-8 and growth factors, such as insulin growth factor, and Akt, also could activate c-Jun NH_2-teminal kinase (JNK) pathway followed by up-regulation of p21, which is one of the cell cycle regulators, resulting in cell growth arrest (Arbel-Goren et al., 2005). Galectin-3 can also bind to $\alpha3\beta1$ integrin on the cell surface of endothelial cells (EC), and these proteins form a complex with NG2 proteoglycan, which is expressed by microvascular pericytes in newly formed blood vessels. This complex may potentiate transmembrane signaling by means of an $\alpha3\beta1$ integrin, which is responsible for EC motility and morphogenesis (Fukushi et al., 2004).

Galectin-1 can interact with the $\beta1$ subunit of integrin on vascular smooth muscle cell (SMC), and incubation of SMCs with exogenous galectin-1 transiently increased the amount of the active form of $\beta1$ integrin followed by up-regulation of tyrosine phosphorylation of FAK (Moiseeva et al., 2003). In a more recent study, it was reported that galectin-1 interacted with the $\alpha5\beta1$ integrin to restrict a variety of carcinoma cells' growth by induction of p21 and p27. These antiproliferative effects result from inhibition of the Ras-MEK-ERK pathway and consecutive transcriptional induction of p27 after the binding of exogenous galectin-1 to $\alpha5\beta1$ integrin (Fischer et al., 2005).

Taken together, several extracellular galectins can modify specific integrin signaling of distinct cell types through conjugation with integrins in a carbohydrate-dependent manner.

Galectins and Growth Factor Receptor-Related Signaling

A large family of cytokines including transforming growth factor (TGF)-β, epidermal growth factor (EGF), platelet-derived growth factor (PDGF), or insulin growth factor (IGF) binds to their specific receptors and can exert a wide range of biological effects on a large variety of cell types. Many of them have also important functions during embryonal development in pattern formation and tissue specification (Heldin et al., 1997). These receptors normally have many N-glycosylation sites like integrins; therefore, galectins have the possibility to bind to them, and, in fact, several galectins have been reported to engage to those receptors, resulting in signal transduction changes.

Galectin-3 can bind to the carbohydrate structures of EGF and TGF-β receptors, which are modified by the specific Golgi enzyme, $\beta1$, 6-N-acetylglucosamnyltransferase V (GnT-V or Mgat5) (Partridge et al., 2004). The multivalence of galectin-3 can make a lattice formation of those receptors, resulting in delays of receptor removal by constitutive endocytosis.Therefore, the lattice ensures up-regulation of surface receptors and increased sensitivity to growth factors. For example, mammary tumor cells from Mgat5-deficient mice, in which lattice formations by galectin-3 are impossible, display a global loss of sensitivity to cytokines judging from the nuclear translocation rates of phosphorylated Erk1/2 and Smad2/3. Similar functional defects are exhibited in the macrophages derived from galectin-3–deficient mice (Hsu et al., 2000). Meanwhile, exogenous galectin-1 and galectin-3 stimulated the proliferation of hepatic stellate cells (HSCs) by the mitogen-activated protein kinase/extracellular signal-regulated kinase (MEK) 1/2 - ERK1/2 signaling pathway similar to PDGF stimulation, whereas those binding partners were undetermined. Interestingly, galectin-3, but not galectin-1, used protein kinase C (PKC) and A (PKA) to induce this process, suggesting each galectin can stimulate a different intracellular signaling pathway in HSCs (Maeda *et al.*, 2003).

Collectively, among the galectin family, galectin-3 can efficiently regulate the cytokine receptors by a lattice formation because of its multivalence.

Galectins and TCR Signaling

T cells constantly sample their environment using T-cell receptors (TCRs) that possess both a low affinity for major histocompatibility complex (MHC) molecules and a highly diverse set of CDR3 regions. The decision of T-cell activation is dependent on TCR interaction, resulting in cell proliferation or apoptosis (Germain and Stefanova, 1999). Galectin-1 and -3 can modify this signaling through their carbohydrate binding properties.

First, it was reported that galectin-1 specifically modulated TCR signaling to enhance TCR apoptosis but inhibit interleukin (IL) -2 production and proliferation. This galectin-1 stimulation is shown to synergize with TCR engagement to specifically enhance ERK2 activation, although it does not uniformly enhance TCR-induced tyrosine phosphorylation (Vespa *et al.*, 1999). Subsequently, it was revealed that galectin-1 could also induce phosphorylation of partial TCR-ζ chain and the generation of inhibitory pp21ζ, to antagonize TCR signal transduction processes (Chung *et al.*, 2000). Meanwhile, galectin-3 is also associated with the TCR complex dependent on the interaction with Mgat5, in which the mechanism is similar to the pattern of cytokine receptors as mentioned previously (Demetriou *et al.*, 2001). However, contrary to the cytokine receptors, a galectin-3/glycoprotein lattice of TCR clustering restricts TCR recruitment to the site of antigen presentation, resulting in negative regulation of TCR signaling and T-cell activation.

Taken together, galectins can negatively modulate TCR signaling in T cells, suggesting that galectins can play an important role in several immune responses.

The Role of Intracellular Galectins in Signal Transduction

Galectins and Ras Signaling

Ras proteins (H-Ras, K-Ras, N-Ras), which are members of a large superfamily of small GTPases, function as core signaling molecules that are activated by convergent signaling pathways after extracellular stimuli. Activated Ras, in turn, regulates a diversity of downstream cytoplasmic signaling cascades and can induce different biological consequences such as cell proliferation, senescence, survival, or death, depending on the cellular context (Mitin *et al.*, 2005).

Galectin-1 is the first reported galectin protein that can directly bind to H-Ras(G12V), which is the constitutively active GTP-bound H-Ras mutant and regulate Ras signaling as an escort protein (Paz *et al.*, 2001). Overexpression of galectin-1 in 293T cells increased membrane-associated Ras, Ras-GTP, and active ERK, resulting in cellular transformation. Subsequently, more detailed Ras signaling modified by galectin-1 was revealed. It was shown that H-Ras/galectin-1 or K-Ras4B/galectin-1 co-transfectants exhibit enhanced and prolonged EGF-stimulated increases in Ras-GTP, Raf-1 activity, and active ERK compared with just Ras transfectants. Conversely, Ras and galectin-1 co-transfection inhibited the EGF-stimulated increase in PI3K activity, but not Raf-1 activity. Thus, galectin-1 binds active Ras and diverts its signal to Raf-1 at the expense of PI3K (Elad-Sfadia *et al.*, 2002). To support the previous phenomena, analysis of Ras localization in microdomains at the plasma membrane revealed that galectin-1 stabilized the association of activated H-ras with non-raft microdomains, in which activated Ras proteins were predominantly distributed (Prior *et al.*, 2003). On the other hand, galectin-3 acts as a selective binding partner of activated K-Ras. Co-transfectants of K-Ras/galectin-3 in HEK-293 cells, but not of H-Ras/galectin-3, exhibited enhanced and prolonged EGF-stimulated increases in Ras-GTP, Raf-1 activity, and PI3K activity. However, contrary to galectin-1, ERK activity was attenuated and PI3K activity was augmented in K-Ras/galectin-3 co-transfectants (Elad-Sfadia *et al.*, 2004). In a more recent study, it was demonstrated that overexpression of galectin-3 in BT-549 human breast cancer cells coincided with a significant increase in K-Ras-GTP coupled with loss in N-Ras-GTP, whereas the nononcogenic galectin-3 mutant proteins (S6E and G182A) failed to induce the Ras isoform switch. Only wild-type galectin-3 protein could activate K-Ras with radical alterations in the Ras signaling pathway, whereby the Akt and Ral activities were suppressed and shifted to ERK activity (Shalom-Feuerstein *et al.*, 2005).

Taken together, specific galectins are involved in controlling the duration and selectivity of the specific Ras signal by direct binding independent of CRD.

Galectins and TNF-Related Signaling

TNF-related signaling, which is activated by death receptors of the tumor necrosis factor (TNF) family such as Fas, TNFR1, or the TNF-related apoptosis-inducing ligand (TRAIL) receptors DR4 and DR5, is generally implicated in apoptosis pathways in diverse diseases. Binding of death factors to their receptors can recruit the adaptor protein, Fas-associated death domain (FADD) in the cytoplasm, which in turn activate the cascade of specific caspases and induce apoptosis (Thorburn, 2004).

Galectin-3 has been shown to have anti-apoptotic activity in several cell types against distinct apoptotic stimuli including TNF-related molecules. Recently, it was reported that overexpression of galectin-3 in J82 human bladder carcinoma cells rendered them resistant to TRAIL-induced apoptosis. This anti-apoptotic effect in galectin-3–transfected J82 cells was resulted from constitutively activated Akt, which is the key downstream signaling for TRAIL-induced apoptosis (Oka et al., 2005). On the other hand, galectin-3-transfected BT549 human breast carcinoma cells were more sensitive to TRAIL-induced apoptosis compared with control vector-transfected cells. In these cells, galectin-3 overexpression inactivated Akt by dephosphorylation (Lee et al., 2003). The mechanism of these opposite effects of galectin-3 against TRAIL has not yet been determined, and the changes of upstream signaling of Akt should be explained. Meanwhile, galectin-3 can also regulate the other TNF-related signaling, CD95 (APO-1/Fas) apoptotic signaling, by binding to CD95 in T-cells. There are two primary CD95 apoptotic signaling pathways, one regulated by the large amount of active caspase-8 (type I) formed at the death-inducing signaling complex and the other by the apoptogenic activity of mitochondria (type II). Transfection of galectin-3 in type II cells (galectin-3 null cells) resulted in converting them into type I apoptotic cells, in which galectin-3 is expressed (Fukumori et al., 2004).

Galectins and NF-κB Signaling

NF-κB (nuclear factor-κB) belongs to a family of heterodimeric transcription factors that play a key role in inflammatory and stress responses as well as in tumor cell resistance to apoptosis (Delhalle et al., 2004). Several issues have been reported about interactions between galectins and NF-κB signaling, but almost all were related to changes of galectin expression through NF-κB signaling. For example, the expression of galectin-3 in human T lymphotropic virus-1 (HTLV-1) infected T-cells was increased by the galectin-3 promoter through interaction with NF-κB (Hsu et al., 1996). Inhibition of NF-κB by specific proteasomal inhibitor attenuated the induction of galectin-3 in glioblastoma cells (Dumic et al., 2000). In addition, down-regulation of NF-κB by nuclear protein, one of the apoptosis-related molecules, decreased the expression level of galectin-3 mRNA/protein (Liu et al., 2004). Conversely, thioglycollate elicited inflammatory cells derived from galectin-3–deficient mice exhibited significantly lower levels of NF-κB response compared with cells from wild-type mice (Hsu *et al.*, 2000).

Galectins and Wnt Signaling

Wnt/Wingless signaling transduction pathway plays an important role in both embryonic development and tumorigenesis. Basically, β-catenin can form complex with the TCF/LEF family of transcription factors and activates transcription of Wnt target genes (Akiyama, 2000). It has been reported that galectin-3 might be involved in this signaling through direct binding to both β-catenin and Axin, regulator proteins of Wnt signaling (Shimura *et al.*, 2004, 2005). Analysis of the human galectin-3 sequence reveled a structural similarity to β-catenin because it also contains the consensus sequence for glycogen synthase kinase-3β (GSK-3β) phosphorylation, and galectin-3 can serve as its substrate (Shimura *et al.*, 2005).

Galectin-12 and Adipogenic Signaling

Galectin-12 is preferentially expressed in peripheral blood leukocytes and adipocytes. It was reported that this protein is required for specific signal transduction in mouse preadipocyte that conveys hormone stimulation to the induction of adipogenic factors essential for adipocyte differentiation. Down-regulation of endogenous galectin-12 expression by RNA interference greatly reduced the expression of the adipogenic transcription factors such as CCAAT/enhancer-binding protein (C/EBP)-β, α, and peroxisome proliferator-activated receptor (PPAR)-γ, and severely suppressed adipocyte differentiation as a result of defective adipogenic signaling including phosphorylation of Akt, ERK, and cAMP response element-binding protein (CREB) (Yang *et al.*, 2004).

Galectin-3 and Phosphorylation

Signal transduction described previously is generally performed by phosphorylation of tyrosine, serine, or threonine residue of the key phosphoproteins in a specific signaling cascade. It has been revealed that galectin-3 is also a phosphoprotein, and this phosphorylation status is critical for the activity of galectin-3 in its biological functions such as cell-cycle regulation, anti-apoptotic activity, or even K-Ras signaling (Shalom-Feuerstein *et al.*, 2005; Yoshii *et al.*, 2002). Therefore, phosphorylation of galectin-3 might be a key signaling factor associated with several signal transductions. In addition, Ser[6] in galectin-3 is phosphorylated with casein kinase I and in turn dephosphorylated with protein phosphatase type 1, and this phosphorylation status can modulate its protein-carbohydrate interactions (Mazurek *et al.*, 2000).

Construction of Expression Vectors

For Extracellular Galectins

To assess the signal transduction changes by extracellular galectins, recombinant galectins are required. Recombinant galectins can be prepared in bacterial expression systems. One such method for expression of recombinant galectin-3 uses the GST gene fusion system (Amersham). The following is based on a previous report (Fukumori *et al.*, 2003). The plasmid encoding GST-galectin-3 is created by subcloning pGEM(7+) vector (Promega) containing human galectin-3 cDNA (Gong *et al.*, 1999) using the restriction enzyme *Eco*R I into the pGEX-6p-3 (Amersham), followed by confirmation of sense direction and appropriate orientation by DNA sequencing.

For Intracellular Galectins

To assess signal transduction changes inside the cell regarding galectins, an overexpression system is used by transfection of galectin cDNA into mammalian cells. In addition, analysis of mutant galectins that can exert certain dominant negative effects is helpful for a more comprehensive understanding. A previous report introduces the procedure for creating the DNA constructs for expression of wild-type and mutant galectin-3 (S6A and S6E) in mammalian cells (Yoshii *et al.*, 2002). Wild-type galectin-3 cDNA was excised from pGEM(7+)/galectin-3 (Gong *et al.*, 1999) with *Eco*RI and inserted into a mammalian expression vector pBK-CMV (Stratagene) at the *Eco*RI site in the sense direction. The proper orientation of the cDNA insert was confirmed by the analysis of restriction enzyme, *Sma*I or *Nco*I. Single mutation within galectin-3 is generated using a Quick Change mutagenesis kit (Stratagene) according to the manufacture's protocol. Ser[6] of galectin-3 is mutated to Ala (S6A) using a sense primer, 5'-CAATTTTGCGCTCCATGAT-3' and an antisense primer, 5'-ATCATGGAGCGCAAAATTG-3', and to Glu (S6E) using a sense primer, 5'-ATGGCAGACAATTTTGACCT-3' and antisense primer, 5'-AGCTCAAAATTGTCTGCCAT-3'. For PCR, pBK-CMV vector containing wild-type galectin-3 cDNA is used as the template. The veracity of the constructs must be verified by DNA sequencing.

Binding Assays for Extracellular Galectins

Preparation of Recombinant Galectin-3

For large-scale preparation, *Escherichia coli* strain BL21(DE3) transformed with expression vector pGEX-6p-3/galectin-3 is incubated overnight in 80 ml of LB medium containing 100 μg/ml ampicillin. On the next day, 20 ml of the cultured bacteria is added into prewarmed 230 ml of LB and then grown for 1–2 h until A_{600nm} reads 0.6. Thereafter, isopropyl-thio-β-D-galactoside (IPTG) is added to a final concentration of 0.1 mM and bacterial cells are incubated for an additional 5 h. Harvested by centrifugation at 1250g at 4°, the cell pellet is suspended in 12.5 ml of phosphate-buffered saline (PBS) containing 1% Triton X-100 and protease inhibitors (0.25 IU/ml aprotinin, 1 μg/ml leupeptin, and 1 μg/ml pepstatin), and lysed by sonication. Lysate is centrifuged at 12,000g for 30 min, and supernatant is incubated with 1 ml of 50% slurry of glutathione-Sepharose 4B (Amersham) equilibrated with PBS for 10 min. After 2–3 h incubation, glutathione-Sepharose is washed with 15 ml of PBS for three to five times. To cut between GST and galectin-3 proteins, the buffer is changed to the cleavage buffer (50 mM Tris-HCl [pH 7.0], 150 mM NaCl, 1mM EDTA, and 1 mM DTT) (480 μl) and incubated with 20 μl of Precision Protease (Amersham) for 5 h in a rotator at 4°. Eluted solution containing wild-type galectin-3 is dialyzed with PBS overnight at 4°, and the recombinant proteins are quantitated using Bio-Rad protein assay reagent (Bio-Rad) and stored at –80° until use.

Cell Surface Binding Assay

The cell surface–binding assay using flow cytometry is useful to confirm that extracellular galectins can bind to some glycoproteins on the cell surface. The following protocol is for the binding of exogenous galectin-3 to the cell surface of T cells. A total of 2×10^5 cells are incubated with 10 μM of recombinant galectin-3 in the presence or absence of 50 mM lactose (the inhibitor for galectin-3 binding through CRD) in a total volume of 200 μl for 2 h at 37°. The cells are washed three times with PBS and incubated with 1:2 diluted anti-galectin-3 mAb (TIB166; ATCC) for 45 min at 4°, followed by 1:200 diluted anti-rat FITC-conjugated secondary Ab for 45 min at 4°. After washing three times with PBS, the stained cells were fixed with 1% formaldehyde in PBS and analyzed for fluorescent intensity using flow cytometry as described previously (Pace *et al.*, 2000).

Binding Analysis of Extracellular Galectins to Specific Ligands

Galectin binding to specific cell surface receptors is detected using immunoprecipitation technique followed by immunoblot analysis. To avoid deviating from the binding partners, cross-linking is generally used before immunoprecipitation. In particular, cell surface proteins are cross-linked using the thiol-cleavable cross-linker reagent, DTSSP (Pierce) at a final concentration of 1 mM in PBS containing 10 mM HEPES (pH 7.4) at 4° for 1 h. The reaction was terminated by the addition of 1 M Tris-HCl (pH 7.4) at a final concentration of 10 mM. After standing for 15 min at 4°, the reaction mixture was aspirated, and the cells were washed three times with ice-cold PBS. Then, cells are lysed in lysis buffer (10 mM Tris-HCl [pH 8.0], 150 mM NaCl, 1% Triton X-100, 0.1% sodium dodecyl sulfate [SDS], 1% sodium deoxycholate, 1 mM EDTA, 1 mM phenylmethylsulfonyl fluoride [PMSF], 10 μg/ml leupeptin, 10 μg/ml aprotinin), and the suspensions are incubated at 4° for 20 min and centrifuged at 10,000g for 15 min at 4°. The supernatants are saved as cell lysates for immunoprecipitation. For immunoprecipitation of galectin-3, cell lysates are precleared by overnight incubation at 4° with 100 μl of 50 % slurry of protein A–Sepharose 6MB (Amersham). The precleared sample is divided into two fresh tubes, one is for addition of 3 μg of purified rabbit whole IgG and the other is for 3 μg of purified rabbit polyclonal anti-galectin-3 Ab (Gong *et al.*, 1999). After incubation at 4° for 3 h, 40 μl of 50% slurry of protein A–Sepharose 6MB is added to each sample and further incubation is performed for additional 1.5 h. The beads are washed five times with lysis buffer, once each with 50 mM Tris-HCl (pH 8.0), and water. SDS sample buffer containing 5% β-mercaptoethanol is added to the washed immunoprecipitates, which are then boiled for 5 min and separated by SDS-PAGE. After the transfer to PVDF membrane, it is blocked with 5% BSA in PBS for 1 h at 25°, and then reacted with properly diluted primary antibody (for galectin-3 detection, TIB166 Ab diluted with 1:500 is used.) for 1–2 h. After washing twice with enough amount of PBS containing 0.1% Tween-20, the membrane is then reacted with the appropriate secondary antibody conjugated with HRP (for TIB166 Ab, HRP-conjugated anti-rat IgG [Sigma]) for 1 h. After washing three times, protein bands reacted with the antibody are visualized by enhanced chemiluminescence (ECL) system (Amersham).

Analyses of Phosphorylated Proteins

Phosphorylation of Galectin-3

To detect the phosphorylation of galectin-3, a metabolic labeling in cultured cells, followed by immunoprecipitation of galectin-3 is used. For labeling *in vivo* with ^{32}P, 8 × 10^5 cells are grown overnight in 35-mm dishes. To deplete endogenous phosphate before labeling, the cells are washed twice with phosphate-free DMEM (Invitrogen) and incubated in phosphate-free DMEM supplemented with 5% dialyzed FBS for 2 h at 37°. The cells are then labeled with 0.2 mCi/ml ^{32}P in phosphate-free DMEM supplemented with 10% dialyzed FBS for 5 h at 37°. After two washes with ice-cold phosphate-free DMEM, the cells are lysed with 400 μl of ice-cold lysis buffer B (50 mM

Tris-HCl [pH 7.5], 120 mM NaCl, 0.5% Nonidet P-40, 2 mM EDTA, 5 μg/ml leupeptin, 10 μg/ml aprotinin, 500 μM PMSF, 200 μM sodium orthovanadate, 100 mM sodium fluoride) and incubated for 30 min at 4°. The cell lysates from two dishes (1.6×10^6 cells) are combined and then centrifuged at 12,000g for 5 min, and each supernatant is adjusted to equal radioactivity and volume. Thereafter, immunoprecipitation of galectin-3 is performed as described previously except the preclear procedure. After SDS-PAGE, galectin-3 phosphorylation state is analyzed by PhosphorImager (Molecular Dynamics, Sunnyvale, CA), and immunoprecipitated galectin-3 is detected by immunoblotting using TIB166 Ab (ATCC).

Phosphorylation of ERK, Akt

To detect the phosphorylation status of ERK and Akt, immunoblot analysis using the specific antibody is useful and commonly performed. Following the objective stimuli, cell lysates are prepared using ice-cold lysis buffer B as described previously. Equal amounts of cell lysates are subjected to SDS-PAGE, followed by immunoblotting with one of the following primary antibodies: 1:2,000 diluted anti-ERK antibody (Santa Cruz Biotechnology); 1:10,000 anti-phospho-ERK antibody (Sigma); 1:1,000 anti-Akt antibody (Cell Signaling); 1:1,000 anti-phospho-Akt antibody (Cell Signaling) (Shalom-Feuerstein et al., 2005).

Kinase Assays

Casein Kinase I Assay

Phosphorylation of galectin-3 at Ser^6 is catalyzed by casein kinase I, and this activity is assayed in vitro with the following protocol. The reaction mixture is prepared to include (50 mM Tris [pH 7.5], 140 mM KCl, 10 mM MgCl$_2$, 5 mM DTT, 0.2 mM [γ-^{32}P]ATP to a final specific activity of 20 mCi/mmol, 500 units of rat recombinant casein kinase I [New England BioLabs], and 20–40 μg of purified recombinant galectin-3) in a final volume of 50 μl. The reaction is performed for 30 min at 30° and terminated by addition of reduced SDS sample buffer. After SDS-PAGE, the phosphorylated galectin-3 can be visualized by autoradiography (Mazurek et al., 2000).

GSK-3β Kinase Assay

To examine whether galectin-3 is phosphorylated by GSK-3β, in vitro kinase assay is available as described previously (Ikeda et al., 1998, Shimura et al., 2005). The reaction mixture is prepared to include

(50 mM Tris-HCl [pH 7.5], 10 mM MgCl2, 1 mM DTT and 50 mM [γ-^{32}P] ATP [500–2000 cpm/pmol], 20–80 units of recombinant GSK-3β [New England BioLabs], and 1-3 μg of purified recombinant galectin-3) in a final volume of 10 μl for 30 min at 30°. During the reaction, the presence of Axin can enhance galectin-3 phosphorylation and in turn LiCl (30 mM), a specific GSK-3β inhibitor, can inhibit it (Shimura *et al.*, 2005). The reaction is stopped by boiling with SDS sample buffer, followed by SDS-PAGE and autoradiography as described previously.

PI3K Assay

To determine the activity of PI3K in the cells, the following protocol reported previously (Pardo *et al.*, 2002) is used with minor modification. Cells are grown in 10-cm dishes to 80% confluence, and then lysed using lysis buffer C (20 mM HEPES-NaOH [pH 7.4], 150 mM NaCl, 1% Triton X-100, 2 mM EDTA, 10 mM sodium fluoride, 10% glycerol, 1 mM PMSF, 5 mM benzamidine, 1 mM N-tosyl-L-lysine chloromethylketone, 20 μM leupeptin, 18 μM pepstatin, 20 μg/ml aprotinin, 1 mM DTT, 2 mM sodium orthovanadate, 10 mM glycerophosphate), followed by preparation of 1 mg protein adjusted to a volume of 500 μl. The enzyme is immunoprecipitated by incubation with rabbit anti-PI3K p85 antibody (Upstate Biotechnology), and its activity is assayed with 0.5 mg/ml phosphatidylinositol and 125 μM ATP and 5 μCi of [γ-^{32}P]ATP. Lipids are extracted with chloroform/ methanol and then separated by thin-layer chromatography (Elad-Sfadia *et al.*, 2002; Pardo *et al.*, 2002). Phospholipid markers are used for the identification of the ^{32}P-labeled phosphatidylinositol 3-phosphate product. The ^{32}P-labeled lipid products are visualized by an overnight exposure on an x-ray film.

References

Ahmad, N., Gabius, H. J., Andre, S., Kaltner, H., Sabesan, S., Roy, R., Liu, B., Macaluso, F., and Brewer, C. F. (2004). Galectin-3 precipitates as a pentamer with synthetic multivalent carbohydrates and forms heterogeneous cross-linked complexes. *J. Biol. Chem.* **279,** 10841–10847.

Akiyama, T. (2000). Wnt/beta-catenin signaling. *Cytokine Growth Factor Rev.* **11,** 273–282.

Arbel-Goren, R., Levy, Y., Ronen, D., and Zick, Y. (2005). Cyclin-dependent kinase inhibitors and JNK act as molecular switches, regulating the choice between growth arrest and apoptosis induced by galectin-8. *J. Biol. Chem.* **280,** 19105–19114.

Barondes, S. H., Castronovo, V., Cooper, D. N., Cummings, R. D., Drickamer, K., Feizi, T., Gitt, M. A., Hirabayashi, J., Hughes, C., Kasai, K., *et al.* (1994). Galectins: A family of animal beta-galactoside-binding lectins. *Cell* **76,** 597–598.

Chung, C. D., Patel, V. P., Moran, M., Lewis, L. A., and Miceli, M. C. (2000). Galectin-1 induces partial TCR zeta-chain phosphorylation and antagonizes processive TCR signal transduction. *J. Immunol.* **165,** 3722–3729.

Cooper, D. N., and Barondes, S. H. (1999). God must love galectins; he made so many of them. *Glycobiology* **9,** 979–984.

Delhalle, S., Blasius, R., Dicato, M., and Diederich, M. (2004). A beginner's guide to NF-kappaB signaling pathways. *Ann. N. Y. Acad. Sci.* **1030,** 1–13.

Demetriou, M., Granovsky, M., Quaggin, S., and Dennis, J. W. (2001). Negative regulation of T-cell activation and autoimmunity by Mgat5 N-glycosylation. *Nature* **409,** 733–739.

Dumic, J., Lauc, G., and Flogel, M. (2000). Expression of galectin-3 in cells exposed to stress-roles of jun and NF-kappaB. *Cell Physiol. Biochem.* **10,** 149–158.

Elad-Sfadia, G., Haklai, R., Ballan, E., Gabius, H. J., and Kloog, Y. (2002). Galectin-1 augments Ras activation and diverts Ras signals to Raf-1 at the expense of phosphoinositide 3-kinase. *J. Biol. Chem.* **277,** 37169–37175.

Elad-Sfadia, G., Haklai, R., Balan, E., and Kloog, Y. (2004). Galectin-3 augments K-Ras activation and triggers a Ras signal that attenuates ERK but not phosphoinositide 3-kinase activity. *J. Biol. Chem.* **279,** 34922–34930.

Fischer, C., Sanchez-Ruderisch, H., Welzel, M., Wiedenmann, B., Sakai, T., Andre, S., Gabius, H. J., Khachigian, L., Detjen, K. M., and Rosewicz, S. (2005). Galectin-1 interacts with the {alpha}5{beta}1 fibronectin receptor to restrict carcinoma cell growth via induction of p21 and p27. *J. Biol. Chem.* **280,** 37266–37277.

Fukumori, T., Takenaka, Y., Oka, N., Yoshii, T., Hogan, V., Inohara, H., Kanayama, H. O., Kim, H. R., and Raz, A. (2004). Endogenous galectin-3 determines the routing of CD95 apoptotic signaling pathways. *Cancer Res.* **64,** 3376–3379.

Fukumori, T., Takenaka, Y., Yoshii, T., Kim, H. R., Hogan, V., Inohara, H., Kagawa, S., and Raz, A. (2003). CD29 and CD7 mediate galectin-3-induced type II T-cell apoptosis. *Cancer Res.* **63,** 8302–8311.

Fukushi, J., Makagiansar, I. T., and Stallcup, W. B. (2004). NG2 proteoglycan promotes endothelial cell motility and angiogenesis via engagement of galectin-3 and alpha3beta1 integrin. *Mol. Biol. Cell* **15,** 3580–3590.

Germain, R. N., and Stefanova, I. (1999). The dynamics of T cell receptor signaling: Complex orchestration and the key roles of tempo and cooperation. *Annu. Rev. Immunol.* **17,** 467–522.

Gong, H. C., Honjo, Y., Nangia-Makker, P., Hogan, V., Mazurak, N., Bresalier, R. S., and Raz, A. (1999). The NH2 terminus of galectin-3 governs cellular compartmentalization and functions in cancer cells. *Cancer Res.* **59,** 6239–6245.

Hadari, Y. R., Arbel-Goren, R., Levy, Y., Amsterdam, A., Alon, R., Zakut, R., and Zick, Y. (2000). Galectin-8 binding to integrins inhibits cell adhesion and induces apoptosis. *J. Cell Sci.* **113** (Pt 13), 2385–2397.

Heldin, C. H., Miyazono, K., and ten Dijke, P. (1997). TGF-beta signalling from cell membrane to nucleus through SMAD proteins. *Nature* **390,** 465–471.

Hirabayashi, J., Hashidate, T., Arata, Y., Nishi, N., Nakamura, T., Hirashima, M., Urashima, T., Oka, T., Futai, M., Muller, W. E., Yagi, F., and Kasai, K. (2002). Oligosaccharide specificity of galectins: A search by frontal affinity chromatography. *Biochim. Biophys. Acta* **1572,** 232–254.

Hsu, D. K., Hammes, S. R., Kuwabara, I., Greene, W. C., and Liu, F. T. (1996). Human T lymphotropic virus-I infection of human T lymphocytes induces expression of the beta-galactoside-binding lectin, galectin-3. *Am. J. Pathol.* **148,** 1661–1670.

Hsu, D. K., Yang, R. Y., Pan, Z., Yu, L., Salomon, D. R., Fung-Leung, W. P., and Liu, F. T. (2000). Targeted disruption of the galectin-3 gene results in attenuated peritoneal inflammatory responses. *Am. J. Pathol.* **156,** 1073–1083.

Hsu, D. K., Zuberi, R. I., and Liu, F. T. (1992). Biochemical and biophysical characterization of human recombinant IgE-binding protein, an S-type animal lectin. *J. Biol. Chem.* **267,** 14167–14174.

Hughes, R. C. (1999). Secretion of the galectin family of mammalian carbohydrate-binding proteins. *Biochim. Biophys. Acta* **1473**, 172–185.

Ikeda, S., Kishida, S., Yamamoto, H., Murai, H., Koyama, S., and Kikuchi, A. (1998). Axin, a negative regulator of the Wnt signaling pathway, forms a complex with GSK-3beta and beta-catenin and promotes GSK-3beta-dependent phosphorylation of beta-catenin. *EMBO J.* **17**, 1371–1384.

Lee, Y. J., Song, Y. K., Song, J. J., Siervo-Sassi, R. R., Kim, H. R., Li, L., Spitz, D. R., Lokshin, A., and Kim, J. H. (2003). Reconstitution of galectin-3 alters glutathione content and potentiates TRAIL-induced cytotoxicity by dephosphorylation of Akt. *Exp. Cell Res.* **288**, 21–34.

Leffler, H., and Barondes, S. H. (1986). Specificity of binding of three soluble rat lung lectins to substituted and unsubstituted mammalian beta-galactosides. *J. Biol. Chem.* **261**, 10119–10126.

Levy, Y., Arbel-Goren, R., Hadari, Y. R., Eshhar, S., Ronen, D., Elhanany, E., Geiger, B., and Zick, Y. (2001). Galectin-8 functions as a matricellular modulator of cell adhesion. *J. Biol. Chem.* **276**, 31285–31295.

Levy, Y., Ronen, D., Bershadsky, A. D., and Zick, Y. (2003). Sustained induction of ERK, protein kinase B, and p70 S6 kinase regulates cell spreading and formation of F-actin microspikes upon ligation of integrins by galectin-8, a mammalian lectin. *J. Biol. Chem.* **278**, 14533–14542.

Liu, F. T., and Rabinovich, G. A. (2005). Galectins as modulators of tumour progression. *Nat. Rev. Cancer.* **5**, 29–41.

Liu, F. T., Patterson, R. J., and Wang, J. L. (2002). Intracellular functions of galectins. *Biochim. Biophys. Acta* **1572**, 263–273.

Liu, L., Sakai, T., Sano, N., and Fukui, K. (2004). Nucling mediates apoptosis by inhibiting expression of galectin-3 through interference with nuclear factor kappaB signalling. *Biochem. J.* **380**, 31–41.

Maeda, N., Kawada, N., Seki, S., Arakawa, T., Ikeda, K., Iwao, H., Okuyama, H., Hirabayashi, J., Kasai, K., and Yoshizato, K. (2003). Stimulation of proliferation of rat hepatic stellate cells by galectin-1 and galectin-3 through different intracellular signaling pathways. *J. Biol. Chem.* **278**, 18938–18944.

Mazurek, N., Conklin, J., Byrd, J. C., Raz, A., and Bresalier, R. S. (2000). Phosphorylation of the beta-galactoside-binding protein galectin-3 modulates binding to its ligands. *J. Biol. Chem.* **275**, 36311–36315.

Mitin, N., Rossman, K. L., and Der, C. J. (2005). Signaling interplay in Ras superfamily function. *Curr. Biol.* **15**, R563–R574.

Moiseeva, E. P., Williams, B., Goodall, A. H., and Samani, N. J. (2003). Galectin-1 interacts with beta-1 subunit of integrin. *Biochem. Biophys. Res. Commun.* **310**, 1010–1016.

Oka, N., Nakahara, S., Takenaka, Y., Fukumori, T., Hogan, V., Kanayama, H. O., Yanagawa, T., and Raz, A. (2005). Galectin-3 inhibits tumor necrosis factor-related apoptosis-inducing ligand-induced apoptosis by activating Akt in human bladder carcinoma cells. *Cancer Res.* **65**, 7546–7553.

Pace, K. E., Hahn, H. P., Pang, M., Nguyen, J. T., and Baum, L. G. (2000). CD7 delivers a pro-apoptotic signal during galectin-1-induced T cell death. *J. Immunol.* **165**, 2331–2334.

Pardo, O. E., Arcaro, A., Salerno, G., Raguz, S., Downward, J., and Seckl, M. J. (2002). Fibroblast growth factor-2 induces translational regulation of Bcl-XL and Bcl-2 via a MEK-dependent pathway: Correlation with resistance to etoposide-induced apoptosis. *J. Biol. Chem.* **277**, 12040–12046.

Partridge, E. A., Le Roy, C., Di Guglielmo, G. M., Pawling, J., Cheung, P., Granovsky, M., Nabi, I. R., Wrana, J. L., and Dennis, J. W. (2004). Regulation of cytokine receptors by Golgi N-glycan processing and endocytosis. *Science* **306**, 120–124.

Paz, A., Haklai, R., Elad-Sfadia, G., Ballan, E., and Kloog, Y. (2001). Galectin-1 binds oncogenic H-Ras to mediate Ras membrane anchorage and cell transformation. *Oncogene* **20**, 7486–7493.

Prior, I. A., Muncke, C., Parton, R. G., and Hancock, J. F. (2003). Direct visualization of Ras proteins in spatially distinct cell surface microdomains. *J. Cell. Biol.* **160**, 165–170.

Shalom-Feuerstein, R., Cooks, T., Raz, A., and Kloog, Y. (2005). Galectin-3 regulates a molecular switch from N-Ras to K-Ras usage in human breast carcinoma cells. *Cancer Res.* **65**, 7292–7300.

Shimura, T., Takenaka, Y., Fukumori, T., Tsutsumi, S., Okada, K., Hogan, V., Kikuchi, A., Kuwano, H., and Raz, A. (2005). Implication of galectin-3 in Wnt signaling. *Cancer Res.* **65**, 3535–3537.

Shimura, T., Takenaka, Y., Tsutsumi, S., Hogan, V., Kikuchi, A., and Raz, A. (2004). Galectin-3, a novel binding partner of beta-catenin. *Cancer Res.* **64**, 6363–6367.

Tamura, M., Gu, J., Tran, H., and Yamada, K. M. (1999). PTEN gene and integrin signaling in cancer. *J. Natl. Cancer Inst.* **91**, 1820–1828.

Thorburn, A. (2004). Death receptor-induced cell killing. *Cell Signal.* **16**, 139–144.

Vespa, G. N., Lewis, L. A., Kozak, K. R., Moran, M., Nguyen, J. T., Baum, L. G., and Miceli, M. C. (1999). Galectin-1 specifically modulates TCR signals to enhance TCR apoptosis but inhibit IL-2 production and proliferation. *J. Immunol.* **162**, 799–806.

Wang, J. L., Gray, R. M., Haudek, K. C., and Patterson, R. J. (2004). Nucleocytoplasmic lectins. *Biochim. Biophys. Acta* **1673**, 75–93.

Yang, R. Y., Hsu, D. K., Yu, L., Chen, H. Y., and Liu, F. T. (2004). Galectin-12 is required for adipogenic signaling and adipocyte differentiation. *J. Biol. Chem.* **279**, 29761–29766.

Yoshii, T., Fukumori, T., Honjo, Y., Inohara, H., Kim, H. R., and Raz, A. (2002). Galectin-3 phosphorylation is required for its anti-apoptotic function and cell cycle arrest. *J. Biol. Chem.* **277**, 6852–6857.

Section VII

Newly Developed Field

[20] *Helicobacter pylori* Adhesion to Carbohydrates

By Marina Aspholm, Awdhesh Kalia, Stefan Ruhl,
Staffan Schedin, Anna Arnqvist, Sara Lindén, Rolf Sjöström,
Markus Gerhard, Cristina Semino-Mora, Andre Dubois,
Magnus Unemo, Dan Danielsson, Susann Teneberg, Woo-Kon Lee,
Douglas E. Berg, and Thomas Borén

Abstract

Adherence of bacterial pathogens to host tissues contributes to colonization and virulence and typically involves specific interactions between bacterial proteins called adhesins and cognate oligosaccharide (glycan) or protein motifs in the host that are used as receptors. A given pathogen may have multiple adhesins, each specific for a different set of receptors and, potentially, with different roles in infection and disease. This chapter provides strategies for identifying and analyzing host glycan receptors and the bacterial adhesins that exploit them as receptors, with particular reference to adherence of the gastric pathogen *Helicobacter pylori*.

Overview

Helicobacter pylori chronically infects the gastric (stomach) mucosa of billions of people worldwide. Infections tend to last for decades once established, despite host defenses such as mucosal shedding and immune and inflammatory responses. Histological inspection of gastric biopsy specimens generally shows most *H. pylori* cells in the thin mucus layer. Mucus matrix is formed by high-molecular-mass oligomeric glycoproteins known as mucins. The mucus layer protects the underlying epithelium from gastric acidity and infection by other microbes. In addition to the many *H. pylori* cells present in the mucin layer, others adhere directly to epithelial cell surfaces, and a few may enter and proliferate within epithelial cells. *H. pylori* is motile and chemotactic for particular host metabolites, including bicarbonate, arginine, and urea, and chemotaxis-driven motility probably allows efficient migration of the pathogen to preferred gastric mucosal sites. In one view, the primary benefit of adherence for *H. pylori* is in escaping clearance by mucosal shedding and peristalsis. Alternately, or in addition, much of adherence's value may come from placing *H. pylori* at or close to epithelial surfaces, for efficient scavenging of nutrients leached from host tissue. Adherence to epithelium also promotes efficient management of host responses by delivery of toxins

METHODS IN ENZYMOLOGY, VOL. 417   0076-6879/06 $35.00
        DOI: 10.1016/S0076-6879(06)17020-2

and other signaling molecules. Therefore, tight adherence may be a mixed blessing for *H. pylori,* however, also potentially exposing it to intense, bactericidal inflammatory responses.

Two of *H. pylori*'s adhesins have been characterized in terms of receptor interactions (BabA, specific for ABO and Lewis b [Leb] antigens; and SabA, specific for sialylated Lewis × [sLex] and sialylated Lewis a [sLea] antigens). Studies of numerous clinical isolates have shown that the BabA and SabA adhesins are diverse in terms of amino acid sequence, glycan specificity, and affinity, and each is expressed by only a subset of *H. pylori* strains. The genes encoding these two adhesins are members of a large multigene family of *H. pylori* outer membrane proteins (HOPs). Some proteins are probably porins, and some members probably encode adhesins, whose activities and regulation of expression have not yet been documented. It is noteworthy in this context that *H. pylori* is extremely diverse genetically as are its gastric mucosal habitats.

Gastric mucosal diversity is illustrated by the differences among people in glycan profiles because of genetic polymorphisms in underlying glycosyl transferase genes and because of changes in a given person's glycosylation patterns caused by infection and host inflammatory response. It is in this context that one sees that *H. pylori* may well benefit from having multiple adhesins, differing in glycoprotein and glycolipid receptor specificities and affinities, and in how their expression is controlled.

Collectively, the constellation of adhesins and cognate receptors is postulated to affect how *H. pylori* resists mucosal shedding and peristalsis and how it signals target cell regulatory circuitry and captures nutrients leached from host tissue.

Characterization of Bacterial Adherence by *In Vitro* Binding to Host Tissue

The use of human biopsy materials and other preserved tissues in *in vitro* tissue binding assays allowed detection of tissue-specific adherence of *H. pylori* and other pathogens and characterization of the receptors used. Early studies showed that many *H. pylori* strains could bind fucosylated Lewis (or ABO histo blood group) antigens and the inflammation-associated sialylated antigens of the gastric epithelium (Aspholm-Hurtig *et al.*, 2004; Borén *et al.*, 1993; Mahdavi *et al.*, 2002); that uropathogenic *E. coli* bound to globoseries glycolipids, which are abundant in kidney tissue (Roberts *et al.*, 1994); and that *Streptococcus pyogenes*–bound protein antigens in human cutaneous tissue (Okada *et al.*, 1994). In typical *in vitro* host tissue binding assays, bacterial cells are cultured, labeled, and overlaid on histotissue sections. The binding patterns observed give insight into expression, localization, and spatial distribution of receptors in target tissues. Further characterization of receptors can emerge from studies using various inhibitors of adherence (e.g., simple glycans to titrate adhesins; glycosidases or to disrupt potential receptors) in this *in vitro* tissue-binding assay. A general flowchart for initial identification of carbohydrate receptor structures is presented in the early review by Falk *et al.* (1994a).

Biopsy specimens from gastroscopy examinations and tissues from gastrointestinal surgery provide valuable material for adherence studies. It is often useful to choose tissues for analysis according to patient ABO and Lewis blood group status and disease condition (e.g., gastritis, peptic ulcer, gastric cancer). Also noteworthy are gender, age, medication, immune status, level of inflammation, and other associated infections, because these may also affect the types and tissue densities of various glycans that *H. pylori* can use as receptors and, thereby, vulnerability to infection or disease. The series of studies referred to previously also illustrates that *H. pylori* clinical isolates differ in types of glycans they can use as receptors and intensity of binding, reflecting diversity in their complement of adhesin genes, as noted earlier.

Tissues can be fixed by standardized formaldehyde treatment and embedded in paraffin blocks for later microtome sectioning. Most pathology laboratories use standardized protocols and robotic fixation and embedding instruments to ensure full reproducibility. Biopsy material can also be snap-frozen and embedded in cryopreservatives for later cryosectioning. There are pros and cons with both methods: Paraffin-embedded/fixed material is usually cut in thinner sections than is cryopreserved material, and thus provides higher resolution images, and fixed material can be stored for years. However, cryosectioned materials retain more of the gastric mucin layer and conformational epitopes, and thereby allow better presentation of certain antigens for analyses of adherence or detection with monoclonal antibodies.

Bacterial cells used for initial *in vitro* tissue adherence studies are often FITC (green fluorochrome) labeled and then overlaid on tissue sections (Falk *et al.*, 1993, 1994b; Borén *et al.*, 1997; see also FITC-labeling in section 2: "Glycoprotein Array" and *8:* "Identification of Bacterial Clones"). Red fluorochromes such as TRITC or Texas-Red can be used equivalently (Van de Bovenkamp *et al.*, 2003), although TRITC tends to leach from bacterial cells and skew adherence profiles unless tissue sections are very thoroughly washed after the bacterial overlay. It is also important to take into consideration that bacterial surface labeling involves modification of basic amino acids and thus may diminish adherence by direct inactivation of an adhesin's binding domain or indirectly through steric hindrance. When this is a concern, fluorochrome-labeled bacterial cells can be tested by RIA or ELISA assays if defined receptors are available (see the next section). If no binding activity of fluorochrome-labeled *H. pylori* cells is detected, the possible influence of fluorolabeling can be tested using unlabeled cells for binding to tissue sections, followed by staining (1) with DNA intercalating agents such as acridine orange or (2) through enzymatic conversion of precursor substrate and fluorochrome-activation of bacterial cytoplasm (see *Molecular Probes*, http://www.probes.invitrogen.com), or (3) with antibodies against bacterial cells or their surface antigens.

Once *H. pylori* cells are successfully fluorochrome labeled they can be aliquoted and stored frozen at −20° for 6–12 mo. Use of aliquots from the same batch of labeled cells can contribute to consistency (e.g., when many tissues or receptors are being studied) and avoids complications from any possible growth-related variation in adhesin expression and presentation on cell surfaces. This procedure depends on intactness of bacterial cells and is compatible with most *H. pylori* strains. This said, there might be certain strains or species in which intrinsic fragility of bacterial cells or their adhesins, either naturally or during freezing and thawing, makes this protocol less suitable, such as *Neisseria gonorrhoeae* and *N. meningitidis* and various *Borrelia* species. An alternative could be to use minimal levels of DMSO as solvent for the FITC staining reagent. The *in vitro* tissue adherence assay is performed essentially as described by Falk *et al.* (1993, 1994b) and Borén *et al.* (1997), with modifications in duration and frequency of washing of the histotissue section overlaid with bacterial cells to fit bacterial affinity. Slides used to test for adherence by high-affinity interactions can be washed more extensively than those used to detect low-affinity adherence or if there are relatively few adhesin proteins per cell. Identification of optimal washing conditions may sometimes need careful titration and comparison with suitable references, such as derivative strains in which the adhesin gene of interest has been switched off in expression by phase variation or has been deleted.

Alternatively, biopsy specimens used for the *in vitro* tissue adherence assay can alternately be cultured *in vitro* for 2–3 days and used for adherence analyses (i.e., by use of the *in vitro* explant culture [IVEC] technique) (Olfat *et al.*, 2002). This is a most useful application for studies of the impact of adherence in terms of bacterial-host crosstalk, such as cytokine release and cellular signaling in host tissue. In addition, the IVEC technique also complements the use of primary cell cultures, because the cells in biopsies are most similar to the true *in vivo* conditions of the epithelium and in expression of *naive* glycosylation patterns. In addition, IVEC also ensures the integrity of the spatial distribution of cell lineages in the intact biopsy materials. However, use of human gastric mucosa depends on close collaboration with the gastrointestinal surgery departments, especially because the gastric mucosa has to be in good and healthy condition, which excludes the use of most cases of dysplastic and cancer tissue obtained from eradication surgery. Representative useful biopsy material can also sometimes be obtained from patients undergoing surgical removal of stomach tissue because of morbid obesity.

Conclusions

The *in vitro* tissue binding assay is valuable for initial characterization of microbial adherence to specific host cell lineages and receptors that the microbe exploits, and of variant strains, whether generated in the laboratory or recovered during the course of natural or experimental infection.

Glycoprotein Array for Screening and Identification of
   Adhesin Binding Properties

Various overlay techniques on immobilized glycolipids or glycoproteins had been developed on the basis of bacterial overlay on either thin-layer chromatography (Hansson *et al.*, 1985; Karlsson and Stromberg, 1987) or nitrocellulose transfers of protein extracts separated by SDS-PAGE (Prakobphol *et al.*, 1987) to identify molecules carrying carbohydrate receptors for bacterial lectin-like adhesins. Carbohydrate receptors are immobilized in these assays which facilitates detection of low-affinity binding that depends on multivalent interactions between multiple adhesins on bacterial surfaces and clustered receptors on the solid-phase substrate. Soluble receptors that bind only weakly in RIA can be analyzed more robustly using solid-phase presentation on array-membranes. However, a disadvantage of the original array methods, as in the RIA assays (described in the following section "Analyses of Binding Activity"), is a requirement for radioactive labeling. As alternatives, the overlay technique was modified by tagging bacteria with biotin (Ruhl *et al.*, 1996) or fluorescein isothiocyanate (FITC) (Walz *et al.*, 2005). The bacterial overlay technique is helpful in searching for glycoprotein receptors on eukaryotic cell surfaces (Ruhl *et al.*, 2000) and for identifying ligands for bacterial adhesins in complex body fluids (Murray *et al.*, 1992; Ruhl *et al.*, 2004). The method can also be used to determine adhesin-binding specificities using membranes that have been spotted with purified glycoproteins or neoglycoproteins carrying a defined oligosaccharide motif, detailed later. In addition, oligosaccharides on glycoproteins can be enzymatically modified once immobilized on nitrocellulose, before overlay with bacteria, which also makes it very useful for initial screening of adhesin specificities. For more detailed comparisons of binding strengths, serial dilutions of appropriate glycoconjugates can be spotted on the nitrocellulose and then used for bacterial overlay. For exact comparisons, the number and density of bound oligosaccharides *per* protein molecule must also be considered. As with *in vitro* tissue adherence to biopsy material, the glycan array method is most useful for analyzing effects of deletions or other mutations in adhesin genes (construction described later) on binding properties (Walz *et al.*, 2005).

Method for Bacterial Overlay

Materials

Fluorescein-5-isothiocyanate (FITC, Molecular Probes, Eugene, OR)

Solutions

FITC stock solution: dissolve 1 mg FITC in 100 μl DMSO. Prepare freshly before use.

Tris-buffered saline (TBS): Tris-HCl buffer 20 mM, pH 7.6, 150 mM NaCl

Blocking buffer: TBS, 5% bovine serum albumin, immunoglobulin-free, fraction V, 1 mM CaCl$_2$, 1 mM MgCl$_2$.

Wash buffer: TBS, 0.05% Tween-20, 1 mM CaCl$_2$, 1 mM MgCl$_2$

Fluorescence Labeling of H. pylori

1. *H. pylori* J99 and its isogenic derivatives with deletion mutations in the *sabA* and/or *babA* genes are grown for 24–27 h at 37° in a microaerophilic atmosphere on Wilkins-Chalgren agar (Oxoid, Wesel, Germany) containing 10% horse blood, Dent supplement (Oxoid) and 0.4 g KNO$_3$ *per* liter.

2. Harvest bacteria from plates by wiping off the colonies with a sterile cotton swab and wash bacteria twice in 20 mM phosphate-buffered saline, pH 7.2 (PBS).

3. Adjust the bacterial concentration to 10^8 cells/ml (equivalent to an optical density of 1) in PBS and label by incubation with FITC at 100 μg/ml (0.1 ml of FITC stock solution *per* 10 ml of bacterial suspension) for 5-30 min at RT. Please note that bacterial numbers estimated by optical density highly depend on the *H. pylori* strain used (a CFU [colony forming units]) analysis is recommended for calibration)

4. Recover labeled bacteria by centrifugation at 800g for 7 min (10 ml volumes), wash three times with PBS (until supernatant is free of yellow color), and resuspend in 10 ml blocking buffer.

Spotting of Glycoprotein Arrays

1. Spot dry nitrocellulose membranes (Schleicher und Schüll, Protran B85) with 1-μl volumes containing 1 μg of glycoproteins or neo-glycoproteins of choice. Human serum albumin (HSA) or bovine serum

albumin (BSA) should be included as negative (nonglycosylated) controls. Purified natural glycoproteins that can be used include fetuin (Calbiochem, Bad Soden, Germany), asialofetuin (Sigma), glycophorin A (Sigma), asialoglycophorin (Sigma), laminin (from human placenta, Sigma), transferrin (Sigma), fibronectin (from human plasma, Sigma), and lactoferrin (from human milk, Sigma).

2. Allow spots to dry before use in the overlay (membranes can be stored for at least a week at RT in a dry dust-free environment).

Pretreatment of Dot Blot Arrays (Optional)

1. For removal of terminal sialic acids, incubate membranes with 0.1 U/ml of sialidase (from *Clostridium perfringens*, type X, Sigma) in TBS containing 1% BSA (fraction V, Sigma), 1 mM CaCl$_2$, and 0.1% sodium azide at 37°. Wash three times with TBS to remove sialidase before overlay with bacteria.

2. For denaturation of spotted proteins, treat membranes with 0.1% SDS (Merck, Darmstadt, Germany) in PBS containing 50 mM beta-mercaptoethanol (Merck) for 5 min in a sealed plastic bag immersed in a cooking water bath.

3. For N-glycosidase F digestion, add 0.05 U/ml of recombinant Glyko N-glycanase from *Chryseobacterium meningosepticum* (PROzyme, San Leandro, CA) and 0.75% NP-40 (PROzyme) after prior denaturation (see earlier) and incubate overnight at 37°.

4. All enzymatic pretreatments of membranes should be performed in sealed plastic bags for purity and to save reagents.

Bacterial Overlay on Nitrocellulose Membranes

1. Block unspecific binding sites on membranes with TBS containing 5% BSA (fraction V, Sigma), 1 mM CaCl$_2$, and 1 mM MgCl$_2$ for 2 h at 4°.

2. Add fluorescence-labeled bacteria to the membranes in a final concentration of 5×10^7 organisms *per* ml of blocking buffer (membranes should be covered with at least 0.6 ml of bacterial suspension *per* cm^2 of nitrocellulose membrane).

3. Incubate overlaid membranes for 30 min at 4° without mixing to allow bacterial binding.

4. Wash three times for 5 min on a rotary shaker with TBS containing 0.05% Tween-20, 1 mM CaCl$_2$, and 1 mM MgCl$_2$ to remove unbound bacteria.

5. The fluorescence of bound bacteria can be detected by a fluorescence scanner (Typhoon 9200, GE Healthcare Biosciences, Freiburg, Germany).

Analyses of Binding Capacity Based on of RIA and Scatchard
   Affinity Assays a Nonradioactive Alternative Based on
   Fluorescent Glycoconjugates

As described previously, the *in vitro* tissue adherence analyses (Falk *et al.*, 1993) and /or glycan arrays can be used to delimit the range of possible host receptor candidates. Radio immunoanalysis (RIA) can then be used for quantitative analyses of an adhesin's or cell's binding capacity and affinity, which together make up its binding activity. Once receptor structures have been better characterized, similar or related glycan substances or conjugates can be obtained or synthesized and analyzed by RIA. The main difference between the RIA and *in vitro* tissue adherence methods is the use of soluble receptors in RIA analyses vs. immobilized "solid-phase" receptors in histo tissue sections.

Method for RIA Assay Based on ^{125}I-Labeled Receptor Conjugates

In RIA analysis, 1 ml of bacteria $OD_{600} = 0.1$ is allowed to react with 300 ng of radiolabeled glycoconjugate for 2 h at RT, and percent binding is calculated. For "strong binders," most binding sites are occupied, and the percent binding corresponds to the total bacterial binding capacity. For "weak binders", the percent binding will also reflect the affinity constant.

Solutions

   $10\times$ PBS (phosphate-buffered saline): 250 mM phosphate, 850 mM
      NaCl
   (6.81 g KH_2PO_4, 34.84 g K_2HPO_4, 49.67 g NaCl, 971 g H_2O)
   PBS-Tween: 25 mM phosphate, 85 mM NaCl, 0.05% Tween-20, pH 7.4
   (106 g $10\times$ PBS, 900 g H_2O, 0.5 ml Tween-20)
   Blocking buffer: 1% BSA in PBS-Tween-20, filter through 0.22-μm filter.
   Phosphate buffer: 50 mM phosphate, pH 7.4
   (1.66 g KH_2PO_4, 6.59 g K_2HPO_4, H_2O to 1 l)

[125]I Labeling of Glycoconjugate (Hunter and Greenwood, 1962)

To 2–10 μg of glycoconjugate diluted to 50 μl with phosphate buffer, 0.2 mCi [125]I (7.4 MBq, carrier free) is added. The reaction is started by adding 20 μl of 0.6 mg/ml Chloramine-T. After 40 sec, the reaction is stopped by adding of 100 μl of 1 mg/ml sodium meta-bisulfite. Then, 200 μl of 9.6 mg/ml KI is added, and the sample is allowed to stand for at least 10 min before the radiolabeled glycoconjugate is separated from unbound ^{125}I on a PD-10 column (GE Healthcare, Uppsala, Sweden). The column is equilibrated with 5 ml blocking buffer for 30 min, washed with PBS-Tween, and eluted with PBS-Tween, and 0.5 ml fractions are collected. The radioactivity of fractions is monitored with a GM counter and fractions that elute in the very first peak ("void fraction") are pooled and kept frozen until used. It should be used within 2 mo, because iodinated material tends to self-destruct with concomitant release of free iodide (and loss of signal). If there is need to use iodinated conjugate of higher specific activity (labeling), less conjugate (<1 µg) is added to the labeling mixture for more focused ^{125}I-labeling.

[125]I-labeled Conjugate Cocktail: 300 ng Conjugate/10 µl, 20.000 CPM/10 µl

Thirty micrograms of unlabeled "cold" conjugate and approximately 2,000,000 cpm of ^{125}I-labeled "hot" conjugate are mixed and diluted to 1 ml with blocking buffer.

RIA Analysis

Ten microliters of cocktail is mixed with 1 ml bacteria (optical density at $A_{600nm} = 0.1$). After 2 h incubation on a rocking table at RT, the bacterial cells are pelleted by a 5–15 min-centrifugation at 20,000g, and the supernatant and pellet are counted separately in a gamma scintillation counter. After subtracting the background count, the percent binding is calculated.

Applications

The RIA method is very useful for rapid tests of mutants, for scoring binding properties in collections of clinical isolates, and for quantitative analyses of variations in binding activities in response to culture conditions, such as limitation in nutrients, temperatures, and oxygen levels.

A Nonradioactive Alternative to Test for H. pylori Binding Properties in Solution: Fluorochrome-labeled Glycoconjugates

Materials

Fluorescein 5(6)-isothiocyanate (FITC, Sigma-Aldrich St Louis, MO)
HSA-Glycoconjugate (Isosep, Tullinge, Sweden)

Solutions

Carbonate buffer: 0.15 M NaCl, 0.05 M carbonate, pH 9.0

FITC Labeling of Glycoconjugates

Dissolve 0.5 mg HSA-glycoconjugate in 0.1 ml carbonate buffer. Add 0.015 mg FITC solubilized in DMSO (do not store FITC dissolved: instead use fresh solution each time). Incubate the vial protected from light for 2 h with agitation. Remove excess of FITC by washing once with 0.5 ml carbonate buffer followed by three washes in PBS in a Microcon YM-50 centrifugal filter device (Millipore, Bedford, MA). The FITC-labeled glycoconjugates can be stored at $-20°$ for >1 y. Aliquot the labeled glycoconjugate into single-use vials to avoid freeze–thaw cycles.

An alternative that provides options for use of multiple fluorochromes is to label the conjugates with the Alexa Fluor Monoclonal Antibody Labeling Kits (*Molecular Probes*, http://www.probes.invitrogen.com).

H. Pylori *Binding Assay Using Fluorescent Glycoconjugates*

Incubate 100 ul *H. pylori* ($OD_{600} = 1.0$) with 500 ng conjugate for 30 min in PBS containing 0.5% human serum albumin (HSA) and 0.05% Tween-20 in a 96-well round bottom plate. Wash three times by centrifugation ($2200g$) in 200 ul PBS containing 0.05% Tween-20 and then read fluorescence.

Several glycoconjugates and strains can be tested simultaneously, although some precaution is advised. Leakage of signal from one well to another can be avoided when using clear-walled microtiter plates by leaving an empty well between samples or better, using black-walled wells. It is also necessary that a negative control with no fluorochrome be added) because *H. pylori* itself can emit autofluorescence.

Affinity Analysis by RIA According to Scatchard

The binding of receptor conjugates is often affected by the multivalent presentation of receptors on carrier molecules, as is seen with Leb-HSA (human serum albumin) conjugates (IsoSep), which carry 15–20 Lewis b antigen (Leb)—oligosaccharides attached to each HSA molecule. This allows low-affinity-interactions to be detected by adding saturating levels of soluble receptor conjugate to the bacterial suspension. However, the results obtained with weak-binding strains can also be easily misinterpreted as "good/strong" binding to receptors. Scatchard analysis can be used for sensitive determination of binding-affinity (association constant $[K_a]$) and also total binding capacity (R_T). Scatchard analysis is performed in a manner similar to that of RIA analysis, with minor modifications: For each sample, a number of test tubes with the same concentration of bacteria are prepared, but with different concentrations (a dilution series) of conjugate. In presenting data from a full series of binding experiments, conjugate bound to bacterial cells versus conjugate free in solution is depicted on the Y-axis; and total conjugate bound is depicted on the X-axis. The Y-axis' highest value is that obtained with only low levels of added conjugate (^{125}I-labeled "hot" conjugate only, with no dilution with unlabeled "cold" carrier conjugate). In comparison, the other end of the curve, with saturating levels of receptor conjugate will eventually cross the X-axis. This value provides information about the maximal number of receptor conjugates bound to each bacterial cell and, thereby, allows the number of cognate adhesins per bacterial cell to be estimated, see Fig. 1A (Scatchard, 1949; Rosenthal, 1967).

A constant number of bacterial cells (typically range of 10^7–10^9 cells/ml) is added in 900 μl to each of eight test tubes containing approximately 10,000 cpm of ^{125}I-labeled conjugate and unlabeled conjugate (in 100 μl) with the relative concentrations 0, 4, 9, 14, 20, 30, 50, and 100. The bacterial numbers are adjusted so that the highest binding is approximately 50% (i.e., bound/free approximately $= 1$ on the Y-axis), and conjugate concentrations are selected so that the lowest binding is 10–20%. The samples are mixed and put on a rocking table until the reaction has reached equilibrium (17 h for binding *H. pylori* BabA adhesin to Leb-HSA-conjugate) and then the bacteria are pelleted by a 5–15-min centrifugation at 20,000g. The supernatant and the bacterial pellet are counted separately in a gamma scintillation counter. After correction for background cpm/counting, the bound/free (pellet/supernatant) ratio and concentration of bound conjugate (conjugate concentration * percent binding/100) is calculated for each sample and plotted as Y and X values, respectively. If all binding sites have the same affinity constant, this plot—a Scatchard plot—should produce a straight line, the slope of which is $-K_a$ and the X-intercept represents R_T (total binding capacity).

$$\text{Theory}: K_a = \frac{RL}{R * L} \quad R_T = R + RL \quad \frac{RL}{L} = K_a * R_T - K_a * RL$$

Practical Considerations

For detailed calculation of affinity, it is essential to achieve conditions of low unspecific binding. Ideally, a bacterial strain lacking all adhesins (e.g., an isogenic deletion derivative) would be the best negative reference. However, the cognate adhesin usually is not known, and thus an unrelated "nonbinding" strain is typically used as control to detect background or nonspecific binding. In the case of ABO/Leb-antigen binding, nonspecific binding is kept to a minimum using BSA/Tween blocking buffer. Alternatives could include complex protein mixtures such as serum, which also includes antibodies, or milk samples (bovine or human) that are rich in glycosylated conjugates and glycans. Typical biochemical parameters such as pH, ion strength, and temperature also need to be considered when optimizing a particular binding assay.

As described previously, bacterial suspensions should be diluted so that the highest binding is approximately 50%. If binding is much higher, most

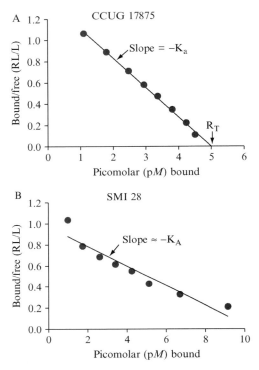

FIG. 1. Scatchard affinity analyses according to Scatchard, essentially as described previously (Rosenthal, 1967; Scatchard, 1949). The equilibrium of glycoconjugate bound to bacterial cells versus conjugate free in solution (bound/free, i.e. pellet/supernatant) is depicted on the Y-axis, whereas total glycoconjugate bound is depicted on the X-axis. The maximal Y-axis value is achieved with only low levels of conjugate. The negative value of the slope derived by linear regression provides fairly good estimation of average K_A. Thus, in such a Scatchard plot, the slope of the straight line is $-K_a$ (i.e., the affinity constant). For bacterial strains with strong binding properties, high (saturating) levels of glycoconjugate make the line graph cross the X-axis and, by so doing, provides information about the total binding capacity of the glycoconjugate tested (R_T). Thus, R_T also tells us about the number of cognate adhesins per bacterial cell. (A) The affinity analysis is illustrated with *H. pylori* strain CCUG 17875 which produces an almost ideal linear slope, with an affinity (K_a) of $2.7 \times 10^{11} M^{-1}$. In contrast, *H. pylori* clinical isolate SMI 28 produces a curved slope, with steeper slope and thus higher K_a during limiting conditions, followed by a more relaxed slope when the receptor glycoconjugate is added in excess (B). This gradual shift in K_a likely reflects heterogeneities in the adhesin complexes alternatively a mix of adhesin activities (proteins) that differ in binding properties and affinities for the receptor. Please note that in a curved slope, the total binding capacity of glycoconjugate (R_T) cannot be deduced from the point where the line eventually crosses the X-axis but needs to be estimated by approximation (on the basis of the slope in the first part of the graph). Alternately, the total binding capacity can be estimated more accurately by performing the RIA binding assay with receptor conjugate added in excess (i.e., RIA during saturating conditions). Fortunately, most BabA adhesins demonstrate high-affinity binding properties that allow for estimations of full binding capacity for interpretation of the corresponding numbers of bacterial adhesin molecules.

receptor conjugates are bound. This includes conjugates of lower multivalency, those damaged by iodination and/or irradiation, or contaminants from receptor purification or synthesis.

Despite these considerations, Scatchard diagrams are sometimes concave, with a steeper curve, indicating stronger affinity at low than at high receptor concentrations (see Fig. 1B). One likely explanation invokes two distinct adhesins that differ in affinity for the receptor used. An alternative explanation invokes heterogeneity in adhesin structure (e.g., if the adhesin consists of a supramolecular complex that also includes a variable number of regulatory or modulatory subunits).

ELISA Analysis of *H. pylori* Binding to Immobilized
  Lewis b Glycoconjugate

Similar to assays using fluorescent glycoconjugates (described previously), ELISA can be used as an alternative to RIA for characterizing clinical strains, if routine use of ^{125}I-labeling is not favored. The difference being that both RIA assays and assays using fluorescent glycoconjugates are based on the use of glycoconjugates free in solution, whereas ELISA assays are based on use of immobilized (solid-phase presented) glycoconjugates. For ELISA assays, freshly isolated bacterial strains are labeled with digoxigenin before testing them for adherence to Lewis b antigen (Leb)–coated microtiter plates. Covalently bound bacteria can then be detected using an anti-DIG-HRP antibody by measuring extinction in a microplate reader. This method allows simultaneous examination of many strains, and yields highly reproducible results.

ELISA Methods

  Materials

  Digoxigenin-3-0-succinyl-aminocaproic acid-*N*-hydroxy-succinimide ester (DIG-NHS) (Roche Diagnostics, Mannheim, Germany).

  Solutions

  Carbonate buffer: mix 10 ml 0.2 *M* Na_2CO_3 with 90 ml 0.2 *M*
    $NaHCO_3$: pH 9.2
  PBST: PBS + 0.5% Tween-20
  Blocking buffer: PBST + 0.5% non-fat dry milk

  SURFACE LABELING OF *H. PYLORI* BACTERIAL CELLS

  1. Grow clinical isolates for 24–48 h, harvest from agar plates with a cotton wool stick, and resuspend in 1 ml carbonate buffer.

2. Wash 2× in carbonate buffer, collect bacteria by centrifugation at 5000 rpm, 5 min. Surface label bacteria by incubation with 100 μg/ml Dig-NHS (freshly prepared in DMSO) in 1 ml carbonate buffer for 10–60 min at room temperature (RT) in the dark. Labeled bacteria are washed twice in PBST and collected by centrifugation (s.a.), dilute in PBST to a density of 1 OD A_{600} and freeze at $-20°$ in 100-μl aliquots until use.

ELISA BINDING ASSAY. For the binding assay we used 96-well Universal Covalent microtiter-plates (Corning Costar, Cambridge, MA), which contain photoactivatable linker used to covalently immobilize biomolecules by abstractable hydrogen using UV illumination resulting in a carbon–carbon bond. These plates are coated with human serum albumin (HSA) conjugated Leb or Lea (obtained from IsoSep AB, Tullinge, Sweden).

1. Dilute antigens at 50 ng/well in 0.2 M carbonate buffer to 1 ng/μl, add 50 μl of the solution (or 50 μl of buffer for controls) to each well.
2. Incubate for 1 h in the dark and remove liquid by carefully pipetting of the supernatants.
3. Expose plates to UV light for 30 sec in a Stratalinker (Stratagene, Germany) to immobilize the glycoproteins.
4. Block remaining binding sites on the plates by adding 100 μl blocking buffer to each well. After incubation for 1 h at RT, decant plates without washing.
5. Add 50 μl of bacterial suspension (diluted 1:1 in blocking buffer, including 10% FCS) and incubate for 1 h at RT in the dark with gentle agitation (100 rpm) to reduce nonspecific binding.
6. Remove unbound bacteria by vacuum aspiration using yellow micropipet tips.
7. Carefully wash each well three times with 200 μl PBS.
8. For detection, dilute anti-DIG-HRP-antibody 1:7000 in 2% BSA/ 50 mM TRIS.
9. Incubate for 30 min at RT and wash as above.
10. Add 50 μl ABTS-solution (Roche Diagnostics) to each well and incubate at 37° in the dark. After 15–30 min (depending on staining intensity versus the controls), extinction can be quantified by a microplate reader (Bio-Rad, Munich, Germany) at 405 nm and normalized to controls (uncoated wells).

Typically, all strains are tested in two Leb-HSA– or Lea-HSA–coated wells and two control (HSA-(nonglycosylated) coated wells). The extinction ratio is calculated from the mean of antigen coated/control values. Strains were considered positive if the ratio of $Ex_{Lewis}/Ex_{control}$ is >1.5.

This method was used to survey *H. pylori* strains from a German population for binding to fucosylated blood group antigens such as Leb (Gerhard *et al.*, 1999). Most strains were readily assigned to either a "binder" or "nonbinder" group, and this classification correlated well with *babA* genotyping. "False-positive" binding (i.e., strains that lacked a functional *babA* gene) was not observed. However, some strains showed no binding to immobilized Leb despite presence of an apparently functional *babA* gene, which suggested allelic variation in binding specificity or mechanisms of gene regulation. In a further analysis with more geographically diverse strain set (Olfat *et al.*, 2005), fewer of the strains from southern than from northern Europe exhibited strong Leb-binding. These findings indicate that BabA expression can be regulated in individual strains.

Receptor Activity–directed Affinity Tagging (Retagging) Technique for Adhesin Protein Identification

Receptor activity–directed affinity tagging (retagging) is particularly valuable for selective adhesin labeling when cognate glycan receptors are available in conjugate form. For example, to identify carbohydrate-binding adhesins, the retagging technique can use albumin glycoconjugates with multivalently linked glycans. Retagging technique was developed and implemented for identification of *H. pylori*'s blood group antigen binding adhesin (BabA) (Ilver *et al.*, 1998) and further refined for identification of its sialic acid binding adhesin (SabA) (Mahdavi *et al.*, 2002). The tri(multi)-functional Sulfo-SBED cross-linker was used for specific biotin (Re)tagging of the BabA and SabA adhesins. This cross-linker contains three reactive groups, an amine-reactive sulfo-NHS-ester, a photo reactive phenyl azide group, and a biotin moiety, each on a separate arm of the structure. The receptor glyco (albumin) conjugate is first labeled through Sulfo-SBED's sulfo-NHS-ester, which reacts with primary amines in the albumin molecule (procedure detailed below). Bacterial cells are then mixed with the cross-linker labeled glycoconjugate to allow the bacterial adhesins to bind the glycan conjugate receptors, which brings the cross-linker in close proximity to the adhesin. The cross-linker is then activated by UV irradiation to photo cross-link the glycoconjugate and adhesin protein. A disulfide bond in the cross-linker's Sulfo-NHS-ester arm is cleaved with DTT or other reducing agents, a reaction that also removes cross-linker containing glycoconjugate but leaves a biotin-tag on the *H. pylori* cells, in particular their adhesins. The biotin group then provides a handle for streptavidin binding, applicable both for visualization and purification of adhesin protein (Fig. 2).

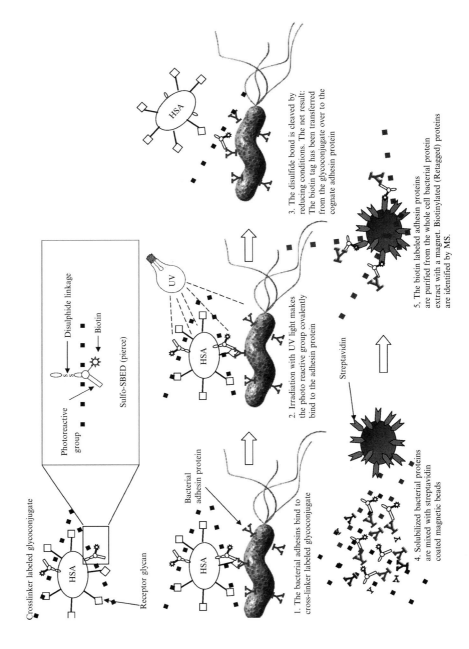

Crosslinker labeled glycoconjugate

Photoreactive group

Sulfo-SBED (pierce)

Disulphide linkage

Biotin

Receptor glycan

Bacterial adhesin protein

HSA

UV

Streptavidin

1. The bacterial adhesins bind to cross-linker labeled glycoconjugate

2. Irradiation with UV light makes the photo reactive group covalently bind to the adhesin protein

3. The disulfide bond is cleaved by reducing conditions. The net result: The biotin tag has been transferred from the glycoconjugate over to the cognate adhesin protein

4. Solubilized bacterial proteins are mixed with streptavidin coated magnetic beads

5. The biotin labeled adhesin proteins are purified from the whole cell bacterial protein extract with a magnet. Biotinylated (Retagged) proteins are identified by MS.

Preparation of Cross-linker–Labeled Glycoconjugate

Solutions

Phosphate-buffered saline (PBS): 150 m*M* NaCl, 10 m*M* phosphate, pH 7.6.
Wash buffer: PBS + 0.05% Tween-20
Tris-buffered saline (TBS): 150 m*M* NaCl, 10 m*M* Tris-HCl, pH 7.4
TBST: TBS + 0.05% Tween-20

Materials

Dimethylsulfoxide (DMSO)
Bovine serum albumin (BSA)
Glycoconjugate: Many glycoconjugates is available from Isosep AB, Tullinge, Sweden.
Sulfo-SBED cross-linker (Pierce, Rockville, IL)
PD10 column (Amersham Pharmacia Biotech, Sweden)
Horseradish peroxidase (HRP)–streptavidin

Methods

The cross-linker is light sensitive, and the following steps should be carried out in reduced-light conditions.

1. Dissolve 1 mg of cross-linker in 100 μl DMSO (i.e., 10 μg/μl).
2. Apply 20 μg (2 μl) cross-linker solution per 100 μg of glycoconjugate in 1 ml PBS.
3. Incubate the reaction on a rotary table for 1 h at RT (cover sample with aluminum foil to protect it from light). In the meantime, prepare a PD10

FIG. 2. Retagging and adhesin purification, essentially as described (Ilver *et al.*, 1998; Mahdavi *et al.*, 2002). The multifunctional Sulfo-SBED cross-linker is chemically attached to the protein core of the cognate receptor glycoconjugate by the NHS group. The cross-linker–labeled glycoconjugate is next mixed with *H. pylori* and the bacterial adhesin proteins bind the glycoconjugate (left panel). The bacterial cells with bound cross-linker–tagged glycoconjugate are subjected to UV irradiation (mid panel), and the photoreactive azide group forms a covalent bond to structures in the immediate surrounding area (usually the adhesin protein). By addition of reducing conditions, the disulfide bond located in the cross-linker structure is cleaved, and the freed glycoconjugate can then usually be washed away (right panel). The retagged bacterial cells are fully solubilized with a harsh detergent to extract the adhesins, and retagged adhesin proteins are purified by adsorption to streptavidin-coated magnetic beads. The isolated adhesin protein is next subjected to MS analyses and identified by its unique peptide composition.

column by washing it with 20 ml of wash buffer. Incubate it for 1 h with PBS-0.05% Tween-20 + 1% BSA and then wash it again with 20 ml of wash buffer.

4. Pass the glycoconjugate-Sulfo-SBED reaction through the PD-10 column to remove any excess cross-linker. Elute 0.5 ml flow-through fractions with wash buffer and use 25 μl from each fraction for analysis of protein content.

5. Pool fractions with the highest protein content (1.5–2 ml in total) and store them protected from light at $-20°$ until use.

6. To assay success of cross-linker labeling, electrophorese 1 μl of cross-linker–labeled glycoconjugate in a sodium dodecyl sulfate polyacrylamide gel (SDS-PAGE) (7.5% gel) and blot proteins to a PVDF membrane (Bio-Rad, Hercules, CA). The PVDF membrane is blocked with 5% BSA in TBST over night at 4° and then incubated for 1 h with HRP-streptavidin diluted 1:10.000 in TBST + 1% BSA. After washing six times with TBST, the biotin-labeled glycoconjugate is visualized by enhanced chemiluminescence using ECL reagents (Amersham Biosciences, Uppsala, Sweden) according to manufacturer's instructions. The cross-linker–labeled conjugate is positive on the HRP-streptavidin immunoblot and has also increased somewhat in molecular mass, because of the SULFO-SBED substitutions.

Retagging Solutions

Wash buffer: PBS-0.05% Tween-20, protease inhibitors (1 mM EDTA and 10 mM benzamidine)

Reducing buffer: PBS-0.05% + Tween-20 + protease inhibitors (1 mM EDTA and 10 mM benzamidine) and 50 mM dithiothreitol (DTT)

Laemmli buffer containing 5% beta-mercaptoethanol but without bromophenol blue staining

Materials

Sulfo-SBED cross-linker labeled glycoconjugate prepared as described previously.

UV-lamp, model UVL-56 (366 nm), Upland, CA.

Method. The following protocol describes conditions used for photo-cross-linking and transfer of biotin tags to the *H. pylori* BabA adhesin. For use of retagging as an effective method for labeling other adhesin proteins, conditions such as the amount of glycoconjugate and incubation time may need to be optimized. This is illustrated by listing protocol modifications that led to successful isolation of the SabA adhesin, whose affinity is generally weaker than that of BabA for their cognate receptors. An albumin conjugate containing an oligosaccharide that is not bound by the bacterial adhesin can serve as a negative control in this retagging procedure, although a better negative control would be a glycoconjugate that is bound by a different type of adhesin.

1. Mix bacteria (A600 nm = OD 1.0) with cross-linker–labeled glycoconjugate and incubate the reaction on a rotary table for 2–4 h. For labeling the BabA adhesin, approximately 1 μg (about 80 μl) of cross-linker labeled glycoconjugate was used per ml bacteria A_{600} nm = OD 1.0.

2. Wash bacterial cells twice with wash buffer to remove unbound glycoconjugate.

3. Resuspend bacterial cells in wash buffer in a volume that covers the surface of a regular petri dish or other tray to be used for UV irradiation (300–366 nm) of the sample. A high surface/volume ratio increases cross-linking efficiency but also allows more fluid evaporation during UV irradiation.

4. Place the tray with the bacterial suspension 5–10 cm from the UV lamp. Irradiate the sample for up to ~15 h (over night). Titration of UV dose may be useful to optimize extent of cross-linking. Incubate in RT or in the cold room, depending on intensity of UV irradiation and heat output.

5. Wash bacterial cells twice for 5 min with reducing buffer to break the cross-linker's disulfide bond, leaving the biotin moiety on the adhesin.

6. Solubilize bacterial cells in Laemmli buffer, containing 5% beta-mercaptoethanol and boil the sample for 5 min.

7. Analyze a small sample of the crude cell extract by SDS-PAGE, blot to a PVDF membrane, and visualize the biotinylated proteins by probing the membrane with streptavidin-HRP, as described previously.

Materials for Magnetic-bead Enrichment of Biotin-retagged Adhesin

Magnetic Tube Holder

Streptavidin-coated magnetic beads (QIAGEN, Hilden, Germany)

Solutions

Wash buffer: PBS-0.05% Tween-20 + protease inhibitors (1 mM EDTA and 10 mM benzamidine)

Laemmli sample buffer with 5% beta-mercaptoethanol but without bromophenol blue staining, because the color makes it difficult to visualize the beads in the buffer.

Method for Magnetic-bead Enrichment of Biotin-labeled Adhesin

1. Dilute the bacterial crude cell extract (prepared as described earlier) 1:20 with wash buffer to lower the SDS concentration

2. Add streptavidin-coated magnetic beads that have been pretreated with wash buffer to the cell extract, and incubate for 3–4 h at 4° on a rotary table. Invert the tube at 30-min intervals to better disperse sedimented beads.
3. Separate magnetic beads from the bacterial cell extract using a dedicated magnetic tube holder, and wash the beads at least twice with wash buffer.
4. Resuspend the magnetic beads in 40–100 μl of Laemmli sample buffer containing 5% beta-mercaptoethanol, mix the beads gently, and boil them for 5 min to elute bound proteins. Separate the magnetic beads from the protein extract by using the magnetic tube holder. Repeat the procedure once (or several times), and pool the eluted fractions.
5. Separate a small fraction of the eluted proteins by SDS-PAGE, Western-transfer them to a membrane, and visualize the biotinylated proteins with streptavidin-HRP, as described earlier.

Identification of Retagged (Biotinylated) Adhesin Protein

Mass-spectrometry (MS) analysis is used for definite identification of the retagged protein. We use SDS-PAGE gels to separate the protein extract that has been eluted from the magnetic beads. The gels are stained with Coomassie Blue (avoid silver stain, because of lowered efficiency in subsequent MS analyses). With a clean razor blade, excise the band of interest that corresponds to the molecular mass detected with streptavidin-HRP, as described previously. In case your retagging is still not efficient enough to generate material for MS analysis, there is still the alternative to scale up the entire preparative procedure. The SDS gel separation can then be replaced with preparative SDS gel fractionation by the Prep-Cell system (Bio-Rad), which allows for large-scale preparative electrophoresis.

The band is normally digested with Trypsin (seq grade, Promega, USA) by the MS core facility, and peptides are identified by mass spectrometry on the basis of peptide masses and sequences. MALDI-TOF-MS on a Tof-Spec E mass spectrometer and ProFound (http://www.proteometrics.com) are used to match peptide masses to proteins in the NCBI database. Peptide identities can be validated by ESI-MS/MS sequencing on a Q-Tof instrument, using the nanospray source. The Mascot program (http://www.matrixscience.com) is recommended for identification of peptide sequences.

Practical Considerations of Retagging such as Modifications for SabA

For successful identification of adhesin proteins by retagging, proper choice of receptor glycoconjugates is critical, based both on high binding affinity and binding specificity. Thus, for efficient retagging, most receptor conjugates need to be tightly bound to the cognate adhesin protein. In addition, the conjugates should bind preferentially to only one type of adhesin protein. That is, well-targeted binding specificity is essential to minimize complications from unspecific and promiscuous binding and retagging. In essence, the probability of successful (Re)tagging of the adhesin is much enhanced by use of highest possible binding specificity and affinity receptors. However, to compensate for low-affinity binding adhesins, 10–100-fold higher concentrations of receptor conjugates can be used. Conjugate with higher level (density) of attached glycans can be used for better efficiency in binding and retagging, because increased numbers of receptor epitopes will increase the multivalency in binding and avidity, also called as the Velcro-effect.

The retagging UV exposure time can also be extended, because the Sulfo-SBED structure has turned out pretty stable. Thus, UV exposure can sometimes be successfully extended overnight, especially in case the receptor-ligand does not form a tight complex, and targeted transfer of biotin is less efficient. As described previously, retagging and transfer of biotin can be followed by immunoblots (SDS-PAGE separated proteins transferred to a membrane and retagged biotin residues are visualized by streptavidin, as described earlier.

When MS has tentatively identified the adhesin candidate, it is highly recommended to construct deletion mutants to verify the lack of phenotypic binding properties. This is especially important when several similar, often related, proteins are pinned down by the MS analyzes. The mutants are then analyzed by the series of binding assays described in this chapter, such as *in vitro* tissue binding, *in vitro* glycan array binding, RIA, and ELISA.

Phylogenetic Methods for Detecting Adaptive Change

Adaptive molecular evolution is fundamentally important, especially in cases of pathogens that interact with host tissues. Surveys of homologous bacterial protein sequences from bacterial populations often reveal substantial amino acid sequence variation both within and between species. A key question is what mechanisms produce and maintain the changes in amino acid sequences of proteins? It is well established that evolutionarily conserved regions of a protein are functionally "critical" and, therefore, generally refractory to amino acid change, with amino acid change at particular sites in protein being driven by "positive selection." Such diversifying selection can contribute importantly to "fine-tuning" of protein function, in particular, when confronted with a new environment. Such adaptive evolution leading to a new or modified protein function is generally episodic in that it affects only few amino acids at select time points

(Gillespie, 1991). The term "adaptive evolution" in the context of the bacterial adhesins studied here refers to changes in adhesin amino acid sequence or parameters such as rate of adhesin synthesis, stability, or localization. These features were likely selected by aspects of gastric physiology that change during the course of infection or that vary among host individuals. This diversifying selection for amino acids changes at different sites is often also referred to as "diversifying selection" or as "positive selection." Diversifying selection is expected to be common for adhesins, toxins, or other proteins that may be targeted by immune responses or that interact with host cells or factors that may themselves be variable in the population or over time (as exemplified by tissue glycosylation patterns). It should be uncommon in "housekeeping" genes whose encoded proteins act only internally (e.g., catalyzing particular steps in metabolic pathways).

Within proteins subject to diversifying selection, amino acid sequences will be better conserved in some domains than in others. Conserved domains may be critical for function and/or not directly involved in host interaction or subject to immune selection. Analyses of other, less conserved domains or motifs that may sometimes involve only a few amino acids is particularly important as a novel source of insights into protein function in variable or hostile host environments, and thus into mechanisms of infection or disease. Our analyses of *BabA* evolutionary patterns illustrate approaches that may be suitable for any adhesin or other protein involved in pathogen-host interaction.

Adaptive evolution is detected by comparing synonymous (silent; d_S) and nonsynonymous (amino acid-altering; d_N) substitution rates in protein-coding DNA sequences (Nei and Kumar, 2000). The ratio, $d_N/d_S = \omega$ measures the difference between the two rates and provides an indication of selective pressures on a protein. Evolutionary theory predicts that if an amino acid change is functionally neutral (i.e., neither improves nor diminishes protein function), it will be fixed at the same rate as a synonymous change with $\omega = 1$. A change that is deleterious will be fixed at lower rate than a synonymous change ($\omega < 1$), and eventually purged from the population by selection against it (termed "purifying selection"). Only when an amino acid change offers a selective advantage is it fixed at a higher rate than a synonymous mutation ($\omega > 1$). Thus, an ω ratio significantly greater than 1 provides a robust indication of adaptive evolution (Nei and Kumar, 2000).

Estimation of d_N *and* d_S *between Homologous Sequences*

Methods for estimating d_N and d_S between two protein-coding DNA sequences can be broadly classified into two classes.

The first class are those referred to as "approximate methods," which involve the following steps: counting synonymous (S) and nonsynonymous (NS) sites, counting synonymous and nonsynonymous differences between the two sequences, and finally correcting for multiple substitution at the same site. Most models assume that mutations occur independently and at a constant rate and do not account for differences in the rates of transition (t_i) and transversion (t_v) base substitution changes. Synonymous mutations are most often t_is rather than t_vs. Ignoring the t_i/t_v bias can thus lead to underestimation of S and overestimation of NS (Fay and Wu, 2003). Codon usage bias, which may result from mutational bias or selection (e.g., for translational efficiency), often has the opposite affect to that seen with t_i/t_v bias: ignoring codon usage bias can lead to overestimation of S, underestimation of NS, and overestimation of ω (Fay and Wu, 2003). Finally, approximate methods average ω over all amino acid sites in the sequence and over time interval separating the two sequences. Because most amino acids are conserved and adaptive evolution is episodic, these methods lack the power to detect positive selection (Fay and Wu, 2003; Nei and Kumar, 2000).

The second class is the maximum likelihood (ML) method: the ML framework provides an explicit model of codon evolution and incorporates both t_i/t_v rate bias and codon-usage bias. Importantly, unlike approximate methods wherein ω is averaged across the entire protein sequence, the ML framework permits codons to evolve at heterogeneous rates (i.e., different rates at different sites) (Yang, 2004). For example, a fraction of codons may be constrained with $\omega < 1$, a fraction are neutral, $\omega = 1$, and a fraction are under diversifying selection wit $\omega \gg 1$.

Several codon-based models of sequence evolution have been developed for detecting heterogeneous selective pressures in protein coding sequences. These differ in the number of codon classes (codons evolving under neutral, purifying or diversifying selection) incorporated. Diversifying selection is inferred when models that incorporate a class of codons under diversifying selection fit the data better than those that do not (Yang, 2004). A protein may experience different selective pressures in different phylogenetic lineages (often equivalent to different human populations in cases of *H. pylori*). By constructing a phylogeny and inferring the ancestral states of a protein (i.e., reconstruction of protein sequences at internal nodes of the phylogeny), ω can be determined for each lineage of a phylogeny. Models can then be constructed to test whether ω varies significantly among lineages of a phylogeny (Yang, 2004). For example, a model assuming a single ω for all lineages can be compared with a model that permits an independent ω for each lineage (free-ratio model). Bayesian statistics can then be used to determine the probability for each codon being associated with diversifying, neutral, or purifying selection. Models for assessing heterogeneous selective pressures along the length of the gene or in the phylogeny are implemented in the CODEML program of PAML version 3.14 software (http://www.abacus.gene.ucl.ac.uk/software/paml.html).

Molecular Adaptation in H. pylori Adhesin Gene, babA

Analyses of receptor specificities of many *H. pylori* strains from different human populations had distinguished "specialist" BabA adhesins that bound the simple Leb antigen (characteristic of people of blood group O) far better than bulkier ALeb and BLeb (characteristic of blood group A and B, respectively), *vs.* "generalist" BabA adhesins that also bound ALeb and BLeb antigens with high affinity (Aspholm-Hurtig *et al.*, 2004). In this study, the ML method was applied to understand the origin and maintenance of specialist and generalist *babA* alleles in *H. pylori* lineages. A central ~268 amino acid segment of BabA contains the domain that is predicted to make contact with the cognate receptor. The *babA* gene sequences encoding this domain were determined from representative strains, and three specific questions were asked: (1) Are the specialist and generalist *babA* alleles phylogenetically distinct? (2) Is the central variable region of *babA* subject to variable selective pressures? (3) Do BabA variants evolve at different rates in different *H. pylori* lineages?

A *babA* phylogeny reconstructed using ML methods showed that specialist and generalist alleles did not form distinct clusters, which suggested that BabA binding specificity did not depend on large sequence differences. The average ω ranged from 04–.18, implying a relaxed selective constraint. ML estimates of tree length, ω, and average d_N/d_S were fairly homogeneous for all models under varying initialization parameters, thereby suggesting optimal convergence of ML algorithms and negligible impact of initialization parameters on model performance. Models that allowed for diversifying selection (M2, M3, and M8) fit the data better than those that did not (M01, M1, and M7); models M3 and M8 both suggested that nearly 9% of codons in the central variable region of *babA* were under diversifying selection. Model M3 models variable selective pressures using a discrete (statistical) distribution allowing for three codon site classes, each associated with its own ω. M3 parameter estimates suggested that 51.7% of codon sites were highly conserved ($\omega_0 = 0.052$), that 38.7% of sites moderately conserved and nearly neutral ($\omega_1 = 0.869$), and that 9.4% of sites were under diversifying selection ($\omega_2 = 3.598$) (Fig. 3). ω-ratios along individual branches in the *babA* phylogeny were determined using the free ratios model of codon evolution. The free-ratio model assumes an independent ω for each branch in the phylogeny, and was significantly better than the null

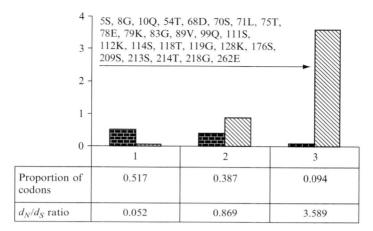

FIG. 3. Histogram showing the frequency distribution of three codon classes and their associated d_N/d_S ratios computed under the M3 model. Arrow indicates codon class under diversifying selection. Codons identified to be under positive selection are shown. *babA* sequence alignment used for this analysis is available from GenBank [gi:49473301].

model (M0), which assumes a single ω for the entire phylogeny (-InL(free-ratio) = 18641.945, -InL(M0) = 18750.405, $\omega_2 = 216.92$, d.f. 129, $p < 0.0001$), thereby suggesting *babA* evolves rapidly in different host lineages. Together, these data provided evidence of selection for amino acid sequence change at particular sites in BabA, which would be likely to affect glycan binding pocket structure. In one view, these sequence changes might simply fine-tune BabA activity for special conditions in different host populations.

A variant model emerged in considering (1) the very similar distribution of receptor affinities of Leb-only specialists and generalists, (2) the many clinical isolates that seemed to naturally lack BabA adhesin function, and (3) a consideration that even though BabA adhesin-receptor interaction must often benefit *H. pylori* it may also be potentially detrimental, for example, at times or sites of intense inflammatory response. Alternating selection pressures for loss and then restitution of adherence would sometimes be satisfied by (1) an amino acid substitution that diminished BabA function, perhaps through change in the binding pocket, and (2) then "suppressor" mutations that restore binding activity through amino acid sequence elsewhere in the protein. Much of the diversifying selection at particular sites in BabA and the rapid emergence of specialist alleles of BabA uniquely in Amerindian Peruvians (peoples that are essentially all blood O) would then be ascribed to cycles of selection for mutations causing loss and restitution of function. Only restitution of Leb adherence would be selected in this population, almost devoid of blood group A or B individuals, and hence generalist binding would be lost by attrition, not direct selection for a tight binding pocket uniquely suited to the streamlined Leb structure.

Knock-out of Adhesin Gene: General Strategy for Genetic Constructions by PCR without Recombinant DNA Cloning

The ability to test the importance of particular proteins or residues in proteins by directed mutation is of immense value in analyses of many traits, bacterial adherence included. Most *H. pylori* strains are easily trans-formable and can undergo homology-based recombination between added DNAs (whether genomic, plasmid or PCR product) and corresponding chromosomal sequences. This makes it feasible to delete genes of interest and to replace one allele with another, especially if the allele is marked with a resistance determinant (e.g., *ery, cat,* or *aph*; resistances to erythro-mycin, chloramphenicol, or kanamycin, respectively).

Traditionally, the needed genetically marked alleles were usually made by recombinant DNA cloning of segments of interest into bacterial plas-mids; digesting the recombinant plasmid with a restriction endonuclease that cleaved the cloned DNA at an appropriate site; ligating the linearized plasmid with a resistance gene-containing restriction fragment; selecting derivative plasmids containing the added resistance determinant by *E. coli* transformation; and further screening to identify plasmids with intended structures and orientation of the resistance determinant. Such genetic engineering, although immensely useful in many studies, remained limited, because useful restriction sites were only rarely exactly where they might be most valuable for high resolution genetic engineering.

PCR-based approaches freed experimentalists from limitations in dis-tributions of useful restriction sites, and thus allowed genetic engineering with unprecedented precision because PCR primers could be designed to fix endpoints of PCR products, and thereby insertion or deletion endpoints of ensuing recombinant DNAs, at any specific template sequences of interest. PCR primers are also easily built with additional un-templated features, such as new or altered regulatory sites, as illustrated by our separation of a naturally occurring gene fusion into two component orfs, each well expressed in *H. pylori* (Raudonikiene *et al.*, 1999).

Plasmid cloning can be laborious, particularly for many genes encoding membrane or secreted proteins or other factors that can be deleterious in *E. coli*. New PCR strategies also allow these problems to be avoided when seeking to engineer *H. pylori* or other transformable species. It had long been known that PCR is a discontinuous process, that DNA synthesis begun on one template can be halted, and then continued in a subsequent cycle on a different template to which the nascent strand's 3′ end is complementary. In one application, this "crossover PCR" was used to map multiple sites of transposon insertion in a gene (Krishan *et al.*, 1991).

Chalker *et al.* (2001), adapted this principle to the high throughput construction of deletion alleles of 42 *H. pylori* genes of interest to them, essentially as diagrammed in Fig. 4. Primers were developed (see list of primers below, primer number is indicated in bold) to amplify *H. pylori* DNA segments that were intended to flank each deletion (primers 1 and 2 for fragment A, and primers 5 and 6 for fragment C, in the nomenclature in this figure) and also for a resistance cassette (primers 3 and 4 for fragment B; see figure). The inside primers for the flanking segments (primers 2 and 5) were designed to allow crossover PCR: added on to the *H. pylori* specific sequences in their 3' halves, were 5' extensions that were complementary to primers 3 and 4 used for amplification of the resistance cassette (indicated by downward angled tails on primers 2 and 5 in Fig. 4). A second round of PCR amplification using a mixture of the left and right flanking DNAs (A, C) plus the resistance cassette (B) with just the exterior primers 1 and 6 usually results in the desired recombinant DNA fragment ("usual route" in Fig. 4).

It is important to recognize that this simple strategy for three-fragment assembly sometimes fails (an outcome that is easily scored by gel electrophoresis and absence of product of expected size) and that such failures can be reproducible. One such case is illustrated in Fig. 5 (the lane marked "??"). The mechanisms underlying such failures have not been examined but likely relate to distributions of alternative potential primer binding sites and/or secondary structures in DNAs to be amplified. More important, in our experience, such difficulty can be circumvented by first making two fragment subassemblies (AB with primers 1 and 4; and BC with primers 3 and 6), and then using these two subassemblies to make the final assembly, as diagrammed in Fig. 4, and with gel data in Fig. 5.

The PCR products generated by either of these assembly routes are well suited for transformation of *H. pylori*. In our experience, all selected transformants contain the desired mutant allele in place of the original wild-type allele, unless the gene in question is essential for viability (Chalker *et al.*, 2001; Tan and Berg, 2004; Tan *et al.*, 2005).

PCR Condition for Initial DNA Segment (A, B, C) Amplification

1. Composition of PCR reaction mix: Template DNA (SS1 or X47 genomic DNA/pRH151 DNA) 1 μl, primer, forward (10 pmole) 1 μl, primer, reverse (10 pmole) 1 μl, 10× NH$_4$ reaction buffer 5 μl, dNTP (2.5 mM each) 5 μl, MgCl$_2$ (50 mM) 1 μl, Biolase DNA polymerase (5 u/μl, Bioline, Germany) 0.4 μl, DW fill up to 50 μl (35.6 μl)
2. Cycling condition: 94°, 40 sec, 55°, 40 sec (50°, 40 sec for amplification of segment B, *ermM*), 72°, 1 min (30 cycles)

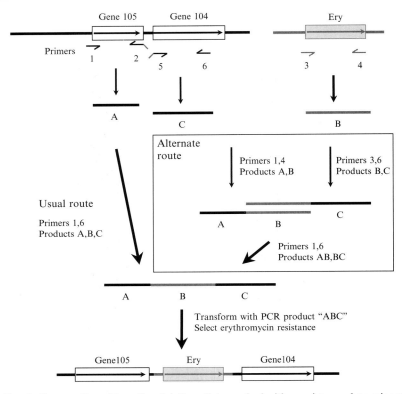

Fig. 4. Construction of insertion deletion allele marked with a resistance determinant by PCR without need for recombinant DNA cloning, essentially as described (Chalker *et al.*, 2001). In general, such constructions entail amplification of DNA segments to left and right of the intended site of insertion/deletion (products A and C, with primers 1 and 2 and with primers 5 and 6, respectively), and also a selectable resistance determinant (here resistance to erythromycin; designated fragment B with primers 3 and 4). The 5' ends of primers 2 and 5 (denoted as downward slanted lines) are complementary to the primers used for amplification of the resistance determinant (fragment B). After mixing of separately amplified fragments A, B, and C, amplification with primers 1 and 6 generally results in an assembled PCR product containing the desired insertion/deletion allele as diagrammed at bottom, on the basis of "crossover PCR" at sites of overlap because of complementarity of 5' ends of primers 2 and 3 and of primers 4 and 5. In some cases this simple three-fragment assembly fails as illustrated in Fig. 5 (gel lane marked ?? at bottom). In such cases, an alternate strategy (boxed) that entails separate two fragment assemblies of A and B, and of B and C, and then a final assembly of subassembly AB and BC often gives the desired product (rightmost experimental lane marked ABC at bottom). Gene105 and gene104 are arbitrary designations for orfs in the *H. pylori* genome. For detailed design of primer sequences for use with erythromycin and other resistance markers, see Tan and Berg (2004) and Tan *et al.* (2005).

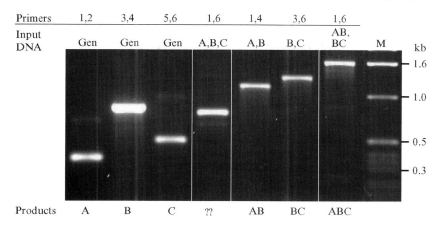

FIG. 5. Agarose gel electrophoresis of products from construction of insertion/deletion allele diagrammed in Fig. 4. Primers 1 through 6 and simple PCR products A, B, C, and assembled products AB, BC, and ABC all correspond to those diagrammed in Fig. 4. Input DNA "gen" indicates genomic DNA. The product designated ?? is a PCR-generated deletion variant of the desired ABC assembly that came to predominate in the PCR product population because of its smaller size and, thus, replication advantage. It has not been characterized further. M indicates marker DNA, with sizes of characteristic fragments indicated at right.

PCR Condition for DNA Fragment Assembly

1. Composition of PCR reaction mix: Template DNA (purified segments, 1 μl each) 2 or 3 μl. Primer, forward (10 pmole) 1 μl, primer, reverse (10 pmole) 1 μl, 10 μl NH$_4$ reaction buffer 5 μl, dNTP (2.5 mM each) 5 μl, MgCl$_2$ (50 mM) 1 μl, Biolase DNA polymerase (5 u/μl, Bioline, Germany) 0.4 μl, DW fill up to 50 μl (33.6 or 34.6 μl)
2. Cycling condition
a. Three fragments assembly (A + B + C): 94°, 40 sec, 55°, 40 sec (or 50°, 40 sec), 72°, 2 min 30 sec (25 cycles)
b. Two fragments assembly (A + B or B + C) 94°, 40 sec, 55°, 40 sec, 72°, 2 min (30 cycles)
c. Final assembly (AB + BC): 94°, 40 sec, 55°, 40 sec, 72°, 2 min 30 sec (25 cycles)
*Agarose gel: 1.2% (with 1-kb ladder DNA marker)

Primers (numbers in parentheses refer to primers used in Figs. 4 and 5)

(**1.**) CR2   5'-GGAGTTAAAAACATGAAAACACC 23mer
(**2.**) CF3   5'-<u>TACTGCAATCTGATGCGATTATTG</u>TCAAACCCC CACTTCAGACCAC 46mer

(**3**.) ermMF   5′-<u>CAATAATCGCATCAGATTGCAGTA</u> 24mer
(**4**.) ermMR   5′-<u>TTACTTATTAAATAATTTATAGCTATT GAA</u> 30mer
(**5**.) CR3   5′-<u>TTCAATAGCTATAAATTATTTAATAAGTAAGT</u> GGTCTGAAGTGGGGGTTTGA 52mer
(**6**.) CF2   5′-ATCCTTGTCAAGCCGTTATTGG 22mer

Identification of Bacterial Clones with Rare and/or Induced Receptor Binding Properties by Use of Glycoprobes

H. pylori is extremely diverse genetically both because of both mutation and recombination, which can operate between different strains or between duplicate and divergent genes within a given lineage (Kersulyte *et al.*, 1999; Suerbaum *et al.*, 1998). Some of *H. pylori*'s genes are particularly prone to genetic change, resulting in metastability or heterogeneity in bacterial phenotypes. For *H. pylori*, the genetic flexibility provides the prerequisites to adapt to certain environmental changes. Included among metastable genes are those for BabA and SabA adhesins, leading to ideas that an ability to cycle between adherent and non-adherent phenotypes contributes importantly to persistent *H. pylori* infection. FITC-labeled bacteria and semisynthetic glycoconjugates (*i.e.*, glycoprobes) were used to study metastability in *H. pylori* Leb antigen binding phenotype (Bäckström *et al.*, 2004). As one indication of metastability, individual cells proficient in binding were visualized by fluorescence microscopy of pure cultures of strains that had been found to have a nonbinding phenotype by RIA, albeit often at frequencies of only 1/50,000 bacterial cells or fewer "Bio-panning" was used to isolate rare clones with adherent phenotypes from these cultures (Aspholm-Hurtig *et al.*, 2004; Bäckström *et al.*, 2004). To accomplish this, bacterial cells were mixed with biotinylated receptor conjugate, and those proficient in binding were enriched for by use of streptavidin-coated magnetic beads. Individual clones that express adhesin protein and readily bind the receptor conjugate were then identified by membrane colony screening technique; biotinylated receptor conjugate was added to the membranes, and colonies that had gained proficiency in Leb antigen binding were detected with streptavidin staining.

Biotinylation of Receptor Conjugate

Materials

Biotin-XX-NHS (Calbiochem, San Diego, CA)
PD10 (Sephadex G-25) column (GE Healthcare Bio-sciences AB, Sweden)

Solutions

  Dimethylsulfoxide (DMSO)
  Carbonate buffer: 0.15 M NaHCO$_3$, pH 9.0
  1 M Tris-HCl, pH 9.0
  PBS (phosphate-buffered saline): 150 mM NaCl, 10 mM phosphate,
    pH 7.6
  Blocking buffer: PBS + 0.5% Tween-20 + 1% BSA
  Elution buffer: PBS + 0.05% Tween-20 + 0.2 M Tris-HCl, pH 9.0

One milligram Biotin-XX-NHS (Calbiochem, San Diego, CA) dis-solved in 50 μl dimethylsulfoxide (DMSO) is mixed with 10 μg of glyco-conjugate (for example, Leb-HSA, Isosep, Tullinge, Sweden) dissolved in carbonate buffer in a total volume of 1 ml. The sample is incubated with end-over-end mixing for 3 h at 23°; 260 μl 1 M Tris-HCl, pH 9.0, is added followed by further incubation for 30 min. Biotinylated receptor conjugate is purified on a column. Before addition of the sample, the PD10 column is washed with PBS + 0.05% Tween-20, blocked with blocking buffer for 30 min and washed with elution buffer. Samples are eluted with elution buffer and collected in aliquots of 0.5 ml. The void fraction, approximately 1.5 ml, is pooled, and the gel filtration is repeated once more to further purify the labeled Leb conjugate from the molar excess of biotin-XX-reagent.

FITC-Labeling of Bacterial Cells

  Materials

  FITC (Fluorescein isothiocyanate), (Sigma, St. Louis, MO)

  Solutions

  PBS (phosphate-buffered saline)
  0.2 M NaHCO$_3$
  Washing buffer: PBS + 0.05% Tween-20
  Blocking buffer: PBS + 0.5% Tween-20 + 1% BSA

Bacteria were grown for 24–36 h on Brucella agar medium supplemen-ted with 10% bovine blood and 1% Iso Vitox (Svenska LABFAB, Ljusne, Sweden) at 37°, under 10% CO$_2$ and 5% O$_2$. A 1-ml bacterial sample (A$_{600}$ 1.0) is washed in washing buffer, resuspended in 1 ml 0.2 M NaHCO$_3$ and incubated for 30 min at 23°; 0.1 mg FITC, dissolved in 15 μl DMSO is added to the bacterial mixture and incubated end-over-end for 8 min. Samples are washed 4 × 8 min (or until the supernatant is clear) in washing buffer and finally resuspended in blocking buffer such that a cell density of A$_{600}$ 0.2 is achieved. Use the fluorolabeled bacterial cells directly or store them in aliquots at −20°. To get rid of lysed cells after storage, thawed samples are washed in blocking buffer followed by adjustment of cell density to A_{600} 0.2.

Fluorescence Microscopy to Identify Clones of Rare Receptor Binding Phenotype

For fluorescence microscopy, a 1-ml sample of FITC-labeled bacteria (A_{600} 0.2) is incubated with blocking buffer for 1 h, incubated with 2 μg of biotinylated receptor conjugate for 2 h at 23°, washed by centrifugation, resuspended in blocking buffer, and incubated with Cy-3-streptavidin (1:10.000) (Sigma) for 1 h at 23°. Samples are washed in washing buffer and resuspended in 200 μl blocking buffer. A 1–10-μl sample is analyzed by fluorescence microscopy (magnification ×400). The total number of cells in the sample is analyzed by counting cells in a Petroff-Hausser chamber (CA Hausser and Son, Philadelphia, PA).

Bio-panning for Clones with Certain Receptor Binding Phenotypes

A 2-ml ($A_{600} = 0.5$) bacterial suspension is incubated with 10 μg of biotinylated receptor conjugate for 2 h at 37° in growth medium on a slow shaker. The bacterial-conjugate mixture is washed at least twice in growth medium, mixed with 100 μg of streptavidin magnetic beads (Qiagene, Inc. Hilden, Germany) for 2 h at 37° or other relevant growth conditions. Bacteria bound to the beads are recovered with a magnet and cultured on agar medium. Progeny bacterial cells can be used for additional rounds of enrichment.

Colony Blotting with Glycoprobes

Materials

Nitrocellulose membranes
Streptavidin-peroxidase (Roche, Stockholm, Sweden)
4-chloro-1-naphtol tablets (Sigma, St. Louis, MO)

Solutions

Washing buffer PBS + 0.05% Tween-20
TBS buffer: 50 mM Tris, pH 7.4, 150 mM NaCl + 0.05% Tween-20
Buffer I: TBS + 1.5% gelatin + 1% BSA

Bacteria are spread at a density of approximately 500 CFU per plate, incubated at appropriate growth conditions for single cell colonies. Bacteria are transferred to nitrocellulose membranes. Membranes must be organized such that positive clones can be identified on the master plate. Membranes are baked for 1 h at 70°, soaked and washed repeatedly in washing buffer, and blocked in buffer 1 overnight at 4°. The following day, 2 μg of biotinylated receptor conjugate in 5 ml buffer 1 is added to membranes, and the membranes are incubated for 2 h at 23°, washed in TBS buffer three × 10 min and thereafter incubated with streptavidin-peroxidase (Roche, Stockholm) (1:5.000) in buffer 1 for 1 h at 23°. Finally, membranes are washed in TBS buffer 3 × 10 min and exposed to 4-chloro-1-naphtol as recommended by the manufacturer (Sigma, St. Louis, MO). Receptor binding colonies are identified by blue color.

Characterization of Spatial Expression of *H. pylori* Virulence Products in Gastric Mucosa by *In Situ* Hybridization

Introduction

Transcripts from specific bacterial genes made during infection can be detected, localized, and quantified by *in situ* hybridization and a point-counting stereological method (Semino-Mora *et al.*, 2003). This method is exquisitely sensitive and specific and is suitable for archived formalin-fixed specimens and biopsies. The specificity of the method could be further increased by use of multiprobe fluorescence *in situ* hybridization (FISH).

Methods

Biopsies. Biopsies are fixed in Z-fix (10% paraformaldehyde + 1% ionized Zn, Anatech LTD, Battle Creek, MI) within 1 min of harvesting, then dehydrated, and embedded in paraffin within the next 2 days. This timing minimizes risks of RNA degradation. Serial 5-μm sections are stained with hematoxylin–eosin or Genta stain (Genta *et al.*, 1994) for grading of gastritis and atrophy according to the Sydney system (Dixon *et al.*, 1997) (0 = none, 1 = mild, 2 = moderate, and 3 = marked). Additional serial sections are prepared for in situ hybridization as described later.

Oligonucleotide PROBES for H. pylori

Specific RNA and cDNA probes were designed using a sequence database for *H. pylori 16S rRNA* (Semino-Mora *et al.*, 2003) or were as published in the case of *cagA* (Tummuru *et al.*, 1993). The 5' end of the oligonucleotides is labeled with either biotin or digoxigenin-3-0-methylcarbonyl-e-aminocaproic acid-*N*-hydroxy-succinimide ester (DIG-NHS ester) (Roche Diagnostics, Indianapolis, IN). In initial experiments, it is important to use both RNA and cDNA probes concurrently and to confirm that the two probes can detect the same structures. In addition, it is critical to demonstrate that detection is abolished by RNase but not DNase (to demonstrate that these probes specifically target mRNA). Having validated the cDNA antisense probes in one's system, RNase and DNase should be included as controls in each run. Once the method is specific, the use of RNA probes can be discontinued, especially because they are more susceptible to degradation and more delicate and costlier to prepare.

In Situ *Hybridization.* All procedures are performed as described in the following for paraffin-embedded tissues (Semino-Mora *et al.*, 2003; Wilkinson, 1998) and at RT except as specified:

1. *Section pretreatment:* Sections are first deparaffinized in xylene, rehydrated by a series of washes in graded ethanol (100%, 95%, 80%, 70%, and 50%), and finally washed in DEPC-treated water and PBS. To improve subsequent probe penetration, sections are treated with proteinase K (Roche Diagnostic, Indianapolis, IN; 10 μg/ml in 100 mM Tris-HCl buffer, pH 8.0, containing 5 mM EDTA) for 10 min, and then washed twice for 5 min each with 1× standard saline citrate solution (SSC, Quality Biological, Inc, Gaithersburg, MD).

2. *Pre-hybridization:* Sections are covered for 5 min with hybridization mixture prepared as follows: 200 μl Denhardts solution (2% bovine serum albumin [BSA], 2% Ficoll, and 2% polyvinylpyrrolidone [PVP] in water), 5 ml formamide (Roche Diagnostic, Indianapolis, IN), 600 μl 5 M NaCl, 2 ml dextran sulfate (0.8 g/8 ml) heated at 65° for 20 min, 20 μl EDTA (0.29 g/10 ml), 2.5 ml hybridization solution (phenol-chloroform extracted salmon testis DNA and SSC), 20 μl 1.0 M Tris-HCl buffer, pH 7.5, and 660 μl DEPC-treated water). Sections are then covered for 20 min with blocking solution (0.2 g BSA fraction V, 0.5 ml normal horse serum [Jackson ImmunoResearch, Inc.]), and 10 ml of immuno-Tris buffer (50 ml of 1 M Tris-HCl buffer, pH 7.5, and 25 ml of 5 M NaCl in 1000 ml water, pH 7.5).

3. *Hybridization:* The probes listed in "DNA probes used" are denatured by heating for 10 min at 65° and subsequent chilling on ice for 3 min. Each unstained section is layered with 70–100 μl of the probe solution (4 pmol/μl), cover-slipped, and placed horizontally on filter paper soaked with water in an airtight box wrapped with aluminum foil. Separate boxes are used for different probes and for negative controls. Each box is incubated overnight at 37°.

4. *Post-hybridization:* After 18 h of incubation, the unbound probe is removed by successive washes in decreasing concentrations of SSC (2× SSC 30 min, 1× SSC 10 min, 0.5× SSC 10 min at RT, and 0.1× SSC 15 min at 60°).

Detection of H. pylori *Gene Expression*

1. Bright-field ISH: Anti-digoxigenin antibody-conjugated or streptavidin-conjugated alkaline phosphatase (AP; Roche Diagnostic, Indianapolis, IN) is

used to detect the digoxigenin-labeled (*cagA*) or the biotin-labeled (*16S rRNA* or *babA2*) probe, respectively (see, DNA probes used). Sections are incubated for 2 h with anti-digoxigenin-AP or streptavidin-AP respectively, 1:500 dilution in blocking solution. Unbound antibody-alkaline phosphatase is washed gently with immuno-Tris buffer (3 times, 5 min each), using a different Coplin jar for each probe and negative controls. Bound alkaline phosphatase is detected by covering the slides with a chromogenic substrate (5-bromo-4-chloro-3indolyl phosphate/nitroblue tetrazolium BCIP/NBT kit, Vector Labs, Burlingame, CA) and keeping them in the dark for 20 min. After washing in distilled water for 3 min, slides are counterstained with Nuclear Fast Red (Vector, Burlingame, CA) for 15 min, washed in water, dehydrated, cleared in xylene, briefly air dried, and mounted with Permount (Biomedia Corporation, Foster City, CA).

2. Fluorescence ISH (FISH) dual localization (Jaju *et al.*, 1999; Semino-Mora *et al.*, 2003; Wilkinson, 1998): The hybridization mixture contains both a biotin-labeled probe recognizing *H. pylori* 16S *rRNA* (Semino-Mora *et al.*, 2003) and a digoxigenin-labeled probe recognizing *cagA* (see, DNA probes used). After incubation as described previously, the slides are covered with the following three layers: first layer: 1 μl avidin-Texas red + 1.5 μl mouse monoclonal anti-digoxigenin; second layer: 10 μl biotin-anti-avidin + 1 μl rabbit anti-mouse-IgG-fluorescein isothiocyanate (FITC) conjugated; third layer: 1 μl avidin-Texas red + 10 μl monoclonal anti-rabbit-FITC conjugated. All antibodies are prepared in 1 ml of blocking solution. Slides are incubated with each antibody layer for 30 min at 37°, in a humid chamber, and protected from the light and washed 3 times 5 min each. After air-drying, the sections are mounted in Vectashield (Vector Labs, Burlingame, CA).

Controls

H. pylori pure cultures are streaked onto precleaned microscope slides and used as positive controls. *E. coli* and *S. flexneri* cultures and gastric biopsies from an *H. pylori*–negative patient are used as negative controls. Control for nonspecific binding is performed by using: (1) sense instead of antisense probe; (2) hybridization buffer instead of antisense probe; (3) unlabeled antisense probe; (4) digoxigenin or biotin-labeled probe for scorpion *Buthus martensi* Karsch neurotoxin sequence [5'-GGC CAC GCG TCG ACT AGT AC-3'] (Lan *et al.*, 1999); (5) RNase A pretreatment (Roche); (6) DNase I pretreatment (Roche); and (7) RNase + DNase I pretreatment.

DNA probes used (RNA probes are identical except for substitution of U for T):

16S rRNA	antisense: 5'-TACCTC TCC CACACTCTAG
(Semino-Mora *et al.*, 2003)	AATAGTAGTTTCAAATGC-3'
	sense: 5'-CTA TGA CGG GTA TCC GGC-3'
cagA (Tummuru *et al.*, 1993)	antisense: 5'-CTG CAA AAG ATT GTT
	TGG CAG A-3'
	sense: 5' GAT AAC AGG CAA
	GCT TTT GAG G-3'

Application of Laser Tweezers Technique for Analyzes of Binding Strength

Optical tweezers (OT) is a laser-based tool for nonintrusive manipulation of particles on the micrometer scale, including living cells and bacteria. It consists basically of a microscope objective with high numerical aperture in which a continuous laser beam is focused. A small object, like a micrometer-sized bead with a refractive index larger than that of the surrounding medium, experiences a restoring force in all spatial directions in the focal region, thus becoming trapped in the laser beam focus. The OT-technique is being used as a sensitive way of measuring weak molecular interactions in biological systems (forces typically in the lower picoNewton range) (Ashkin, 1992; Ashkin, *et al.*, 1990; Simmons *et al.*, 1996). The object under study (e.g., a bacterial cell) is approached and allowed to bind to a bead trapped in the focus. By slowly retracting the cell from the trapped bead, a force is exerted on the system, and the position of the bead is shifted. The shift of the bead in the trap constitutes a measure of the exerted force. We recently developed an assay for *in situ* force measurements of specific interactions between the *H. pylori* BabA adhesins and the blood group antigen Lewis b (Leb) presented in solid (surface) phase (Björnham *et al.*, 2005). Moreover, physical properties of *E.coli* P pili, associated with stretching and contracting of the pilus structure, were experimentally assessed by OT force measurements. Three characteristic elongation regions were identified while loading the pili under constant rate (Jass *et al.*, 2004). It was found that pili extension is fully elastic, because the pilus structure exhibited equivalent reciprocal forces during unloading and loading (Fällman *et al.*, 2005). In the following, a short review is presented of recent OT experiments for probing the *H. pylori* BabA-Leb interaction.

Materials and Methods

Instrumentation. The optical tweezers system is constructed around an inverted microscope (Olympus IX70) with a high numerical aperture oil-immersion objective (Zeiss CP-Achromat 100× N.A. = 1.25). An argon-ion laser–pumped continuous-wave titanium-sapphire laser, operating at a wavelength of 810 nm, is used for optical trapping. A complete description of the optical system for controlling the laser beam that forms the optical trap and modifications made to the microscope to realize the OT system is reported in Fällman and Axner (1997) and Fällman et al. (2004). The displacement of a small bead held in the optical trap constitutes a measure of the force that is exerted on the bead. To monitor the displacement, a weak probe laser beam (HeNe) is conveyed into the microscope. Figure 6 illustrates the optical trap and the position detection principle. The probe beam is focused a short distance below the trapped bead, whereby the trapped bead acts as a lens and refracts the beam, in such a way that a well-defined light spot is produced in the far field. A position sensitive detector (PSD) is used to monitor the lateral position of this spot. If an object (e.g., a bacterial cell) bound to the trapped bead is slowly moved (in the horizontal direction), the trapped bead will be shifted a short distance from the focal point, resulting in a shifted position of the probe laser spot on the PSD. The output voltage signal of the PSD is directly proportional to the distance the trapped bead is shifted. A calibration procedure based on the Brownian motion of the trapped bead in the liquid is used to transform the position data to force data in units of pN (Fällman et al., 2004).

Biological Model System

The model system for probing the Leb-BabA interaction contained the following components suspended in phosphate-buffered saline (PBS): *H. pylori* cells, a functionalized large bead that served as immobilization tool for the bacterium and a small Leb-coated bead that served both as a surface hosting the receptors and as a force indicator for the OT. The *H. pylori* strains used were the 17875 wild type (Leb binding) strain and the 17875*babA1*::kan*babA2*::cam (denoted the *BabA*-mutant; non-binding) (Mahdavi et al., 2002). Bacteria were grown for 40–45 h on *Brucella* agar medium supplemented with 10% bovine blood and 1% Iso Vitox (Becton Dickinson, US) at 37°, under 10% CO_2 and 5% O_2, and suspended in a density of $\sim$50·10^6 cells per ml. The large functionalized polystyrene beads with carboxyl surface groups (Interfacial Dynamics Corp, USA) had a diameter of 9.6 μm and were immobilized on the coverslip. The carboxyl surface group beads were activated to allow covalent attachment of bacteria by use of a two-step protocol (Staros et al., 1986). The small carboxyl polystyrene beads (Interfacial Dynamics Corp, USA), with a diameter of 3.2 μm, were coated with Leb HSA (human serum albumin) conjugate (IsoSep AB, Tullinge,

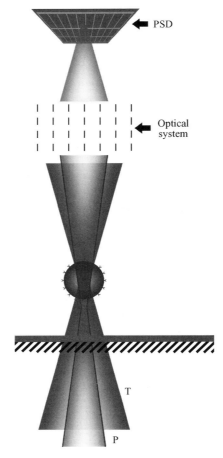

FIG. 6. Position detection of a bead confined in the optical trap. T, trapping laser beam (Ti-sapphire); P, probe laser beam (HeNe); PSD, position sensitive detector.

Sweden) (Björnham *et al.*, 2005). The sample mixture consisted of a $\sim 1 \cdot 10^6$ bacterial cells, large beads with activated surface carboxyl groups ($\sim 500 \cdot 10^6$ per ml) and small Leb-beads ($\sim 50 \cdot 10^6$ per ml), all suspended in a droplet of ~ 25-μl PBS. The sample mixture was placed in a sealed chamber, which was mounted on a piezo stage (Physik Instrumente, GmbH) that allowed precise translation of the sample at a controlled speed (0.5 μm/s) relative to the microscope objective.

Measurement Procedure

A typical force measurement is illustrated in Fig. 7. A free-floating individual bacterial cell was first trapped by the OT, approached to the large bead, and allowed to bind covalently to the surface. Next, the large bead with the bacterium was brought nearer to the small Leb bead held in the optical trap. The initial bacterium-bead contact allowed specific binding between the BabA-adhesins and the Leb-receptors. After ~5 sec, the bacterium was retracted from the Leb bead by a moving the piezo stage at constant rate (0.5 μm/s) (Fig. 7A). The force in the system increased until the binding suddenly breached (Fig. 7B), whereby the Leb bead returned instantaneously to the center of the trap. The last position of the Leb bead before breakage is thus a measure of the maximal binding force.

Conclusions

A typical force-versus-distance graph from a measurement on an individual bacterial cell of the 17875Leb strain is displayed in Fig. 8A. The graph can be seen as consisting of two characteristic regions. The first region consists of a linear force-versus-distance dependence, which demonstrates adhesion and originates from the stress response of the biological system under elongation. The negative force data at the beginning demonstrate an initial contact between the bacterial cell and the Leb bead. The force increases with elongation until the binding breaks, whereby the Leb bead detaches from the bacterium and returns instantaneously to the center of the optical trap. The second part of the data, which constitutes of a zero force region, indicates thereby that the binding has ruptured. The distribution of rupture forces from in total 61 measurements (performed in eight series) on the 17875Leb strain is displayed in the form of a histogram (Fig. 8B). The histogram intervals are chosen as multiples of 12.5 pN, with a width of $\pm$6.25 pN. Apparently, most measurements (46 measurements of 61) gave rise to rupture forces that appear in the even intervals marked with filled bars, which occurs at multiples of 25 pN. A substantially smaller fraction of all measurements (15 measurements) gave rise to forces that appear in the odd intervals (marked with unfilled bars). Thus, a pronounced high-low staggering distribution of the histogram is apparent. Such a distribution of forces suggests the existence of an elemental force in the system, F_{Leb}. The even intervals represent, thereby, multiple bindings with a force of NF_{Leb}, with N being an integer, whereas the odd intervals correspond to intermediate force intervals, centered around $(N - 0.5)F_{Leb}$. It was found that the strongest high-low staggering appearance

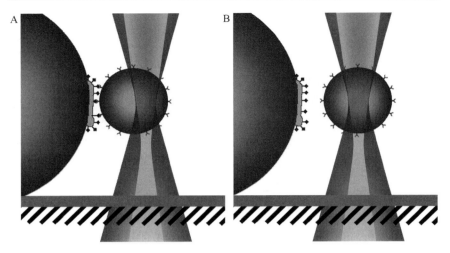

FIG. 7. Illustration of the measurement procedure. (A) The large bead with the bacterium pulls the small Leb-coated bead away from the center of the optical trap. (B) The force in the system increases until the binding breaks, whereby the Leb bead returns to the center.

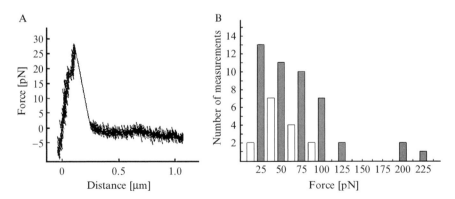

FIG. 8. (A) Typical force-versus-distance graph from a measurement on a single bacterium of the 17875 Leb strain. The force increases linearly force until the binding breaks. (B) Histogram of the distribution of rupture forces from in total 61 measurements on the 17875 Leb strain. The intervals are centered at multiples of 12.5 pN with a width of ±6.25 pN. Most of the rupture forces appear in the even intervals (marked with filled bars). This suggests the existence of an elemental force in the system of $\sim$25 pN, which is interpreted as the rupture force of a single BabA-Leb binding.

occurred at an elemental force of $F_{Leb} = 25$ pN. Thus, the measurement data suggest a rupture force of a single BabA-Leb binding that amounts to $\sim$25 pN.

Negative control measurements on the BabA-mutant strain were performed to test the inferred specific BabA-Leb binding. It was found that the rupture force of BabA-mutants was 75% lower than that of 17875Leb. This difference showed specific BabA-Leb binding, in agreement with previous findings (Borén *et al.*, 1993). Also, the data from the BabA-mutant do not exhibit a high-low staggering feature when plotted in a histogram, which, in contrast to the 17875Leb, suggests the absence of an elemental force in the control system.

Further investigations will include assessment of the rupture force under different pH conditions (acidic compared with neutral), different loading rates, and influence of shear forces (shear flow), which all are assumed to be important factors for a general characterization of the adhesion properties of *H. pylori*.

Analyses of Nonopsonic Bacterial–Cell Interactions by Chemiluminescence Measurements of Released Oxidative Burst Metabolites

Role of Adhesins in Activation of Human Neutrophils

Certain *H. pylori* strains have a nonopsonic (i.e., nonantibody mediated) neutrophil activating capacity and are found significantly more in patients with peptic ulcer disease than gastritis only (Rautelin *et al.*, 1993). To test whether this *H. pylori*–induced neutrophil activation involves lectino-phagocytosis, *H. pylori* strains were incubated with oligosaccharides, and effects on the oxidative burst of subsequently challenged neutrophils were measured by luminol-enhanced chemiluminescence (CL) and flow cytometry (FC) (Teneberg *et al.*, 2000). By both methods a reduced response was obtained by incubation of *H. pylori* with sialic acid–terminated oligosaccharides, whereas lactose had no effect. This suggested that this *H. pylori*–induced neutrophil activation entails bacterial adhesin recognition of sialylated glycoconjugates on neutrophil cell surfaces, leading to phagocytosis and the oxidative burst reaction.

After identification of SabA and characterization of sialyl-dimeric-Le[x] as a high-affinity receptor (Mahdavi *et al.*, 2002), variant gangliosides structurally related to the sialyl-dimeric-Le[x] nonaglycosylceramide were tested in binding studies (Roche *et al.*, 2004). SabA also bound sialylα3-neolacto-tetraosyl-ceramide, sialylα3-neolacto-hexaosyl-ceramide, sialylα3-neolacto-octaosyl-ceramide, the VIM-2 ganglioside, and sialyl-dimeric- Le[x] glycosphingolipids, complex gangliosides also present in human neutrophils (Müthing et al., 1996; Stroud et al., 1996a,b). This led to investigation of SabA's contribution to human neutrophil activation (Unemo et al., 2005). Wild-type and mutant strains that were deficient in SabA demonstrated no neutrophil-activating capacity, showing that H. pylori binding to sialylated receptors on neutrophils contributes to bacterial adherence and phagocytosis.

Methodological Considerations

CL measures both activation of plasma membrane NADPH oxidase (usually referred to as the external oxidative burst) and intracellular NADPH oxidase (internal oxidative burst). The external burst can be checked by quenching (and chemical elimination) with catalase, and the internal burst can be checked by quenching with horseradish peroxidase and azide or by use of isoluminol instead of luminol (Dahlgren and Karlsson, 1999; Dahlgren and Lock, 1988). Flow cytometry of hydroethidine-loaded neutrophils measures only the internal burst (Danielsson and Jurstrand, 1998; Teneberg et al., 2000). A biphasic response is usually obtained when H. pylori–induced neutrophil activation is measured by luminol enhanced CL. Most H. pylori strains induce a strong initial phase within the first minutes, mostly because of activation of the plasma membrane NADPH oxidase. The second phase, which develops in the following 5–30 min, represents activation of both plasma membrane and intracellular NADPH oxidases.

There is variation among strains that express SabA in the ability to induce the oxidative burst manifested by differences in peak values (mV), time to reach peak (min), and absence of a recognizable biphasic response. Luminol-enhanced CL offers some advantages in this respect, because it can be used to measure kinetics, as well as detecting external and internal oxidative bursts.

Isolation of Human Neutrophil Granulocytes

Neutrophils are prepared from heparinized blood from healthy blood donors by Ficoll-Paque (Amersham Biosciences Ab, Uppsala, Sweden) centrifugation as described (Böyum, 1974), with slight modifications (Rautelin et al., 1993). For each series of experiments on a particular day neutrophils were prepared and pooled from three blood donors of the same blood group (A rh+ or O rh+). Neutrophils were thus obtained from different blood donors at each experiment. They were suspended in PBS supplemented with $MgCl_2$, $CaCl_2$, glucose, and gelatin (PBS-GG) (Rautelin et al., 1993). The purity and viability of the neutrophils exceeded 95%.

Culture Conditions

The *H. pylori* strains were grown in a microaerophilic atmosphere at 37° on GC agar plates (GC II agar base; BBL, Cockeysville, MD) supplemented with 1% bovine hemoglobin (BBL), 10% horse serum, and 1% IsoVitaleX enrichment (BBL), without antibiotics for the wild-type strains, and with appropriate antibiotics for mutant strains. After 48 h, bacteria were collected in PBS and used in CL experiments.

Chemiluminescence (CL) Experiments

The oxidative burst of neutrophils challenged with nonopsonized whole and live *H. pylori* organisms was measured with luminol-enhanced CL as previously described (Rautelin *et al.*, 1993). To each test tube (LKB, Bromma, Sweden) was added 300 μl of PBS supplemented with $MgCl_2$ and $CaCl_2$, 100 μl of neutrophils (5×10^6/ml), 50 μl of 10^{-5} M luminol (Sigma, St. Louis, MO), and finally 50 μl of nonopsonized *H. pylori* (5×10^8/ml).

For oligosaccharide inhibition experiments, 50 μl of *H. pylori* (5×10^8/ml) were mixed with 50 μl of oligosaccharides at final concentrations of 0.1–1.0 mM in the CL, for 15 min at 37°, and 100 μl of the mixture was transferred to a test tube for CL measurement.

The oxidative bursts of the neutrophils were measured as luminol-enhanced CL with a luminometer (LKB Wallac 1251, Turku, Finland), starting within 1 min of adding the bacterial suspension. The assays were performed at 37°, and CL from each sample was measured at 60–90-sec intervals during a period of 30–60 min. Data were stored in a computer for computerized calculations. Strain *H. pylori* NCTC 11637 (*sabA*+: provides a strong and rapid CL response) and C-7050 (*sabA*-: inducing no CL response) (Rautelin *et al.*, 1993) were included in each series of experiments as positive and negative controls, respectively.

Acknowledgments

This work was supported by the Umei University Biotechnology Fund, the County Council of Västerbotten, the JC Kempe and Seth Kampe Foundation (T.B.), the Swedish Research Council (T.B., S.T., A.A., S.S.), Cancer Fonden (T.B., S.T., A.A.), SSF programs "Glycoconjugates in Biological Systems" (T.B., S.T.) and Infection and Vaccinology (M.A., T.B.), the Swedish Medical Society (S.T.), the Swedish Society for Medical Research (S.T., S.L.), Åke Wiberg Foundation (A.A.), Nanna Svortz Foundation (A.A.) EMBO Long-Term Fellowship (M.A.), The Research Foundation of Örebro County Council (M.U., D.D.), and N.I.H. (CA82312) (A.D.), R01DK063041 (D.E.B.), Ralph E. Powe Junior Faculty Enhancement Award and University of Louisville Research Incentive Grant (A.K.), Deutsche Forschungs gemeinschaft (DFG), SFB 585/B5 (S.R.).

References

Ashkin, A., Schutze, K., Dziedzic, J. M., Euteneuer, U., and Schliwa, M. (1990). Force generation of organelle transport measured *in vivo* by an infrared laser trap. *Nature* **348**, 346–348.

Ashkin, A. (1992). Forces of a single-beam gradient laser trap on a dielectric sphere in the ray optics regime. *Biophys. J.* **61**, 569–582.

Aspholm-Hurtig, M., Dailide, G., Lahmann, M., Kalia, A., Ilver, D., Roche, N., Vikström, S., Sjöström, R., Linden, S., Bäckström, A., Lundberg, C., Arnqvist, A., Mahdavi, J., Nilsson, U. J., Velapatino, B., Gilman, R. H., Gerhard, M., Alarcon, T., Lopez-Brea, M., Nakazawa, T., Fox, J. G., Correa, P., Dominguez-Bello, M. G., Perez-Perez, G. I., Blaser, M. J., Normark, S., Carlstedt, I., Oscarson, S., Teneberg, S., Berg, D. E., and Borén, T. (2004). Functional adaptation of BabA, the *H. pylori* ABO blood group antigen binding adhesin. *Science* **305**, 519–522.

Bäckström, A., Lundberg, C., Kersulyte, D., Berg, D. E., Borén, T., and Arnqvist, A. (2004). Metastability of *Helicobacter pylori bab* adhesin genes and dynamics in Lewis b antigen binding. *Proc. Natl. Acad. Sci. USA* **101**, 16923–16928.

Björnham, O., Fällman, E., Axner, O., Nilsson, U., Borén, T., and Schedin, S. (2005). Measurements of the binding force between the *Helicobacter pylori* adhesin BabA and the Lewis b blood group antigen using optical tweezers. *J. Biomedical Optics.* **10**, 044024-1–0444024-9.

Borén, T., Falk, P., Roth, K. A., Larson, G., and Normark, S. (1993). Attachment of *Helicobacter pylori* to human gastric epithelium mediated by blood group antigens. *Science* **262**, 1892–1895.

Borén, T., Wadström, T., Normark, S., Gordon, J. I., and Falk, P. G. (1997). Methods for the Identification of *H. pylori* Host Receptors. *In* "Helicobacter pylori Protocols" (C. L. Clayton and H. L. T. Mobley, eds.), Chapter 8 pp. 205–224. Humana Press Inc, Totowa, NJ.

Böyum, A. (1974). Separation of blood leukocytes, granulocytes and lymphocytes. *Tissue Antigens* **4**, 269–274.

Chalker, A. F., Minehart, H. W., Hughes, N. J., Koretke, K. K., Lonetto, M. A., Brinkman, K. K., Warren, P. V., Lupas, A., Stanhope, M. J., Brown, J. R., and Hoffman, P. S. (2001). Systematic identification of selective essential genes in *Helicobacter pylori* by genome prioritization and allelic replacement mutagenesis. *J. Bacteriol.* **183**, 1259–1268.

Dahlgren, C., and Lock, R. (1988). The limitation of the human neutrophil chemilumines-cence response by extracellular peroxidase is stimulus dependent: Effect of added horseradish peroxidase on the response induced by both soluble and particulate stimuli. *J. Clin. Lab. Immunol.* **26**, 49–53.

Dahlgren, C., and Karlsson, A. (1999). Respiratory burst in human neutrophils. *J. Immunol. Methods* **232**, 3–14.

Danielsson, D., and Jurstrand, M. (1998). Nonopsonic activation of neutrophils by *Helicobacter pylori* is inhibited by rebamipide. *Dig. Dis. Sci.* **43** (9 Suppl.), 167S–173S.

Dixon, M. F., Genta, R. M., Yardley, J. H., and Correa, P. (1997). Histological classification of gastritis and *Helicobacter pylori* infection: An agreement at last? The International Workshop on the Histopathology of Gastritis. *Helicobacter* **2** (Suppl. 1), S17–S24.

Falk, P., Borén, T., and Normark, S. (1994a). Characterization of microbial host receptors. *Methods Enzymol.* **236**, 353–374.

Falk, P., Borén, T., Haslam, D., and Caparon, M. (1994b). Bacterial adhesion and colonization assays. *Methods Cell. Biol.* **45**, 165–192.

Falk, P., Roth, K. A., Borén, T., Westblom, T. U., Gordon, J. I., and Normark, S. (1993). An *in vitro* adherence assay reveals that *Helicobacter pylori* exhibits cell lineage-specific tropism in the human gastric epithelium. *Proc. Natl. Acad. Sci. USA* **90**, 2035–2039.

Fällman, E., and Axner, O. (1997). Design for fully steerable dual-trap optical tweezers. *Appl. Opt.* **36**, 2107–2113.

Fällman, E., Schedin, S., Jass, J., Andersson, M., Uhlin, B. E., and Axner, O. (2004). Optical Tweezers based force measurement system for quantitating binding interactions: System design and application for the study of bacterial adhesion. *Biosensors Bioelectronics* **19**, 1429–1437.

Fällman, E., Schedin, S., Jass, J., Uhlin, B. E., and Axner, O. (2005). The unfolding of the P pili quaternary structure by stretching is reversible, not plastic. *EMBO Repo.* **61**, 52–56.

Fay, J. C., and Wu, C. I. (2003). Sequence divergence, functional constraint, and selection in protein evolution. *Ann. Rev. Genomics Hum. Genet.* **4**, 213–235.

Genta, R. M., Robason, G. O., and Graham, D. Y. (1994). Simultaneous visualization of *Helicobacter pylori* and gastric morphology: A new stain. *Hum. Pathol.* **25**, 221–226.

Gerhard, M., Lehn, N., Neumayer, N., Borén, T., Rad, R., Schepp, W., Miehlke, S., Classen, M., and Prinz, C. (1999). Clinical relevance of the *Helicobacter pylori* gene for blood-group antigen-binding adhesin. *Proc. Natl. Acad. Sci. USA* **96**, 12778–127783.

Gillespie, J. H. (1991). The Causes of Molecular evolution Oxford University Press, NY.

Hansson, G. C., Karlsson, K. A., Larson, G., Strömberg, N., and Thurin, J. (1985). Carbohydrate-specific adhesion of bacteria to thin-layer chromatograms: A rationalized approach to the study of host cell glycolipid receptors. *Anal. Biochem.* **146**, 158–163.

Hunter, W. M., and Greenwood, F. C. (1962). Preparation of iodine-131 labelled human growth hormone of high specific activity. *Nature* **194**, 495–496.

Ilver, D., Arnqvist, A., Ögren, J., Frick, I. M., Kersulyte, D., Incecik, E. T., Berg, D. E., Covacci, A., Engstrand, L., and Borén, T. (1998). *Helicobacter pylori* adhesin binding fucosylated histo-blood group antigens revealed by retagging. *Science* **279**, 373–377.

Jaju, R., Haas, O. A., Neat, M., Harbott, J., Saha, V., Boultwood, J., Brown, J. M., Pirc-Danoewinata, H., Krings, B. W., Muller, U., Morris, S. W., Wainscoat, J. S., and Kearney, L. (1999). A new recurrent translocation, t(5;11)(q35;p15.5), Associated with del(5q) in childhood acute myeloid leukemia. *Blood* **94**, 773–780.

Jass, J., Schedin, S., Fällman, E., Ohlsson, J., Nilsson, U., Uhlin, B. E., and Axner, O. (2004). Physical properties of *Escherichia coli* P Pili measured by optical tweezers. *Biophys. J.* **87**, 4271–4283.

Karlsson, K. A., and Strömberg, N. (1987). Overlay and solid-phase analysis of glycolipid receptors for bacteria and viruses. *Methods Enzymol.* **138**, 220–232.

Kersulyte, D., Chalkauskas, H., and Berg, D. E. (1999). Emergence of recombinant strains of *Helicobacter pylori* during human infection. *Mol. Microbiol.* **31**, 31–43.

Krishnan, B. R., Kersulyte, D., Brikun, I., Berg, C. M., and Berg, D. E. (1991). Direct and crossover PCR amplification to facilitate Tn5supF-based sequencing of lambda phage Au C202 clones. *Nucleic Acids Res.* **19**, 6177–6182.

Lan, Z.-D., Dai, L., Zhuo, X.-L., Feng, J.-C., Xu, K., and Chi, C. W. (1999). Gene cloning and sequencing of BmK AS and BmK AS-1, two novel neurotoxins from the scorpion *Butus martensi* Karsch. *Toxicon.* **37**, 815–823.

Mahdavi, J., Sonden, B., Hurtig, M., Olfat, F. O., Forsberg, L., Roche, N., Ångström, J., Larsson, T., Teneberg, S., Karlsson, K. A., Altraja, S., Wadström, T., Kersulyte, D., Berg, D. E., Dubois, A., Petersson, C., Magnusson, K. E., Norberg, T., Lindh, F., Lundskog, B. B., Arnqvist, A., Hammarström, L., and Borén, T. (2002). *Helicobacter pylori* SabA adhesin in persistent infection and chronic inflammation. *Science* **297**, 573–578.

Murray, P. A., Prakobphol, A., Lee, T., Hoover, C. I., and Fisher, S. J. (1992). Adherence of oral streptococci to salivary glycoproteins. *Infect. Immun.* **60**, 31–38.

Müthing, J., Spanbroek, R., Peter-Katalinic, J., Hanisch, F. G., Hanski, C., Hasegawa, A., Unland, F., Lehmann, J., Tschesche, H., and Egge, H. (1996). Isolation and structural characterization of fucosylated gangliosides with linear poly-N-acetyllactosaminyl chains from human granulocytes. *Glycobiology* **6**, 147–156.

Nei, M., and Kumar, S. (2000). Molecular Evolution and Phylogenetics. Oxford University Press, NY.

Okada, N., Pentland, A. P., Falk, P., and Caparon, M. G. (1994). M protein and protein F act as important determinants of cell-specific tropism of *Streptococcus pyogenes* in skin tissue. *J. Clin. Invest.* **94**, 965–977.

Olfat, F. O., Näslund, E., Freedman, J., Boren, T., and Engstrand, L. (2002). Cultured human gastric explants: A model for studies of bacteria-host interaction during conditions of experimental *Helicobacter pylori* infection. *J. Infect. Dis.* **186**, 423–427.

Olfat, F. O., Zheng, Q., Oleastro, M., Voland, P., Boren, T., Karttunen, R., Engstrand, L., Rad, R., Prinz, C., and Gerhard, M. (2005). Correlation of the *Helicobacter pylori* adherence factor BabA with duodenal ulcer disease in four European countries. *FEMS Immunol. Med. Microbiol.* **44**, 151–156.

Prakobphol, A., Murray, P. A., and Fisher, S. J. (1987). Bacterial adherence on replicas of sodium dodecyl sulfate-polyacrylamide gels. *Anal. Biochem.* **164**, 5–11.

Raudonikiene, A., Zakharova, N., Su, W. W., Jeong, J. Y., Bryden, L., Hoffman, P. S., Berg, D. E., and Severinov, K. (1999). *Helicobacter pylori* with separate beta- and beta'-subunits of RNA polymerase is viable and can colonize conventional mice. *Mol. Microbiol.* **32**, 131–138.

Rautelin, H., Blomberg, B., Fredlund, H., Järnerot, G., and Danielsson, D. (1993). Incidence of *Helicobacter pylori* strains activating neutrophils in patients with peptic ulcer disease. *Gut.* **34**, 599–603.

Roberts, J. A., Marklund, B. I., Ilver, D., Haslam, D., Kaack, M. B., Baskin, G., Louis, M., Möllby, R., Winberg, J., and Normark, S. (1994). The Gal(alpha1-4)Gal-specific tip adhesin of *Escherichia coli* P-fimbriae is needed for pyelonephritis to occur in the normal urinary tract. *Proc. Natl. Acad. Sci. USA* **91**, 11889–11893.

Roche, N., Ångström, J., Hurtig, M., Larsson, T., Borén, T., and Teneberg, S. (2004). *Helicobacter pylori* and complex gangliosides. *Infect. Immun.* **72**, 1519–1529.

Rosenthal, H. E. (1967). A graphic method for the determination and presentation of binding parameters in a complex system. *Anal. Biochem.* **3**, 525–532.

Ruhl, S., Sandberg, A. L., Cole, M. F., and Cisar, J. O. (1996). Recognition of immunoglobulin A1 by oral actinomyces and streptococcal lectins. *Infect. Immun.* **64**, 5421–5424.

Ruhl, S., Cisar, J. O., and Sandberg, A. L. (2000). Identification of polymorphonuclear leukocyte and HL-60 cell receptors for adhesins of *Streptococcus gordonii* and *Actinomyces naeslundii*. *Infect. Immun.* **68**, 6346–6354.

Ruhl, S., Sandberg, A. L., and Cisar, J. O. (2004). Salivary receptors for the proline-rich protein-binding and lectin-like adhesins of oral actinomyces and streptococci. *J. Dent. Res.* **83**, 505–510.

Scatchard, G. (1949). The attraction of proteins for small molecules and ions. *Ann. N. Y. Acad. Sci.* **51**, 660–672.

Semino-Mora, C., Doi, S. Q., Marty, A., Simko, V., Carlstedt, I., and Dubois, A. (2003). Intracellular and interstitial expression of *H. pylori* virulence genes in gastric precancerous intestinal metaplasia and adenocarcinoma. *J. Infect. Dis.* **187**, 1165–1177.

Simmons, R. M., Finer, J. T., Chu, S., and Spudich, J. A. (1996). Quantitative measurements of force and displacement using an optical trap. *Biophys. J.* **70**, 1813–1822.

Staros, O., Wright, R. W., and Swingle, D. M. (1986). Enhancement by N-hydroxysulfosuccinimide of water-soluble carbodiimide-mediated coupling reactions. *Anal. Biochem.* **156**, 220–222.

Stroud, M. R., Handa, K., Salyan, M. E. K., Ito, K., Levery, S. B., Hakomori, S.-i., Reinhold, B. B., and Reinhold, V. N. (1996a). Monosialogangliosides of human myelogenous leukemia HL60 cells and normal human leukocytes. 1. Separation of E-selectin binding from nonbinding gangliosides, and absence of sialosyl-Lex having tetraosyl to octaosyl core. *Biochemistry* **35**, 758–769.

Stroud, M. R., Handa, K., Salyan, M. E. K., Ito, K., Levery, S. B., Hakomori, S.-i., Reinhold, B. B., and Reinhold, V. N. (1996b). Monosialogangliosides of human myelogenous leukemia HL60 cells and normal human leukocytes. 2. Characterization of E-selectin binding fractions, and structural requirements for physiological binding to E-selectin. *Biochemistry* **35**, 770–778.

Suerbaum, S., Smith, J. M., Bapumia, K., Morelli, G., Smith, N. H., Kunstmann, E., Dyrek, I., and Achtman, M. (1998). Free recombination within *Helicobacter pylori*. *Proc. Natl. Acad. Sci. USA* **95**, 12619–12624.

Tan, S., and Berg, D. E. (2004). Motility of urease-deficient derivatives of *Helicobacter pylori*. *J. Bacteriol.* **186**, 885–888.

Tan, S., Fraley, C. D., Zhang, M., Dailidiene, D., Kornberg, A., and Berg, D. E. (2005). Diverse phenotypes resulting from polyphosphate kinase gene (ppk1) inactivation in different strains of *Helicobacter pylori*. *J. Bacteriol.* **187**, 7687–7695.

Teneberg, S., Jurstrand, M., Karlsson, K. A., and Danielsson, D. (2000). Inhibition of *Helicobacter pylori*-induced activation of human neutrophils by sialylated oligosaccharides. *Glycobiology* **10**, 1171–1181.

Tummuru, M. K., Cover, T. L., and Blaser, M. J. (1993). Cloning and expression of a high-molecular-mass major antigen of *Helicobacter pylori*: Evidence of linkage to cytotoxin production. *Infect. Immun.* **61**, 1799–1809.

Unemo, M., Aspholm-Hurtig, M., Ilver, D., Borén, T., Bergström, J., Danielsson, D., and Teneberg, S. (2005). The sialic acid binding SabA adhesin of *Helicobacter pylori* is essential for nonopsonic activation of human neutrophils. *J. Biol. Chem.* **280**, 15390–15397.

Van de Bovenkamp, J. H., Mahdavi, J., Korteland-Van Male, A. M., Buller, H. A., Einerhand, A. W., Boren, T., and Dekker, J. (2003). The MUC5AC glycoprotein is the primary receptor for *Helicobacter pylori* in the human stomach. *Helicobacter.* **8**, 521–532.

Walz, A., Odenbreit, S., Mahdavi, J., Borén, T., and Ruhl, S. (2005). Identification and characterization of binding properties of *Helicobacter pylori* by glycoconjugate arrays. *Glycobiology* **15**, 700–708.

Wilkinson, D. G. (1998). "*In Situ* hybridization. A Practical Approach," 2nd Ed., pp. 23–66. Oxford University Press, B.D., Hames.

Yang, Z. (2004). Inference of selection from multiple species alignments. *Curr. Opin. Genet. Dev.* **12**, 688–694.

[21] *Caenorhabditis elegans* Carbohydrates in Bacterial Toxin Resistance

By Brad D. Barrows, Joel S. Griffitts, and Raffi V. Aroian

Abstract

The major virulence factor produced by the bacterium *Bacillus thuringiensis* (Bt) is a pore-forming toxin called crystal (Cry) toxin, which targets and kills insects and nematodes. To understand how this bacterial toxin interacts with its invertebrate hosts, a genetic screen in *C. elegans* for nematodes resistant to Bt toxin was carried out. Four of the five genes that mutated to toxin resistance encode glycosyltransferases. These genes were found to participate in the biosynthesis of *C. elegans* glycosphingolipids. These glycolipids in turn were shown to directly bind Bt toxin. Thus, resistance to Bt toxin in *C. elegans* can develop as a result of loss of glycolipid receptors for the toxin. Here we describe the isolation of Bt toxin resistance mutants in *C. elegans*, isolation of *C. elegans* glycolipids, and their separation by thin-layer chromatography, overlay assays to demonstrate direct binding of Bt toxin to glycolipids, and the purification of specific *C. elegans* glycolipid species.

Overview

The free-living, soil-dwelling, nematode *Caenorhabditis elegans* has shed important insights into many different areas of research. *C. elegans* has a short generation time, a mapped, sequenced genome, well-known forward and reverse genetics, and can be grown into large, genetically homogeneous populations for biochemical purifications. *C. elegans* has been increasingly used in the study of glycobiology (Altmann *et al.*, 2001; Cipollo *et al.*, 2005; Griffitts *et al.*, 2005; Haslam *et al.*, 2002; Schachter, 2004). One recent example involves the interaction of *C. elegans* with a bacterial toxin, the crystal protein Cry5B made by the bacterium, *Bacillus thuringiensis* (*Bt*). Cry5B is a pore-forming toxin (PFT) that intoxicates and kills *C. elegans* and other nematodes on ingestion and binding to receptors on the intestine (Griffitts *et al.*, 2001; Marroquin *et al.*, 2000; Wei *et al.*, 2003). Mutations in five genes were identified that result in *C. elegans* resistant to intoxication by Cry5B—that is, loss of any one of these genes results in animals that are much healthier than wild-type animals when ingesting the

METHODS IN ENZYMOLOGY, VOL. 417   0076-6879/06 $35.00
                                              DOI: 10.1016/S0076-6879(06)17021-4

PFT (Marroquin *et al.*, 2000). It was subsequently demonstrated that four of these genes (*bre-2*, *bre-3*, *bre-4*, and *bre-5*) have significant homology to glycosyltransferases and are part of a single biosynthetic pathway involved in biosynthesis of carbohydrate chains on glycosphingolipids (Griffitts *et al.*, 2001, 2003, 2005). The glycolipid carbohydrates were shown to be intestinal receptors for the PFT. Thus, the resistance phenotype associated with loss of any one of these genes can be attributed to the loss of toxin-binding sites. Research into understanding the molecular mechanisms of resistance to Bt crystal toxins is important for protecting this invaluable natural resource. Bt crystal proteins are increasingly used around the world as topical sprays and in transgenic plants to control insects that cause damage to crops or act as vectors for spreading disease (Crickmore, 2005; Griffitts and Aroian, 2005). For example, in the year 2005, 8.5 million hectares of transgenic cotton expressing a Bt crystal protein were grown, accounting for ~24% of all the cotton in the world (James, 2005).

In this chapter, we will discuss the key techniques used in studying the *bre* mutants, namely isolation of *C. elegans* Cry5B resistance mutants, isolation of *C. elegans* glycolipids, analysis by thin layer chromatography (TLC), overlay binding experiments with labeled Cry5B toxin on TLC-resolved *C. elegans* glycolipids, purification of specific glycolipid species, and enzymatic separation of the oligosaccharide from glycolipids.

Maintenance of C. elegans

All worm strains were maintained on 60-mm NG agar plates, which were seeded with approximately 50 μl of saturated *Escherichia coli* OP50 (Brenner, 1974). Strains were maintained by picking individual L4 stage worms to seeded plates every 2–3 days. Worm plates are incubated at a constant temperature of 20° (Hope, 1999). The *C. elegans* strain N2 Bristol is the standard wild-type laboratory strain (Brenner, 1974).

C. elegans *Chemical Mutagenesis*

Ethyl methanesulfonate (EMS) mutagenesis was the method used to generate mutations in *C. elegans* N2 animals. The use of EMS to isolate mutants in *C. elegans* was used in Brenner's initial genetics studies of *C. elegans* mutants (Brenner, 1974). EMS genetic screens have been shown to be capable of recovering one null mutation for every 2000 copies of a specific gene analyzed in any given screen (Jorgensen and Mango, 2002). EMS is a very efficient mutagen, which typically results in the generation of G/C → A/T point mutations (Anderson, 1995).

Procedure for EMS Mutagenesis

Section 1: Preparing a Synchronized Population of Worms

A P_0 generation of N2 worms synchronized at the L4 stage is typically used for mutagenesis (Sulston, 1988). A large population of worms is obtained first by using a sterile spatula to cut and transfer (like a piece of pie) 1/4 of a starved 60 mm plate of N2 worms onto a 100-mm ENG plate (see recipe in Section 3), which have been seeded the day before with 100 μl of a saturated culture of OP50. A total of three such plates are generated. These three plates are then allowed to grow 2–3 days until the worms have nearly consumed all the bacteria and become gravid adults. The adults are then washed from each ENG plate twice with 4 ml of H_2O and transferred to a 15-ml conical tube (Corning #430052). They are then centrifuged at 500g for 45 sec in a swinging bucket clinical centrifuge and the supernatant aspirated; 5 ml of water is then added and spun again to wash the pellet of worms and remove excess OP50. After aspirating the supernatant, 4 ml of bleach solution is added (for 5 ml bleach solution, 3.5 ml dd$H_2$0, 1.0 ml 5% NaOCl, 0.5 ml 5 N KOH). The worms are then continually mixed by inversion in the bleach solution while monitoring their lysis on a dissecting microscope. After the worms have lysed and eggs are released, the eggs are centrifuged, the bleach solution aspirated, and 8 ml of sterile double-distilled water is added. The eggs are centrifuged and aspirated as before and washed with water one more time and then one time with 8 ml of M9 buffer (22 mM KH$_2$Po$_4$, 42 mM Na$_2$HPO$_4$, 85.5 mM NaCl, 1 mM MgSO$_4$). Because the eggs do not pellet as well in M9 buffer, increase the spin to 750g for 75 sec. Carefully pipette off the supernatant and repeat the M9 wash. After the washes are completed, the eggs are suspended in 2 ml M9 and placed on a rotary platform to hatch overnight at room temperature (Huffman *et al.*, 2004).

After hatching the embryos, synchronized L1 larvae are plated onto 1–10 100-mm ENG plates previously seeded with OP50. The concentration of L1s in the tube can be obtained by counting the number of worms in several 5 μl amounts on an NG plate using a dissecting microscope. Each plate is inoculated with 10,000–20,000 worms. The worms are allowed to grow to the L4 stage (36–42 h at 25°).

Section 2: EMS Mutagenesis

After the P_0 generation has reached the L4 stage, the worms are harvested from the plates with 4 ml of M9 and transferred to a 15-ml conical tube (Corning #430052). The worms are then centrifuged for 30 sec at 200g. The supernatant is then aspirated, and the worm pellet is washed with an additional 8 ml of M9 to remove any residual bacteria. After spinning and aspirating, the worms are resuspended in 2 ml of M9. At this point a 2× solution (0.06 M) of EMS (Sigma # M-0880) can be prepared in a fume hood. Because of EMS being volatile and a recognized carcinogen, all steps involving the use of EMS are performed with great care in the fume hood while using skin and eye protection (e.g., double-gloved hands that are frequently changed). In addition to this, all items that come in contact with the EMS (pipettes, pipette tips, tubes) must be decontaminated with 1 M NaOH before they can be safely discarded. The 2× EMS solution is made by adding 12 μl EMS to 2 ml of M9 solution and mixing well to completely suspend the EMS (which is oily). This will produce a 60-mM EMS solution. The 2× EMS solution is then added to the 2 ml worm suspension. The tube with worms in EMS is capped, sealed with Parafilm, and then placed on a gentle rocker in the fume hood for 4 h. After the 4 h of incubation, the worms are pelleted by centrifuging for 30 sec at 200g. The supernatant is removed and discarded into a waste bucket containing 1 M NaOH. The worms are then washed two times with 4 ml of M9, and the supernatant is discarded into a waste bucket containing 1 M NaOH. At this point the worms are resuspended in 0.5 ml M9 and plated onto two 100-mm ENG plates seeded with OP50. The lids are left slightly open in the fume hood to allow evaporation of any remaining EMS. Once the plates are dry, the mutagenized P_0 worms are grown to gravid adults by incubating overnight at 20°.

Section 3: Resistant Mutant Genetic Screen

After growing the P_0 worms to gravid adults, they are collected and bleached as described in Section 1. After hatching the F1 embryos, the synchronized F1 larvae are plated on 100-mm OP50 seeded ENG plates at 5000–10,000 worms per plate. These worms are allowed to grow to gravid adults at 20°, at which point they are bleached one final time to obtain F2 embryos. The synchronized F2 L1 animals are challenged by *Bt* crystal toxin in an effort to identify possible resistant mutants. The screening for resistance can be done on plates (below and Marroquin *et al.* (2000) or in wells (Marroquin *et al.*, 2000).

ENGIC Plate Method for Cry5B-resistant Mutants. In this method, a toxin-expressing *E. coli* strain functions as both a toxin source and a food source for the animals being screened. The JM103 strain of *E. coli* is transformed with a plasmid containing (QIAGEN) crystal toxin gene under the control of an isopropyl-β-d-thiogalactopyranoside (IPTG)-inducible *T5* promoter (vector pQE9 from QIAGEN). Overnight cultures of two bacteria (a strain with Cry5B cloned into pQE9 vector and a strain with empty pQE9 vector) are spread separately onto 100-mm ENGIC plates (ENGIC media

Wild type   Mutant

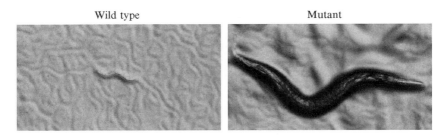

Fig. 1. A wild-type animal (left) and a *bre-3*(*ye28*) (right)–resistant animal grown from the L1 stage on Cry5B-expressing *E. coli* shown at the same magnification. This figure illustrates the ease with which a resistant mutant can be visualized. See Marroquin *et al.* (2000) for more on EMS screening for resistance mutants.

recipe can be found below). After spreading 100 μl of culture on each plate, they are dried overnight at 25°. It is good to set up many Cry5B toxin plates and a few empty vector plates.

F2 populations of mutagenized worms are then plated at the L1 stage (see preceding) at approximately 3500 worms per plate. As a control, 3500 animals are also plated on a plate spread with only empty vector. In addition, 20–30 non-mutagenized L1 worms are plated on one toxin plate and one empty vector plate (controls). After approximately 3 days, the toxin plates with mutagenized worms are scanned for rare (~1/1000–1/5000) healthy well-developed, resistant worms that seem unaffected by the toxin (Fig. 1). In contrast, the nonresistant worms will be severely underdeveloped. The putative resistant mutants are collected onto OP50 (nontoxin plates) and allowed to multiply before their progeny are retested to confirm heritable resistance.

Plate Media

ENG media

5 g Bacto-Peptone
1 g Yeast extract
3 g NaCl
20 g Granulated agar
1000 ml H_2O
Autoclave for 45 min
When cool enough to touch, add:
1 ml 5 mg/mL cholesterol in 100% EtOH
1 ml of 1 M $CaCl_2$
1 ml of 1 M $MgSO_4$
25 ml of 1 M KPO_4

After mixing, distribute the media to 100-mm plates at a volume of 30 ml per plate. Allow the media to dry for 2 days at room temperature before inoculating with a fresh bacterial culture.

For ENGIC plates, in addition add 1 ml of 50 mg/ml carbenicillin and 200 μl of 0.5 M IPTG after cooling, for final concentrations of 0.1 mM IPTG and 50 μg/ml carbenicillin.

C. elegans *Glycosphingolipids*

C. elegans possesses a host of genes involved in glycosylation similar to those found in mammals (Schachter, 2004). Because of the ease with which a researcher can carry out *C. elegans* genetic experiments, the animal provides an excellent system for studying the function of specific glycosylation genes and specific disorders that may result from the lack of these genes (Schachter, 2004). Studies of *C. elegans* Bt resistance mutants has led to the explanation and detailed characterization of intestinal glycolipids (Griffitts *et al.*, 2005). It was demonstrated that the carbohydrate portion of arthroseries glycolipids found in the *C. elegans* intestine are essential for intoxication and bind Cry5B—that is, they act as toxin receptors. The sequence of carbohydrates and their linkages were delineated for three Cry5B-binding glycolipids, with the most complex one containing 11 sugar residues. Further work demonstrated that terminal galactose residues play a particularly important role in the binding of Cry5B to glycolipids (Griffitts *et al.*, 2005). The techniques used in this research demonstrate the value of *C. elegans* in studying glycobiology. Following are the procedures used to isolate and characterize glycolipids from *C. elegans*.

Isolation of Glycosphingolipids from *C. elegans* by Svennerholm Partitioning

This method of purification involves using chloroform, methanol, and water to purify glycolipids from homogenized worm pellets. This is achieved by extracting lipid material into two chemical phases (Svennerholm and Fredman, 1980). The upper phase, which is more hydrophilic, attracts glycolipids with more polar carbohydrate structures. The hydrophobic lower phase collects simple glycolipids and other nonpolar lipids, which have less complexity or no sugar structure. After separation, the upper phase solution is depleted of any nonlipid contaminants by applying the samples to a small hydrophobic column (Schnaar, 1994). The samples purified off of the column are dried down and resuspended in methanol for further analyses (Schnaar and Needham, 1994). For example, the glycolipids from a wild-type *C. elegans* population can be compared with the glycolipids from a population lacking the *bre-2* gene. After these comparative studies, specific glycolipid species (e.g., those absent in the mutant animals but present in wild-type animals) may be subjected to further purification and analyses. Processing of a 600-μl worm pellet will result in an upper phase volume of 300 μl and a lower phase volume of 600 μl.

Section 1: Preparing Worms

To prepare a useful volume of sample, six 100-mm ENG plates should be made for each strain to be processed. Each plate should be spread with 200 μl of fresh overnight culture of *E. coli* strain JM103. These plates should be allowed 48 h to dry and grow a healthy lawn at room temperature.

The six seeded ENG plates (see preceding) are inoculated with approximately 2000 mixed life stage worms from the specific strain of interest. These 2000 worms can be collected from typical 60-mm maintenance plates. The worms are washed from the maintenance plates with 1.5 ml of H$_2$O and transferred to a 15-ml conical tube (Corning #430052). The residual OP50 is washed from the worms by performing three washes (see Notes below). Each wash consist of adding 8 ml of water to the worms, gently inverting several times, centrifuging at 500g for 45 sec in a swinging bucket clinical centrifuge, and aspirating the supernatant. The density of worms is estimated by counting the number of worms in several 5-μl amounts on an NG plate using a dissecting microscope. An appropriate amount is then pipetted to place 2000 worms on each of the 100-mm plates.

Notes: Mixed life stages are used here to ensure that glycosphingolipids that may only be produced during specific life stages are not excluded. Completely starved worms are avoided because of the possibility of altered glycosphingolipid expression resulting from starvation.

After inoculation of a 100-mm ENG plate with 2000 mixed stage worms, the plates are grown for approximately 4 days or more at 20°, depending on the growth rate of the specific strain, until the worm population is nearly starved (see Notes). Plates that are close to starving can be easily identified by a visible wave of worms forming at the edge. After the wave forms, the plate will starve within approximately 8 h at 20° as the wave slowly closes in on the center of the plate. After the plate has reached a point near starvation, the worms can be washed from the plates (see Notes). Using 3 ml of water for each plate, soak the worms for 10 min. The suspended worms can be transferred to a 15-ml conical tube (Corning #430052). Spin and wash the worms three times with chilled water as described for the initial inoculation. After the final spin, remove as much of the supernatant as possible and flash freeze the pellet in liquid nitrogen (see Notes). The pellets can be stored at –80° until the time of purification. The typical size of the worm pellet is between ~0.5 and ~1 ml.

Notes: Do not try to hasten the growth by incubating at 25°, because the increase in temperature may also contribute to variances in glycolipid expression. Worms are harvested at a point near starvation to maximize the amount of worm material collected while minimizing the quantity of bacteria to wash away before purification. Even if you are performing the purification on the same day as the worm collection, you should flash freeze the pellets in liquid nitrogen to aid in weakening worm cuticle.

Section 2: Lipid Extraction

Thaw the previously frozen pellets at room temperature. When purifying glycolipids from more than one strain, it is helpful to adjust all pellets to equal volumes for steps involving centrifugation. The volume is adjusted by removing a small amount of material from slightly larger pellets. After adjusting, add three pellet volumes of sterile double distilled water (ddH_2O) to each pellet and transfer it to a 10-ml glass conical vial (Pierce # 13225). Sonicate the samples for 30 sec at 8 watts (power dial between 4 and 5) using a microtip on a sonicator (Fisher model #60). Keep the samples on ice during the sonication. Verify complete disruption of the sample by tilting the vial and looking for whole worms under a dissecting microscope. If whole worms are still visible, repeat the sonication.

After the sonication, confirm the pellet volumes and adjust all samples to be equal by adding sterile ddH_2O. After confirming the volume of each sample, add 2.67 homogenate volumes of methanol to each sample and briefly vortex (see Notes). At this point it is no longer necessary to keep the sample on ice. Allow the samples to incubate at room temperature for 5 min. After the incubation, add 1.33 homogenate volumes of chloroform and mix by gently inverting several times (see Notes). This mixture contains a final ratio of 4:8:3 (chloroform/methanol/water).

After adding the chloroform and methanol, the samples are incubated at 37° for 2 h. During the 2 h of incubation, mix the samples by inverting several times every 15 min.

On completion of the incubation, centrifuge the samples at 1400g for 5 min using a swinging bucket clinical centrifuge. Using a 25-ml glass pipette, transfer the supernatant to a new 30-ml glass centrifuge tube (Kimble HS No. 45600-30, Cap No. 24-400). Be careful not to disturb the pellet when transferring the supernatant. Add 0.173 volumes of water to each of the supernatants, cap, and mix by inverting several times (see Notes). After mixing the samples, centrifuge them at 1400g for 5 min. At this point, two distinct layers should be visible in the sample tubes. The top layer (upper phase) contains glycolipids with polar oligosaccharide structures, which are soluble in the more aqueous solution, whereas the bottom layer contains less substituted lipids, phospholipids, and cholesterol (Schnaar, 1994).

Notes: Methanol added = 2.67 × (original wet pellet volume + 3 pellet volumes of water). Chloroform added = 1.33 × (original wet pellet volume + 3 pellet volumes of water). After adding the chloroform and methanol, the samples are stable enough to be left at 4° until time permits the completion of the protocol. Water added = 0.173 × (the volume transferred to the 30-ml glass tube).

Section 3: Purification

To remove remaining contaminants from the upper phase glycolipids, it is necessary to perform reversed-phase chromatography. This process removes nonlipid contaminants while concentrating glycolipids (McCluer *et al.*, 1989; Schnaar, 1994; Williams and McCluer, 1980). This process can be performed with a Sep-Pak + cartridge (Millipore Corp Product #–WAT036810) and a 5-ml glass syringe (BD #512471) for applying sample to the cartridge. After applying the sample, it is washed, eluted, and dried down to concentrate the sample. Three solvent mixtures are required for this protocol: 2:43:55 (chloroform/methanol/water), methanol, and 1:1 (chloroform/methanol).

After the 30-ml tubes have been centrifuged to produce two distinct layers, they may remain at room temperature during the purification. Transferring of sample and all solvents should be performed with glass pipettes to reduce product loss. After adding the first solution to the cartridge (see Notes), air must not be passed over the cartridge during the procedure. Flow rate should be a fast drip. One Sep-Pak + column contains enough substrate to efficiently purify approximately 8 ml of upper phase glycolipids. Collect all waste solution in a large beaker in a fume hood.

The purification cycle begins by pushing 4 ml of 1:1 chloroform/methanol over the column at the pace of a fast drip using a 5-ml syringe (approximately 2 drops/sec). At the end of every solution addition, be sure to stop before air is pushed into the column to maintain substrate homogeneity. Continue the equilibration by adding 4 ml of methanol to the column. After the methanol has been pushed through the column, add 4 ml of 2:43:55 (chloroform/methanol/water) to finish the equilibration. The next step requires a clean test tube in which your first run of sample can be collected. Every 4-ml portion of upper phase should be run over the column twice to ensure a maximum yield. Add 4 ml of the first upper phase sample. Push the sample through the column while collecting the run through in a clean test tube. Place the first run through aside while proceeding to the first wash step. Wash the sample by applying 4 ml of 2:43:55 chloroform/methanol/water. The run through for this step can be collected in the waste container. The washed sample can now be eluted into a clean 10-ml glass conical vial (Pierce # 13225). Elute the sample by adding 4 ml of 1:1 chloroform/methanol. Allow the first five water-rich drops from the previous wash to run into the waste container. Elute the rest of the solution into the conical vial. Your purified sample should emerge as a white precipitate between the first 10 and 20 drops of eluted solution. This is the one step in which you may push the eluting solution through with air to completely evacuate the column. This completes the first half of one cycle. The second half of the cycle involves following the same procedure, except the sample run-through saved in the test tube is applied to the column. During this half of the cycle, the sample run-through can be collected in the waste beaker. Eluting the remaining sample into the conical vial completes one full cycle. One column is good for purifying 8 ml of raw sample. The columns should be switched when either changing strains or 8 ml of raw sample has been run through the column twice.

Purified upper-phase lipids can be dried down by gently blowing a continuous stream of nitrogen gas over the sample while it sits on a 45° heating block. The lower phase can be collected by drawing it from the bottom of the 30-ml glass tube with a glass pipette to transfer it to a separate conical vial. The lower phase sample should also be dried down using a stream of nitrogen. Samples may be damaged if they are left on the heating block too long after they have completed drying.

Section 4: Suspension of Dried Samples

The upper and lower phase dried samples are suspended in a volume of solvent proportional to the original worm pellet volumes. Upper phase samples are dissolved in a volume of methanol, which is equal to one half the original worm pellet volume. To completely dissolve the sample, it may be necessary to scrape it from the sides using a glass Pasteur pipette. Vortexing or sonicating in a bath-type sonicator in addition to gentle heating (in a 45° water bath) may also help dissolve the sample (Schnaar, 1994). Remaining white precipitate can be discarded, because it likely consists of protein impurities. Lower phase samples can be dissolved in a volume of 1:1 chloroform/methanol, which is equal to the volume of the original worm pellet. Vortexing is typically sufficient to completely dissolve the sample.

Analysis of Glycolipids by Thin Layer Chromatography

Analysis of both upper and lower phase glycolipids can be accomplished by resolving the samples on thin layer chromatography (TLC) plates.

This system involves using solvent solutions to separate different glycolipid species on the basis of polarity (Schnaar and Needham, 1994). The sample is applied to a thin layer of silica substrate. The silica is a polar substrate, which binds samples applied to the TLC plate. Sample resolution is achieved through the use of specifically prepared solvents. These solvents participate in polar interactions with the glycolipids targeted for resolution. The mobility of glycolipids on silica TLC plates is generally inversely proportional to the number of monosaccharide residues in the head group, although non-carbohydrate polar substitutions can change mobility dramatically (Schnaar and Needham, 1994).

After resolving glycolipid samples by TLC, the resolved band pattern produced can be visualized by staining. One common method used for staining glycolipids involves the use of an orcinol-sulfuric acid test (Schnaar and Needham, 1994; Svennerholm, 1956). Orcinol is a general carbohydrate stain, which is capable of detecting neutral sugar residues in quantities as low as 500 pmol (Schnaar and Needham, 1994). In the interest of studying how specific glycolipids may interact with other molecules, fixing the resolved samples to a TLC plate makes it possible to probe them with labeled proteins much like a Far Western blot (Magnani et al., 1987; Schnaar and Needham, 1994). In studies involving the Bt crystal toxin Cry5B, this overlay approach has been used to determine that glycolipids directly bind Cry5B. Overlay methods have also been used in competition studies to evaluate the specific properties of the identified receptors (Griffitts et al., 2005). This information was later used to target specific glycolipids for further analysis and help determine which monosaccharides play an important role in the binding interaction (Griffitts et al., 2005).

Section 1: Resolving Sample by TLC

C. elegans glycolipids can be resolved using Merck glass-backed silica-60 High-Performance Thin Layer Chromatography (HPTLC) plates (Fisher # M5631-5), a 1-μl Hamilton syringe (Fisher # 14824200) for loading sample, solvents, and a developing tank (Fisher # K416180-0000). The solvent used to resolve upper phase glycolipids contains 4:4:1 (chloroform/methanol/ water), and the solvent used to resolve lower phase glycolipids contains 45:18:3 (chloroform/methanol/water). The silica plate can be prepared by lightly marking with a pencil to label and evenly space multiple samples. Sample lines should be approximately 1 cm long and at least 1 cm from the bottom or either edge of the plate. Sample lines should be spaced at least 0.1 cm apart. An additional mark can also be made 4.5 cm from the bottom as a point of termination for solvent migration (Fig. 2). This will permit the

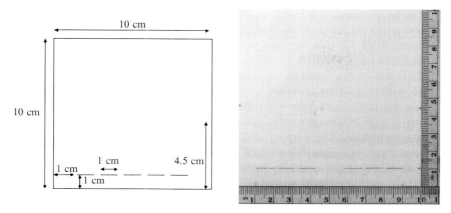

Fig. 2. Pencil mark layout for a 10 × 10-cm TLC plate.

use of two ends of a 10 cm by 10 cm silica plate. If greater sample resolution is required, mark the termination point 1 cm from the top of the plate. After the plate has been labeled, it should be dried in an oven at 125° for 10 min before applying sample (Schnaar and Needham, 1994).

While drying the plate, resolving solutions should be added to the developing tank at a volume of approximately 20 ml (0.5-cm depth). The tank should be allowed to equilibrate with the solvent solution for at least 30 min at room temperature before developing the plate (Schnaar and Needham, 1994). After the silica plate has cooled, thin lines of sample can be added just above the pencil guidelines. A 1-μl Hamilton syringe can be used to apply the sample. The sample should be applied in a thin line to achieve better resolution of closely spaced glycolipid bands. A 5-μl load is typically sufficient to produce a visible band pattern after orcinol staining. After applying the sample, allow the sample to dry on a heat block at 45°. The dried silica plate can then be placed in the equilibrated resolving chamber. Quickly replace the airtight cover and allow the solvent front to travel ~4.5 cm. The plate is then dried on a heat block at 45°. The plate is now ready for staining.

Section 2: Orcinol Staining Resolved Glycolipids

Equipment required for this protocol includes a fine mist sprayer (Fisher # K422530-0125), a sprayer box, a fume hood, and a drying oven. Chemicals required for this protocol include orcinol (Sigma # O-1875), methanol (EMD # MX0485-7), water, and sulfuric acid (Fisher # S79200MF).

The orcinol solution is made in the reservoir portion of the glass sprayer, which should be wrapped with aluminum foil to prevent the degradation of the orcinol by light exposure; 40 ml of orcinol solution can be made as follows: In a fume hood, add 200 mg of orcinol followed by 25 ml of methanol and 5 ml of water directly to the flask sprayer reservoir. Gently mix the three components by swirling the flask. Then 10 ml of sulfuric acid should slowly be added to the mix while gently swirling (taking appropriate precautions when working with the strong acid). The acid must be added slowly because of the heat-producing chemical interaction between the acid and aqueous solutions. After mixing all of the components and allowing the sprayer flask to cool, the sprayer may be assembled. The orcinol solution can be kept in a dark area at 4° for a period of approximately 7 days (Schnaar and Needham, 1994).

The application of the orcinol staining solution should be done in a fume hood while using a sprayer box. Apply air to the sprayer at a pressure that produces a fine and even mist. Evenly spray the plate with multiple passes, but stop before the plate shows signs of visible moisture. Immediately move the plate to the drying oven at a temperature of approximately 125° and allow it to dry for no more than 10 min. After drying, resolved banding patterns should be visible on the surface of the plate (Fig. 3). The plate should immediately be scanned or photographed, because the stain will begin fading soon after being removed from the oven (Schnaar and Needham, 1994).

Section 3: Preparing Protein Probe for Overlay Analysis

Cry5B is labeled by biotinylation before probing glycolipids to test for direct binding. The labeling reaction is performed by combining the protein of interest with a fourfold to sixfold molar excess of N-hydroxysuccinimidyl

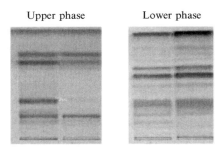

Upper phase   Lower phase

FIG. 3. Thin layer chromatography of upper and lower phase samples from two different strains of *C. elegans*. The solvent used to resolve the upper phase glycolipids consisted of 4:4:1 (chloroform/methanol/water). The solvent used to resolve the lower phase sample consisted of 45:18:3 (chloroform/methanol/water). After samples were resolved, staining was performed with an orcinol-sulfuric acid spray. The two lanes for both plates represent glycolipids from N2 and *bre-2(ye31)* animals, respectively, from left to right.

esters of biotinamidocaproate (Sigma #B-2643). After combining in an appropriate buffer, incubate for 2 h at 20°. Excess label can be removed by desalting with P-6 resin beads (Bio-Rad # 150-0738) equilibrated in the protein's final suspension buffer. A buffer consisting of 20 mM HEPES at a pH of 8.0 was used for labeling and desalting Cry5B (Griffitts *et al.*, 2005).

Biotinylated lectins may also be used as protein probes for overlay analysis. There are many lectins that bind specific sugar residues. These can be used to obtain a general analysis for the exposed structure of glyco-lipid samples resolved by TLC. The only potential difference in protocol for using a lectin versus Cry5B as a probe is the buffer used. In either case, a buffer specific to the protein or lectin may be used to encourage target binding to the glycolipids.

Section 4: Preparing Resolved Samples and Protein Overlay Analysis

Preparation of glycolipid samples for overlay analysis first involves resolving the samples on an HPTLC plate as described in Section 1. The resolved glycolipid samples can be fixed to the plate using a solution of 0.02% polyisobutylmethacrylate (PIBM) (Aldrich # 18154-4) in hexane (Fisher # H303-1) after the running solvent has been completely evapo-rated (Griffitts *et al.*, 2005; Schnaar and Needham, 1994).

All steps involving hexane are performed in a fume hood. Before fixing, the TLC plate is equilibrated in a plastic tray (w × h × d 7.5 × 2.5 × 11 cm) containing 40 ml hexane for a period of 60 sec. The plate is then quickly transferred to a second tray containing 40 ml 0.02% PIBM in hexanes for 60 sec. After 60 sec, move the plate to a 45° heat block where the residual hexane is completely evaporated. The fixed plate is now ready to be blocked in a 0.5% bovine serum albumin (BSA), phosphate-buffered saline (PBS) (136 mM NaCl, 2.6 mM KCl, 10 mM Na$_2$HPO$_4$, 1.8 mM KH$_2$PO$_4$) with 0.02% Tween-20.

The volume of buffer for this protocol was designed for a 5 cm by 10 cm plate in a plastic tray; 200 ml of block solution (180 ml H$_2$O, 20 ml 10× PBS pH with HCl to 7.2, 160 μl 25% Tween-20, 1 g BSA) is sufficient for completing this protocol. The initial block requires 10 ml of blocking buffer and 30 min on a rocker at room temperature. The plate can then be probed with the labeled protein in block solution. When probing with labeled Cry5B, the crystal toxin is added to 10 ml of block solution to a concentra-tion of 11 nM (Griffitts *et al.*, 2005). After adding probe, the plate should be incubated on a rocker for 2 h at room temperature. When probing is complete, remove the solution and add 10 ml block solution and rock for 1 min. Wash the plate a second time with 10 ml of block solution for 5 min. After completing the washes, a streptavidin-linked alkaline phosphatase solution can be added to the plate. Components to make this solution can be obtained from a Vetastain Kit (Vector Labs # AK-5000). The two components of this kit include streptavidin and biotinylated alkaline phosphatase. The kit components are diluted 150 fold when added (40 μl each) to a tube containing 6 ml of block solution. Mix by gently inverting the tube several times before adding it to the plate. Allow the plate to incubate in the solution on a rocker for 1 h at room temperature. After removing the alkaline phosphatase solution, wash the plate three times for 1 min in 10 ml of clean blocking buffer. The specific glycolipids that bind with the labeled protein can be revealed by a chemical color reaction.

The color reaction is initiated by submerging the plate in a solution of 5-Bromo-4-Chloro-Indolyl-Phosphatase (BCIP) and Nitro blue Tetrazolium (NBT) diluted in alkaline phosphatase (AP) buffer (100 mM Tris pH 9.5, 100 mM NaCl, 50 mM MgCl$_2$). Add 30 μl each of BCIP and NBT, in order, to 5 ml of AP buffer. After the washes, move the plate to a new tray and submerge it in 5 ml of the color reaction solution. Allow the plate to rock for 10 min or until stained bands become clearly apparent (Griffitts *et al.*, 2005). After achieving the desired amount of staining, the plate can then be washed with water and dried for scanning or photography (Fig. 4).

Section 5: Polystyrene Microplate Overlay

Semiquantitative analysis of Cry toxin-binding affinity with different *C. elegans* strain upper phase samples can be performed through the use of a polystyrene microplate overlay analysis.

Upper phase glycolipids from mixed life stage worm pellets in which the total number of worms can be quantitated by sampling worm quantities in several small aliquots. The glycolipids are dried down and dissolved in a

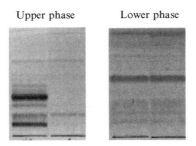

FIG. 4. Cry5B overlay analysis of upper and lower phase glycolipid samples from two different *C. elegans* strains. The two samples were resolved as described in Fig. 3. The plates were fixed with 0.02% PIBM in hexane before being probed with biotinylated Cry5B. The two samples and their order are the same as described in Fig. 3.

solution of 1:1 (methanol/water). Solution volumes, representative of specific numbers of worms, are then transferred to a 96-well polystyrene microtiter plate (Costar #9017, medium binding). This solution is allowed to evaporate at room temperature for a period of 135 min. Any remaining solution is removed and replaced with blocking solution (42 mM Na$_2$HPO$_4$, 85 mM NaCl, 1 mM MgSO$_4$, 0.2% defatted BSA) and allowed to block for 30 min. The wells can then be probed with biotinylated toxin ($\sim$20 nM) in blocking solution for 1 h at room temperature. The wells are then washed twice with blocking buffer and incubated with the alkaline phosphatase solution (Vector Labs # AK-5000), made as described in section 4 , in blocking buffer for 45 min. Wells were then washed twice with BSA-free block solution and once with water. P-nitrophenyl phosphate (PnPP) (Sigma #N9389) is then added at a concentration of 1 mg/ml in PnPP buffer (50 mM HCO$_3$, .5 mM MgCl$_2$, pH 10). After positive control (wild-type glycolipids at the highest concentration) wells reach an OD405 of 1, the color reaction can be stopped by adding 3 M NaOH at a 1:5 ratio to each well. OD405 measurements can then be taken for all wells.

Isolation of Specific Glycosphingolipid Components

To further investigate the structure and function of a specific glycolipid species, it is necessary to separate it from all other isolates collected during the bulk purification described previously. Because any specific glycolipid species will make up only a small portion of the total extracted sample, the purification procedure must be scaled up. The glycolipid extraction protocol for C. elegans should be scaled up to accommodate a 10–20 g worm pellet (approximately 10 150-mm ENG plates). The purification of a specific glycolipid species uses the same principles as resolving glycolipid samples on a TLC plate. This method involves using a silica gel column for separating specific species that are eluted by use of a gradient of solvents. Solvent ratios can be predetermined using TLC to identify a specific solution that is capable of providing adequate resolution for the target species. Because of the possibility that multiple species may have different carbohydrate structures while still migrating very close together, the isolation of one specific glycolipid species cannot be guaranteed. In cases involving species that are difficult to separate, additional methods may be required for completing purification (Schnaar, 1994).

Section 1: Preparation of a 4-ml Silica Gel Column

A silica gel column can be prepared using a 4-ml glass column (Pierce #20055), silica gel (Sigma #227196), and a small amount of clean sand

(Sigma #274739). Approximately 1 g of silica is used for every 10 g of wet worm pellet. The silica should be dried in an oven at 125° before being resuspended in 45:18 (chloroform-methanol); 2 g of the silica slurry can then be added to a column with the provided end cap in place; 2 mm of clean sand is added on top of the silica to help prevent surface disruption when adding solvent and sample to the column. Wash the column with 45:18 (chloroform-methanol) at 30 ml/g of dry silica. To prevent air from entering the column, stop the flow of solvent at a point above the silica at the end of every step.

Section 2: Sample Preparation and Glycosphingolipid Component Purification

Desalted upper phase sample should be dried down and resuspended in an equal volume of 45:18 (chloroform/methanol) before loading it onto the column. The sample should be loaded, using a Pasteur pipette, onto the column at a flow rate of less than 1 ml/min to permit adequate time for substrate binding to occur. To ensure the column is capable of retaining the entire sample, it may be necessary to analyze the flow-through by TLC. After loading the sample, wash the column with several column volumes of 45:18 (chloroform/methanol). After washing the sample, a multistep elution can be performed with multiple solutions with different chloroform/methanol/water ratios. The best ratios for purification of specific glycolipids can be predetermined by TLC analysis. With respect to wild-type *C. elegans*, specific upper phase components can be well separated using a gradient from 30:30:3 to 30:30:8 (chloroform/methanol/water) over 20 column volumes. This is achieved by making six solvent solutions that all differ by 1 part water (30:30:3-8). The amount of each solution used should be split evenly among the 20 column volume fractions ($\sim$8 ml each). Start the gradient with the least polar solvent (30:30:3) and increase by one part water with each additional step in the gradient. All fractions are checked by TLC and pooled by the specific component targeted for isolation. After completing the purification, pooled fractions are dried down and resuspended in methanol (Griffitts *et al.*, 2005; Schnaar, 1994).

Section 3: Enzymatic Release of a Purified Glycolipid Carbohydrate

After a specific glycolipid species has been purified, the carbohydrate portion can be isolated from the lipid by ceramide glycanase (Calbiochem #219484); 1 unit of ceramide glycanase from *Macrobdella decora* is capable of releasing carbohydrate from approximately 100 nmol of purified glycolipid. Purified sample can be prepared for this reaction by drying down and suspending in ceramide glycanase buffer (0.1% taurodeoxycholate, 30 mM

sodium acetate at pH 5.0). Add ceramide glycanase at a ratio of 1 unit per 100 nmol of purified glycolipid. Incubate the reaction at 37° for 24–36 h. TLC can be used to check for complete digestion every 12 h. When carbohydrate has been completely released, a dark band will remain at the origin when resolved with 4:4:1 (chloroform/methanol/water) (see Notes) (Griffitts et al., 2005; Schnaar and Needham, 1994).

When the reaction is complete, the sample can be purified from the lipid portion with a Sep-Pak + cartridge (Millipore Corp Product #– WAT036810). The cartridge is primed by adding, in order, 4 ml chloroform, 4 ml methanol, and 4 ml water. After priming, the sample is added and eluted off the column with water in approximately 12 200-μl fractions. TLC can be used to check for the presence of released carbohydrate in each fraction. All fractions containing carbohydrate can be pooled and dried under nitrogen (Griffitts et al., 2005; Schnaar and Needham, 1994).

Notes: 2:1 (isopropanol/water) can also be used as a resolving buffer for the detection of carbohydrates. The polar sample is capable of further migration and resolution when using a more hydrophilic buffer.

Section 4: Determination of Carbohydrate Concentration

The concentration of a purified carbohydrate sample can be determined using a 2% anthrone solution in ethyl acetate. In a spectrophotometer-compatible glass test tube, add 400 μl of sample diluted in water to 100 μl 2% anthrone and 1 ml sulfuric acid (Fisher #S79200MF) for a total volume of 1.5 ml. Gently swirl until the ethyl acetate is hydrolyzed and a floc of anthrone appears. Swirl more rapidly to dissolve the anthrone and allow the solution to develop for 10 min. After the solution has developed, check the optical density of the sample at 620 nm (Loewus, 1951).

The sample OD can be compared with a series of known standards near a similar concentration. A standard curve can be produced using three concentrations of glucose. A typical carbohydrate sample can be quantitated using a glucose standard curve, which includes 0.4 μg, 1.6 μg, and 4.0 μg in the 400-μl sample portion (Loewus, 1951).

References

Altmann, F., Fabini, G., Ahorn, H., and Wilson, I. B. (2001). Genetic model organisms in the study of N-glycans. Biochimie 83, 703–712.

Anderson, P. (1995). Mutagenesis. Methods Cell Biol. 48, 31–58.

Brenner, S. (1974). The genetics of Caenorhabditis elegans. Genetics 77, 71–94.

Cipollo, J. F., Awad, A. M., Costello, C. E., and Hirschberg, C. B. (2005). N-Glycans of Caenorhabditis elegans are specific to developmental stages. J. Biol. Chem. 280, 26063–26072.

Crickmore, N. (2005). Using worms to better understand how *Bacillus thuringiensis* kills insects. *Trends Microbiol.* **13**, 347–350.

Griffitts, J. S., and Aroian, R. V. (2005). Many roads to resistance: How invertebrates adapt to Bt toxins. *Bioessays* **27**, 614–624.

Griffitts, J. S., Whitacre, J. L., Stevens, D. E., and Aroian, R. V. (2001). Bt toxin resistance from loss of a putative carbohydrate-modifying enzyme. *Science* **293**, 860–864.

Griffitts, J. S., Haslam, S. M., Yang, T., Garczynski, S. F., Mulloy, B., Morris, H., Cremer, P. S., Dell, A., Adang, M. J., and Aroian, R. V. (2005). Glycolipids as receptors for *Bacillus thuringiensis* crystal toxin. *Science* **307**, 922–925.

Griffitts, J. S., Huffman, D. L., Whitacre, J. L., Barrows, B. D., Marroquin, L. D., Muller, R., Brown, J. R., Hennet, T., Esko, J. D., and Aroian, R. V. (2003). Resistance to a bacterial toxin is mediated by removal of a conserved glycosylation pathway required for toxin-host interactions. *J. Biol. Chem.* **278**, 45594–45602.

Haslam, S. M., Gems, D., Morris, H. R., and Dell, A. (2002). The glycomes of *Caenorhabditis elegans* and other model organisms. *Biochem. Soc. Symp.* 117–134.

Hope, I. A. (1999). "*C. elegans.*" Oxford University Press, Oxford.

Huffman, D. L., Bischof, L. J., Griffitts, J. S., and Aroian, R. V. (2004). Pore worms: Using *Caenorhabditis elegans* to study how bacterial toxins interact with their target host. *Int. J. Med. Microbiol.* **293**, 599–607.

James, C. (2005). Global Status of Commercialized Biotech/GM Crops. *In* "ISAAA Briefs No. 34." ISAAA, Ithaca, NY.

Jorgensen, E. M., and Mango, S. E. (2002). The art and design of genetic screens: *Caenorhabditis elegans. Nat. Rev. Genet.* **3**, 356–369.

Loewus, F. A. (1952). Improvement in anthrone method for determination of carbohydrates. *Amer. Chem. Soc. J.* **24**, 219.

Magnani, J. L., Spitalnik, S. L., and Ginsburg, V. (1987). Antibodies against cell surface carbohydrates: determination of antigen structure. *Methods Enzymol.* **138**, 195–207.

Marroquin, L. D., Elyassnia, D., Griffitts, J. S., Feitelson, J. S., and Aroian, R. V. (2000). *Bacillus thuringiensis* (Bt) toxin susceptibility and isolation of resistance mutants in the nematode Caenorhabditis elegans. *Genetics* **155**, 1693–1699.

McCluer, R. H., Ullman, M. D., and Jungalwala, F. B. (1989). High-performance liquid chromatography of membrane lipids: Glycosphingolipids and phospholipids. *Methods Enzymol.* **172**, 538–575.

Schachter, H. (2004). Protein glycosylation lessons from *Caenorhabditis elegans. Curr. Opin. Struct. Biol.* **14**, 607–616.

Schnaar, R. L. (1994). Isolation of glycosphingolipids. *Methods Enzymol.* **230**, 348–370.

Schnaar, R. L., and Needham, L. K. (1994). Thin-layer chromatography of glycosphingolipids. *Methods Enzymol.* **230**, 371–389.

Sulston, J., and H. J. (1988). Methods. *In* "The Nematode *Caenorhabditis elegans*" (W. B. Wood, ed.), pp. 587–606. Cold Spring Harbor Laboratory, Boulder.

Svennerholm, L. (1956). The quantitative estimation of cerebrosides in nervous tissue. *J. Neurochem.* **1**, 42–53.

Svennerholm, L., and Fredman, P. (1980). A procedure for the quantitative isolation of brain gangliosides. *Biochim. Biophys. Acta.* **617**, 97–109.

Wei, J. Z., Hale, K., Carta, L., Platzer, E., Wong, C., Fang, S. C., and Aroian, R. V. (2003). *Bacillus thuringiensis* crystal proteins that target nematodes. *Proc. Natl. Acad. Sci. USA* **100**, 2760–2765.

Williams, M. A., and McCluer, R. H. (1980). The use of Sep-Pak C18 cartridges during the isolation of gangliosides. *J. Neurochem.* **35**, 266–269.

[22]   N-Glycans Are Involved in the Response of Caenorhabditis elegans to Bacterial Pathogens

By Hui Shi, Jenny Tan, and Harry Schachter

Abstract

Caenorhabditis elegans is becoming a popular tool for the study of glycan function particularly as it applies to development. More than 150 C. elegans genes have been identified as homologs of vertebrate genes involved in glycan metabolism. However, only a relatively small number of these genes have been expressed and studied in any detail. Oligomannose N-glycans (Man$_{5-9}$ GlcNAc$_2$Asn), major components of the N-glycans of all eukaryotes including C. elegans, are essential, at least in part, for eukaryote survival, because they play an important role in protein quality control. In addition, vertebrates make hybrid (GlcNAcMan$_{3-5}$GlcNAc$_2$Asn) and complex (XGlcNAc$_{2-6}$Man$_3$ GlcNAc$_2$Asn) but little or no paucimannose (Man$_{3-4}$GlcNAc$_2$Asn)N-glycans, whereas plants, insects, and C. elegans make paucimannose but little or no hybrid nor complex N-glycans. UDP-GlcNAc:α3-D-mannoside β1,2-N-acetylglucosaminyltransferase I (encoded by the gene Mgat1) controls the synthesis of hybrid, complex, and paucimannose N-glycans in all eukaryotes. C. elegans has three genes encoding β1,2-N-acetylglucosaminyltransferase I (gly-12, gly-13, gly-14). To determine the functional requirement for this enzyme in worms, we generated seven worm strains with mutations in these three genes (gly-12, dpy-6 gly-13, gly-14, gly-12 gly-13, gly-14;gly-12, gly-14; dpy-6 gly-13 and gly-14;gly-12 gly-13). Whereas mice and Drosophila melanogaster with null mutations in Mgat1 suffer severe developmental abnormalities, all seven C. elegans strains with null mutations in the genes encoding β1,2-N-acetylglucosaminyltransferase I develop normally and seem to have a wild-type phenotype. We now present evidence that β1,2-N-acetylglucosaminyltransferase I-dependent N-glycans (consisting mainly of paucimannose N-glycans) play a role in the interaction of C. elegans with pathogenic bacteria, suggesting that these N-glycans are components of the worm's innate immune system.

Overview

Glycosylation is one of the most common and complex posttranslational modifications of proteins (Spiro, 2002). Glycans are often branched and composed of several different monomeric sugar components connected by

METHODS IN ENZYMOLOGY, VOL. 417
0076-6879/06 $35.00
DOI: 10.1016/S0076-6879(06)17022-6

a variety of linkages. This type of structure confers on glycans the ability to carry a great deal of information in very compact structures (Laine, 1994) and thereby to mediate many different functions (e.g., cell adhesion, control of the immune system, embryonic development and differentiation) (Fukuda, 2002; Lowe, 2003; Lowe and Marth, 2003; Stanley, 2000; Varki, 1993). Not surprisingly, defects in glycosylation are involved in the etiology of many diseases (e.g., metastatic cancer, congenital disorders of glycosylation, congenital muscular dystrophies, and immune deficiency) (Granovsky *et al.*, 2000; Grunewald *et al.*, 2002; Jaeken and Carchon, 2004; Lowe and Marth, 2003; Schachter, 2005). The common factor responsible for the diverse functions of protein-bound glycans seems to be their ability to interact in a specific manner with a variety of sugar-binding proteins. Humans and mice with mutations in genes involved in protein glycosylation often show severe phenotypic defects (Lowe and Marth, 2003; Schachter, 2005), reflecting the fact that many proteins are affected by the mutation. To understand this phenomenon, it is necessary to identify the proteins targeted by the mutated gene and to determine the role of glycans in the functions of the target molecules. Most observations on the role of glycosylation in metazoan development have been carried out in mice and men (Lowe and Marth, 2003), but *Drosophila melanogaster* and *Caenorhabditis elegans* are becoming popular model organisms (Haltiwanger and Lowe, 2004; Schachter, 2004).

The Biology of *Caenorhabditis elegans*

C. elegans is considered to be the most completely understood metazoan in terms of anatomy, genetics, development, and behavior (Epstein and Shakes, 1995; Riddle *et al.*, 1997; Wood, 1988). The worms have a life cycle of about 3 days and are easily grown on petri dishes. There are two sexes, hermaphrodites and males. Juvenile worms hatch and develop through four larval stages (L1–L4). The mature adult is fertile for 4 days and lives for an additional 10–15 days when feeding on *Escherichia coli* OP50 grown on the standard nematode growth medium (NGM). The anatomy is relatively simple. The transparent body consists of two concentric tubes separated by a space called the pseudocoelom that contains the gonad in the adult. The inner tube is the intestine, and the outer tube consists of a collagenous cuticle, hypodermis, musculature, and nerve cells. The complete cellular architecture and cell lineages of *C. elegans* are known, providing a comprehensive description of the ontogeny of all cells in the organism. Development is invariant; all wild-type adult hermaphrodites and males have 959 and 1031 somatic nuclei, respectively, each of which is traceable to the zygote along distinct cell lineages.

The complete genomic DNA sequence of *C. elegans* is known. The haploid genome contains 10^8 nucleotide pairs on six chromosomes, five autosomes (I–V), and a sex chromosome (X). Extensive and powerful genetic methods has been developed for this organism. Genes can be introduced into the *C. elegans* germ line by microinjection of DNA into the syncytial ovary of the hermaphrodite, where they form functional extrachromosomal tandem arrays. Self-fertilization in the hermaphrodite makes both dominant and recessive genetics relatively simple. Many *C. elegans* genes have counterparts in the mammalian genome. Both classical forward genetics and reverse genetics (by various types of chemical mutagenesis, transposon mutagenesis, or RNA interference) have been used to make *C. elegans* mutants. Many well-characterized mutant strains are available and can be preserved and stored by freezing. Methods for targeted mutagenesis are not available.

The *C. elegans* genome contains DNA sequences with significant similarities to a large number of mammalian genes involved in the metabolism of protein-bound glycans (GlcNAcβ1-*N*-Asn, GalNAcα1-*O*-Ser/Thr, GlcNAcβ1-*O*-Ser/Thr, Manα1-*O*-Ser/Thr, Fucα1-*O*-Ser/Thr, Xylβ1-*O*-Ser), lipid-bound glycans, and chitin (Schachter, 2004). In recent years some of these genes have been cloned and expressed, and worms with mutated genes have become available for study. Structural data have been obtained showing that worms do, in fact, make GlcNAcβ1-*N*-Asn *N*-glycans (Altmann *et al.*, 2001; Cipollo *et al.*, 2002; Fan *et al.*, 2004, 2005; Guerardel *et al.*, 2001; Haslam *et al.*, 2002; Hirabayashi *et al.*, 2002; Kaji *et al.*, 2003; Natsuka *et al.*, 2002; Schachter *et al.*, 2002), GalNAcα1-O-Ser/Thr *O*-glycans (Guerardel *et al.*, 2001), glycosaminoglycans (Bulik and Robbins, 2002; Guerardel *et al.*, 2001; Toyoda *et al.*, 2000; Yamada *et al.*, 1999), glycolipids (Chitwood *et al.*, 1995; Gerdt *et al.*, 1997, 1999; Griffitts *et al.*, 2005; Houston and Harnett, 2004; Lochnit *et al.*, 2000), and chitin (Veronico *et al.*, 2001).

The Structure and Biosynthesis of GlcNAcβ1-*N*-Asn Glycans (*N*-Glycans) in *C. elegans*

More than 200 glycoproteins have been identified in *C. elegans* with Asn-X-Ser/Thr sequons occupied by *N*-glycan structures (Fan *et al.*, 2004, 2005; Hirabayashi *et al.*, 2002; Kaji *et al.*, 2003; Mawuenyega *et al.*, 2003). Fine structure analyses of *N*-glycans, primarily by mass spectrometry (MS) (Altmann *et al.*, 2001; Cipollo *et al.*, 2002; Fan *et al.*, 2004, 2005; Guerardel *et al.*, 2001; Hanneman *et al.*, 2005; Haslam and Dell, 2003; Haslam *et al.*, 2002; Hirabayashi *et al.*, 2002; Kaji *et al.*, 2003; Natsuka *et al.*, 2002; Schachter *et al.*, 2002; Zhang *et al.*, 2003; Zhu *et al.*, 2004), have shown that *C. elegans* contains a predominance of oligomannose $Man_{5-9}GlcNAc_2Asn$ N-glycans identical to those found in vertebrates (Fig. 1). However, the complex and hybrid N-glycans that are highly abundant in vertebrates are absent in *C. elegans* and others are present at very low levels (Cipollo *et al.*, 2002; Haslam and Dell, 2003; Haslam *et al.*, 2002; Natsuka *et al.*, 2002). *C. elegans* contains large amounts of paucimannose $Man_{3-4}GlcNAc_2Asn$ N-glycans (Fig. 1) not usually seen in vertebrates.

Many paucimannose N-glycans are fucosylated (from 1–4 Fuc residues on a single N-glycan) (Fig. 2A). The Asn-linked core GlcNAc residue may be substituted with one or two Fuc residues (Haslam *et al.*, 2002; Natsuka

Manα1-2Manα1-6
  Manα1-6
Manα1-2Manα1-3   Manβ1-4R
Manα1-2Manα1-2—Manα1-3

M9Gn2

Manα1-6
  Manα1-6
Manα1-3   Manβ1-4R
  Manα1-3

M5Gn2

Manα1-6
  Manβ1-4R
Manα1-3

M3Gn2

Manα1-6
  Manα1-6
Manα1-3   Manβ1-4R
GlcNAcβ1-2Manα1-3

GnM5Gn2

Manα1-3/6Manα1-6
  Manβ1-4R
GlcNAcβ1-2Manα1-3

GnM4Gn2

Manα1-3/6Manα1-6
  Manβ1-4R
Manα1-3

M4Gn2

Manα1-6
  Manβ1-4R
GlcNAcβ1-2Manα1-3

GnM3Gn2

GlcNAcβ1-2Manα1-6
  Manβ1-4R
GlcNAcβ1-2Manα1-3

Gn2M3Gn2

GlcNAcβ1-4GlcNAcβ1-N-Asn
GlcNAcβ1-4(Fucα1-3)GlcNAcβ1-N-Asn F^3
GlcNAcβ1-4(Fucα1-6)GlcNAcβ1-N-Asn F^6
GlcNAcβ1-4[(Fucα1-3)(Fucα1-6)]GlcNAcβ1-N-Asn F^3F^6
R groups

FIG. 1. N-glycan structures. The figure shows the major N-glycan structures in *C. elegans*. Most of these structures have been detected by MS but a few (e.g., GnM5Gn2, GnM4Gn2) are based on enzymatic studies. The names assigned to these structures are used in the synthetic schemes shown in Fig. 2 (M, Man; Gn, GlcNAc; F, Fuc). Oligomannose N-glycans have from 5–9 Man residues; only M9Gn2 and M5Gn2 are shown. Oligomannose structures are present in all eukaryotes (vertebrates, insects, *C. elegans*, yeast, plants) but have not been found in prokaryotes. M3Gn2 and M4Gn2 are paucimannose N-glycans; they are major N-glycans in insects, *C. elegans* and plants, but vertebrates make little, if any, of these structures. The remaining structures in the figure all have a GlcNAcβ1-2Manα1-3 moiety and are, therefore, dependent on prior GlcNAcTI action. The R group is defined in the figure; there are four R variants depending on the absence or presence of core α1-3- and α1-6–linked Fuc residues (designated as F^3 and F^6, respectively).

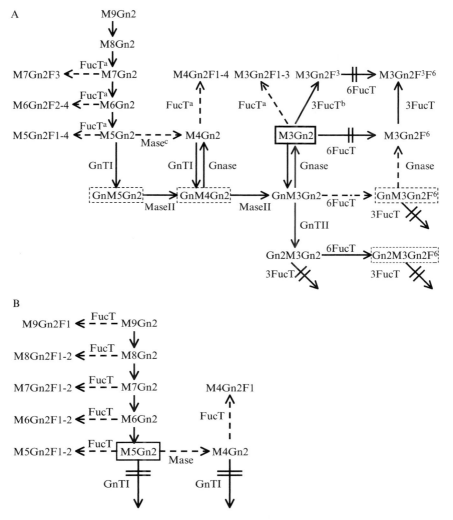

FIG. 2. (A) N-glycan synthesis in wild-type C. elegans. The names of the N-glycans are defined in Fig. 1. Reactions shown by continuous arrows have been established experimentally, whereas the discontinuous arrows are based on indirect evidence (Paschinger et al., 2005). Arrows crossed with double lines indicate reactions that do not occur. The figure shows the conversion of oligomannose N-glycans (M9-5Gn2) to hybrid (GnM5-3Gn2), paucimannose (M4-3Gn2), and complex (Gn2M3Gn2) N-glycans. Structures in boxes with discontinuous lines have not been detected by our group using MS (Zhu et al., 2004; Hanneman et al., 2005) but are included in the figure on the basis of other evidence (Paschinger et al., 2004; 2005). M9Gn2 is converted to M5Gn2 by the sequential action of α1,2-mannosidases (Herscovics, 1999). GlcNAcTI (GnTI in figure) adds GlcNAc in β1-2 linkage to the Manα1-3 arm of

et al., 2002; Hanneman *et al.*, 2005), but there are differences of opinion on whether the non-core Fuc residues are linked to peripheral Man (Altmann *et al.*, 2001; Cipollo *et al.*, 2002; Hanneman and Reinhold, 2003) or peripheral Gal (Haslam and Dell, 2003; Haslam *et al.*, 2002). There have also been reports of protein-bound O-methylated Fuc (Altmann *et al.*, 2001; Haslam *et al.*, 2002), O-methylated Man (Altmann *et al.*, 2001), and phosphorylcholine (Cipollo *et al.*, 2002; Haslam and Dell, 2003; Haslam *et al.*, 2002).

M5Gn2 to form the truncated hybrid *N*-glycan GnM5Gn2 (Schachter, 2000). Two Man residues are removed from GnM5Gn2 by the action of α3,6-mannosidase II (MaseII) to form the truncated hybrid *N*-glycans GnM4Gn2 and GnM3Gn2 (Moremen, 2002). GnM3Gn2 can be acted on by GlcNAcTII (GnTII in figure) to initiate synthesis of complex *N*-glycans. A specific β-*N*-acetylglucosaminidase (Gnase) not present in vertebrates (Zhang *et al.*, 2003) competes with GlcNAcTII for GnM3Gn2 and removes GlcNAc to form M3Gn2 paucimannose *N*-glycans; Gnase also converts GnM4Gn2 to M4Gn2. The major structure in wild-type worms is M3Gn2 (boxed with a continuous line). The sites of action of the two core fucosyltransferases, α1,6FucT (6FucT, FUT-8) and α1,3FucT (3FucT, FUT-1), are shown (Paschinger *et al.*, 2004; 2005). The products of these enzymes are designated as F^6 and F^3, respectively. The fucosylation pathways are restricted by two important properties of the core FucTs: (1) α1,6FucT must act before α1,3FucT to make the difucosylated structure, and (2) α1,6FucT requires the prior action of GlcNAcTI, whereas α1,3FucT acts only on structures with a nonsubstituted Manα1-3 residue on the *N*-glycan core. Wild-type *C. elegans* makes M3-7Gn2 oligomannose *N*-glycans with 1–4 Fuc residues (Zhu *et al.*, 2004) as indicated in the figure. These Fuc residues are attached both to the *N*-glycan core and to peripheral residues (Hanneman *et al.*, 2005) by FucTs (designated FucT[a] in the figure) that have not yet been characterized (Paschinger *et al.*, 2004; 2005). The action of the core α1,3FucT on M3Gn2 has been experimentally established (Paschinger *et al.*, 2005) and is, therefore, shown separately (3FucT[b]) from the other FucTs acting on M3Gn2 (FucT[a]). The fact that GlcNAcTI-deficient worms make M4Gn2 and M4Gn2F1 (Zhu *et al.*, 2004) suggests that there must be a GlcNAcTI-independent mannosidase (Mase[c]) that can act on M5Gn2 upstream of GlcNAcTI. Such mannosidases have been described in other species but not as yet in worms (Francis *et al.*, 2002; Kawar *et al.*, 2001). (B) *N*-glycan synthesis in GlcNAcTI-null *C. elegans*. The pathway was derived from Fig. 2A by removing all reactions dependent on the presence of GlcNAcTI and adding or altering FucT reactions affected by the absence of GlcNAcTI (Zhu *et al.*, 2004). Definitions of *N*-glycan structure names and arrows are described in Figs. 1 and 2A. The major structure in the GlcNAcTI-null worm is M5Gn2 (boxed). It has been shown previously that (1) the presence of an α1-3-linked core Fuc is essential for expression of the horseradish peroxidase (HRP) epitope (Paschinger *et al.*, 2004); (2) the core α1,3FucT (FUT-1) required for synthesis of the HRP epitope does not require prior GlcNAcTI action (Paschinger *et al.*, 2004; 2005); (3) the HRP epitope can be detected by anti-HRP in the GlcNAcTI-null worm (Paschinger *et al.*, 2004). These findings are consistent with the presence of fucosylated M4-9Gn2 *N*-glycans in the GlcNAcTI-null worm (Zhu *et al.*, 2004). It is not known which of the fucosylated *N*-glycans in the GlcNAcTI-null worm has a core α1-3Fuc. Some fucosylated *N*-glycans are made only in GlcNAcTI-null and not wild-type worms, whereas others are in wild-type but not GlcNAcTI-null worms (Zhu *et al.*, 2004). The mechanisms responsible for these observations are not understood, because GlcNAcTI acts downstream of M5Gn2. Possible explanations are loss of feedback controls by downstream metabolites, or the modification of some of the FucTs acting on M5-7Gn2 with functionally important GlcNAcTI-dependent *N*-glycans.

The *C. elegans* N-glycan profile is unique for each developmental stage (L1–L4, Dauer, adult), suggesting a role for these glycans in worm development (Cipollo *et al.*, 2005).

The first and second stages of GlcNAc-β1-N-Asn type N-glycan synthesis are similar in most eukaryotes. The first stage involves the assembly of a lipid-linked precursor Glc$_3$Man$_9$GlcNAc$_2$-pyrophosphate-dolcihol and the oligosaccharyltransferase-catalyzed transfer of the Glc$_3$Man$_9$GlcNAc$_2$-moiety to an Asn residue within an Asn-X-Ser/Thr sequon (Aebi and Hennet, 2001; Burda and Aebi, 1999; Hemming, 1995; Verbert, 1995). The second stage involves the processing, within the lumen of the endoplasmic reticulum and Golgi apparatus, of Asn-linked Glc$_3$Man$_9$GlcNAc$_2$ to Man$_5$GlcNAc$_2$Asn (Herscovics, 1999; Moremen, 2002).

In vertebrates, the third and final stage of the pathway (Schachter, 1991, 1995) takes place in the Golgi apparatus and involves the conversion of Man$_5$GlcNAc$_2$Asn to hybrid and complex N-glycans (Fig. 2A). UDP-Glc NAc:α1-3-D-mannoside β1,2-N-acetylglucosaminyltransferase I (GlcNAc TI) converts Man$_5$GlcNAc$_2$Asn to the hybrid N-glycan GlcNAcMan$_5$Glc NAc$_2$Asn. This is followed by the action of α3,6-mannosidase II to form the hybrid N-glycans (Fig. 2A). GlcNAcMan$_3$GlcNAc$_2$Asn is converted to complex N-glycans by the action of UDP-GlcNAc:α1-6-D-mannoside β1,2-N-acetylglucosaminyltransferase II (GlcNAcTII) and other branching N-acetylglucosaminyltransferases (Schachter, 1986, 2000). Further action by other glycosyltransferases (galactosyl-, sialyl- and fucosyltransferases) on the distal nonreducing ends of the glycan creates a large variety of complex N-glycans.

However, in plants (Vitale and Chrispeels, 1984), insects (Altmann *et al.*, 1995), and *C. elegans* (Zhang *et al.*, 2003), the metabolism of GlcNAc-Man$_3$GlcNAc$_2$Asn takes a different route because of the presence of an unusual β-N-acetylglucosaminidase that removes most of the GlcNAc residues inserted by GlcNAcTI before GlcNAcTII can act (Fig. 2A). This results in the synthesis of large amounts of paucimannose N-glycans at the expense of complex N-glycans. It is important to note that although the synthesis of most of both fucosylated and fucose-free paucimannose N-glycans by *C. elegans* requires the prior action of GlcNAcTI, there are small amounts of Man$_4$GlcNAc$_2$Asn and Man$_4$GlcNAc$_2$Fuc$_1$Asn present in GlcNAcTI-deficient worms (Zhu *et al.*, 2004). A suggested synthetic pathway for these compounds is shown in Fig. 2A, B.

The Function of N-Glycans in *C. elegans*

Oligomannose N-glycans (Fig. 1) are ancient structures present in both unicellular and multicellular eukaryotes and are essential for the viability of all eukaryotes that have been tested (Atienza-Samols *et al.*,

1980; Chang and Korolev, 1996; Finnie, 2001; Hübel and Schwarz, 2002; Marek et al., 1999; Pittet et al., 2006; Lin et al., 1999; Surani, 1979). This is probably due at least in part to the role played by oligomannose N-glycans in endoplasmic reticulum–associated degradation (ERAD) of misfolded or incompletely assembled glycoproteins (Spiro, 2004). RNA interference studies on C. elegans suggest that oligomannose N-glycans may also be essential for C. elegans viability (Schachter, 2004).

GlcNAcTI-dependent N-glycans appeared in evolution at about the same time as multicellular organisms. Embryonic lethality was observed in mice lacking GlcNAcTI (Ioffe and Stanley, 1994; Metzler et al., 1994). Mice (Wang et al., 2001; 2002) and humans (Charuk et al., 1995; Jaeken et al., 1994; Tan et al., 1996) with null mutations in the gene encoding GlcNAcTII, downstream of GlcNAcTI (Fig. 2A) suffer severe developmental abnormalities. It has been shown that GlcNAcTI-dependent N-glycans are also required for the normal development of Drosophila melanogaster (Sarkar et al., 2006). In contrast, GlcNAcTI-null mammalian cells in culture show no obvious phenotypic abnormalities (Chen and Stanley, 2003). It is concluded that, at least in vertebrates and flies, GlcNAcTI-dependent N-glycans serve a major function in cell–cell and cell–environment interactions. This is consistent with the location of N-glycans on membrane-bound and secretory glycoproteins.

C. elegans has three genes encoding enzymatically active GlcNAcTI (gly-12, gly-13, gly-14) (Chen et al., 1999). In sharp contrast to the developmental abnormalities caused by absence of GlcNAcTI in vertebrates and flies, gly-14 (III);gly-12 gly-13 (X) worms with null mutations in all three GlcNAcTI genes develop into apparently normal adults (Zhu et al., 2004). The worms are viable and fertile, although worm extracts have no detectable GlcNAcTI activity and show a highly abnormal N-glycan pattern (Zhu et al., 2004). It is concluded that paucimannose and other N-glycans dependent on prior GlcNAcTI action are not essential for the normal development of C. elegans grown under laboratory conditions. We now present evidence that GlcNAcTI-dependent N-glycans (consisting mainly of paucimannose N-glycans) play a role in the interaction of C. elegans with pathogenic bacteria, suggesting that these N-glycans are components of the worm's innate immune system.

Experimental

General Techniques

General techniques for the handling of nematodes were as previously described (Brenner, 1974; Epstein and Shakes, 1995). Standard molecular biology procedures were used (Ausubel et al., 1993; Sambrook et al., 1989).

Oligonucleotides were synthesized on an automated DNA synthesizer and purified by the cartridge method (Hospital for Sick Children Biotechnology Center, Toronto, Canada). DNA was sequenced in both directions by the double-strand dideoxy method (Sanger *et al.*, 1977).

C. elegans Strains and Culture Method

C. elegans strains N2 (nonclumping) wild-type (Bristol) and *dpy-6(e14)* are maintained in our laboratory. AS270 *gly-12(id47) (X)* (Chen *et al.*, 2003), AS271 *gly-14(id48) (III)* (Chen *et al.*, 2003), *gly-14 (III);gly-12 (X)* (Chen *et al.*, 2003), AS341 *gly-14 (III);gly-12 gly-13 (X)* (Zhu *et al.*, 2004), and *gly-12 gly-13 (X)* (Zhu *et al.*, 2004) have been reported. Strain RB871 *gly-13(ok712) (X)* was obtained from the *C. elegans* Gene Knockout Project at the Oklahoma Medical Research Foundation; this strain had not been backcrossed, and AS338 *dpy-6(e14) gly-13(ok712) (X)* worms were constructed to facilitate this (Zhu *et al.*, 2004). NGM (nematode growth medium) agar plates were prepared as follows: 3 g NaCl, 2.5 g peptone, and 17 g agar were autoclaved in 1 l of water, followed by cooling to 55° and the addition in order of 1 ml cholesterol (5 mg/ml ethanol), 1 ml 1 M CaCl$_2$, 1 ml 1 M MgSO$_4$, and 25 ml 1 M KH$_2$PO$_4$ at pH 6.0 (Brenner, 1974; Lewis and Fleming, 1995). Unless otherwise stated, *C. elegans* were fed *Escherichia coli* strain OP50 grown at 20° on NGM agar plates; growth of the OP50 strain is limited on NGM by its autotrophy for uracil.

Life Span of *C. elegans* Growing on Bacterial Pathogens

A single colony of *Pseudomonas aeruginosa* strain 14 (PA14, provided by Dr. Lori Burrows, McMaster University, Hamilton, Ont.) was added to 2 ml LB (Luria-Bertani) broth (1% w/v tryptone, 0.5% w/v yeast extract, 1% w/v NaCl) and grown at 37° with shaking overnight. Agar bacterial culture plates, 3.5 cm diameter, were prepared with PGS medium (peptone-glucose-sorbitol) using 5 ml of 1% w/v Bacto-peptone, 1% w/v NaCl, 1% w/v glucose, 0.15 M sorbitol, and 1.7% w/v agar per plate (Tan, 2002). PGS plates were stored at 4° for no longer than 7 days after preparation. *E. coli* OP50 or PA14 were spread on the plates, incubated for 24 h at 37°, and kept at ambient temperature for 2 h before worms were added (Tan, 2002). Either *C. elegans* L4 larvae or 1-day-old young adults (30–40 worms/plate) were transferred from a lawn of *E. coli* OP50 on NGM to either a lawn of *E. coli* OP50 or PA14 on PGS. Plates were incubated at 25°, and worms were tested for viability at regular intervals (usually 5, 12, 24, 48, and 72 h). A worm was scored as dead if it did not move in response to touch (Tan, 2002). It is important to note that the only indication of life may be a slight movement of a part of the worm's body.

Two approaches were used to determine whether worms were killed by an infectious process or by a soluble toxin (Gan *et al.*, 2002; Mahajan-Miklos *et al.*, 1999; Tan *et al.*, 1999a). In the filter method, approximately 0.01 ml moist pellet of bacterial culture was spread over a filter (Millipore HVLP, 0.45-μm pores), and the filter was placed on a PGS agar plate. The plate was incubated at 37° overnight. The filter was removed, and 30–40 worms were transferred to the PGS agar plate from a lawn of *E. coli* OP50 on NGM. The worms were incubated on this plate at 25° in the absence of bacteria, and the number of dead worms was scored at regular intervals as described previously. In the heat killing method, 0.010 ml of the bacterial culture was spread directly over the agar surface of a PGS agar plate. The plate was incubated at 37° overnight and placed inside a plastic bag. The bag was sealed and placed in a water bath at 65° for 30 min (Tan *et al.*, 1999a). The heat-treated PGS agar plate was removed from the water bath, and 30–40 worms were transferred to it from a lawn of *E. coli* OP50 on NGM. The worms were incubated at 25° in the presence of heat-killed bacteria, and the number of dead worms was scored at regular intervals as described previously. In the case of the filter method, killing of worms in the absence of bacteria strongly suggests the involvement of a toxin, but a nondiffusible toxin cannot be excluded if worms show normal survival in this assay. Normal survival of worms in the presence of heat-killed bacteria cannot exclude the possibility of a heat-sensitive toxin.

Construction of the gly-14(III);dpy-6 gly-13(X) *Double Mutant*

Male worms were obtained by the heat shock method as follows. Hermaphrodite *gly-14(III)* L3 larvae (6–10 worms) were incubated at 30° for 6 h on small NGM plates. The worms were then allowed to self at 20°. After a few days, the progeny contained about 5–6% male worms. The *gly-14(III)* males (5–6 worms) were mated (Epstein and Shakes, 1995) to *dpy-6 gly-13 (X)* homozygous hermaphrodites (1–2 worms) at 20°. After 24 h, *gly-14/ +(III); dpy-6 gly-13/++(X)* heterozygous hermaphrodites were obtained and allowed to single. Dumpy homozygous hermaphrodite progeny were screened by single worm PCR (Zhu *et al.*, 2004) to obtain the *gly-14;dpy-6 gly-13* double null mutants.

Production of gly-13 *and* gly-14 *Phenotypes by RNA-mediated Interference (RNAi)*

RNA-mediated interference (RNAi) was used to produce worms with phenotypes similar to *gly-13* and *gly-14* mutant worms to obtain confirmatory evidence that the abnormal phenotypes of these mutant worms are, indeed, specific to mutations in the *gly-13* and *gly-14* genes respectively. Drs. Aldis Krizus and James W. Dennis (Samuel Lunenfeld Research Institute, Toronto) provided two *E. coli* clones, each of which contains a plasmid that encodes double-stranded RNA for RNA-mediated interference (Open Biosystems Expression Arrest RNAi); the plasmid DNA sequences encode sections of the wild-type *gly-13* and *gly-14* genes, respectively. The plasmids were isolated, DNA sequences were verified, and aliquots (1 ml) of glycerol stocks were stored at −80°. Bacteria were grown to the mid log phase in LB medium containing ampicillin (0.1 mg/ml) and tetracycline (12.5 μg/ml). Expression of RNA was induced by addition of isopropyl-β-D-thiogalactopyranoside (IPTG, 0.4 mM) followed by incubation at 37° for 1 h. The bacterial culture was centrifuged at 3000 rpm for 20 min in a bench top centrifuge, the supernatant was removed, and a cut pipette tip was used to add about 0.01 ml moist pellet to a 35-mm NGM agar plate containing ampicillin (0.1 mg/ml), tetracycline (12.5 μg/ml), and IPTG (0.4 mM). A round-bottom test tube was used to spread a thin film of concentrated bacteria on the agar surface, leaving a clear area around the edge of the plate to deposit *C. elegans* eggs (30–40 per plate). The plates were incubated at 25° to allow eggs to hatch and develop to the L4 larval stage. Larvae (about 20/plate) were transferred to PGS agar plates covered with either OP50 *E. coli* (*gly-14* RNAi) or PA14 (*gly-13* RNAi), and survival data were obtained.

Rescue of the Mutant gly-13 *Phenotype by a Wild-type* gly-13
Extrachromosomal Array

Rescue experiments were carried out on mutant *dpy-6 gly-13* worms using an extrachromosomal array carrying the wild-type *gly-13* gene. We have previously described the construction of homozygous hermaphrodite +/+; *Ex[rol-6; hs::Myc-gly-13(+)]* rollers that overexpress wild-type GLY-13 under the control of a heat shock promoter (Chen *et al.*, 1999). L3/L4 N2 males (10 worms) were obtained by the heat shock method (described previously) and mated with 1 or 2 L3 +/+; *Ex[rol-6; hs::Myc-gly-13(+)]* hermaphrodites. F1 +/+; *Ex[rol-6; hs::Myc-gly-13(+)]* L3/L4 male rollers were prepared from the progeny and mated with homozygous dumpy *dpy-6 gly-13* hermaphrodite L2 larvae at a ratio of 10–20 males to 1 hermaphrodite. Young hermaphrodite larvae and a high proportion of males to hermaphrodites are required in crosses with male rollers. F2 *dpy-6 gly-13/+ +; Ex[rol-6; hs:: Myc-gly-13(+)]* heterozygous hermaphrodite rollers were obtained and allowed to self. F3 homozygous dumpy *dpy-6 gly-13* hermaphrodites were selected. It is not possible to distinguish *dpy-6 gly-13* worms that carry the

Ex[rol-6; hs::Myc-gly-13(+)] array from those that do not, because the dumpy phenotype suppresses the roller phenotype that identifies the array (Mello and Fire, 1995).

Single worm PCR was carried out on all F3 dumpy worms to identify homozygous *dpy-6 gly-13; Ex[rol-6; hs::Myc-gly-13(+)]* worms. The PCR primer pair (forward, 5'-GTGGCATGAACTAGAACCAA-3'; reverse, 5'-CAATCTGGATTATATGCTGA-3') was designed to distinguish the wild type *gly-13* sequence in *Ex[rol-6; hs::Myc-gly-13(+)]* (528 nt PCR band) from the wild-type *gly-13* sequence in the genome of wild-type and heterozygous worms (1055 nt PCR band). Each F3 worm was placed in a separate PCR tube containing 5 μl lysis buffer (10 mM Tris-HCl, pH 8.0, 50 mM KCl, 2.5 mM MgCl$_2$, 0.45% v/v Igepal CO-630, 0.45% v/v Tween-20, 0.01% gelatin, 0.1 % w/v proteinase K) (Zhu *et al.*, 2004). The tube was kept at $-80°$ for 1 h, 65° for 1 h to activate proteinase K, and 95° for 15 min to inactivate proteinase K. PCR was carried out in 25-μl reaction volumes containing 2.5 μl 10× Invitrogen PCR buffer, 1.0 μl 50 mM MgCl$_2$, 2.5 μl dNTP mixture (1 mM of each nucleotide), 0.5 μl 10 μM forward primer, 0.5 μl 10 μM reverse primer, 2.5 μl single worm extract, 0.2 μl Invitrogen Platinum Taq DNA Polymerase (5 units/μl), and distilled water to a final volume of 25 μl. PCR conditions were 94° for 1.5 min, followed by 34 cycles at 94° for 30 sec, 51° for 30 sec, and 68° for 1 min. PCR products were analyzed by electrophoresis in a 2.0% agarose gel in TBE buffer (0.089 M Tris-borate, pH 8.3, 0.025 M disodium EDTA) and visualized with UV light. The extrachromosomal array was detected in 6 of 13 F3 worms analyzed by PCR.

Before PCR analysis, the dumpy F3 worms were placed on NGM agar plates, one worm per plate, and allowed to lay eggs for 2–3 h at 20°. The F3 parent was then removed for PCR analysis as described previously. Eggs were incubated for a further 6 h at 20°. After hatching, the agar on the plate was cut into two equal pieces. One half of the larvae ($\sim$10–20/plate) was subjected to heat shock at 33° for 1 h. Heat shock was continued at 33° for 1 h at 12-h intervals until either the L4 larval or adult stages (Chen *et al.*, 1999). The other half of the larvae ($\sim$10–20/plate) was incubated at 20° over the same time period to serve as non-heat shocked controls. The F4 progeny (heat-shocked and non-shocked controls) of F3 parents identified by PCR as homozygous *dpy-6 gly-13; Ex[rol-6; hs::Myc-gly-13(+)]* worms were tested for survival on *P. aeruginosa* 14 as described previously. Inheritance of extrachromosomal arrays varies from $\sim$10 to $\sim$90%, with an average of approximately 50% in each generation (Mello and Fire, 1995). Suppression of the roller phenotype in the dumpy F4 progeny prevents detection of worms that have undergone spontaneous loss of the extrachromosomal array, and, therefore, it is to be expected that some of the F4 worms will not be rescued.

Results

Effects of E. coli *OP50 on PGS Medium on the Survival of* C. elegans

When wild-type L4 larvae or adult worms are grown on *E. coli* OP50 on NGM, there are no or very few deaths during the 72-h time period used in our survival experiments (Garsin *et al.* [2001] and data not shown); the worms eventually die on NGM after 10–15 days (Sifri *et al.*, 2005). However, both wild-type L4 larvae (Fig. 3A) and adult worms (Ruiz-Diez *et al.* [2003] and Fig. 3B) are killed at a significant rate on *E. coli* OP50 on PGS medium during the 72 h after loading on the plates. It has been suggested that worms are killed by *E. coli* OP50 on media such as PGS (Ruiz-Diez *et al.*, 2003) and brain heart infusion (Garsin *et al.*, 2001) because these enriched media allow *E. coli* to synthesize virulence factors that the bacteria cannot make in the attenuated

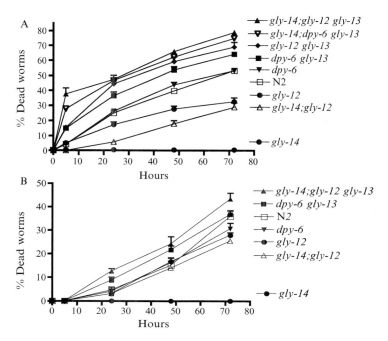

FIG. 3. Survival of *C. elegans* on *E. coli* OP50 on PGS medium. Worms (30–40/plate) were loaded on *E. coli* OP50/PGS plates at (A) the L4 larval or (B) the adult stage. From three to six plates were analyzed for each worm strain. The number of dead worms at various times after loading is plotted as a percent of worms loaded on the plate. Reproducibility between runs was excellent as indicated by the short error bars (standard error of the mean). The worm strains are identified by the symbols on the figure.

NGM (Garsin *et al.*, 2001). The killing rates of N2 and *dpy-6* worms on *E. coli* OP50 on PGS are identical (Fig. 3A, B), showing that the presence or absence of the *dpy-6* mutation in worms with the *gly-13* mutation does not affect the survival data.

L4 larvae in which there is a mutation in *gly-13* (*dpy-6 gly-13, gly-12 gly-13, gly-14;dpy-6 gly-13* and *gly-14;gly-12 gly-13*) die more rapidly in the first 5 h than wild-type and other mutant L4 larvae (Table I, Fig. 3A), but the killing rate slows down after 5 h. This abrupt change at 5 h is attributed to the fact that most L4 larvae have developed into adults at that time. To confirm that the reduction in killing rate reflects the resistance of adults relative to L4 larvae, we have shown that loading *dpy-6 gly-13, gly-12 gly-13, gly-14;dpy-6 gly-13* and *gly-14;gly-12 gly-13* adults on the plate does not lead to the early rapid killing observed for the larval forms (compare Fig. 3A, B). Furthermore, the rates of killing for all the worm strains studied are the same when larvae (after 5 h, Fig. 3A) and adults (from time zero, Fig. 3B) are loaded on the plate.

Table I shows a statistical analysis of survival rates of L4 larvae at 5 h after loading. The data are presented in three groups that show the effects of *gly-12, gly-13,* and *gly-14* null mutations, respectively. In each group of four worms, L4 larvae with the mutation under study (strain B, mutation in bold type) are compared with L4 larvae (strain A) that differ from strain B only in the absence of the mutation under study.

Insertion of the *gly-12* mutation has no significant effect on the survival of any of the four L4 larvae tested (Table I). In contrast, insertion of the *gly-13* mutation results in a highly significant increase in the rate of killing of all four larvae tested (Table I). The situation in adult worms is quite different. Insertion of the *gly-12* mutation into adult wild-type, *dpy-6 gly-13* and *gly-14;dpy-6 gly-13* worms, and of the *gly-13* mutation into adult wild-type, *gly-12* and *gly-14;gly-12* worms has no effect on the killing slopes (Fig. 3A, B). However, insertion of either the *gly-12* or *gly-13* mutation into adult *gly-14* worms causes a significant increase in the slope of the killing plot (Fig. 3A, B).

Remarkably, insertion of the *gly-14* mutation into L4 larvae results in 100% protection against killing both in the absence and presence of the *gly-12* mutation (Table I). The presence of the *gly-13* mutation suppresses the protective effect of the *gly-14* mutation in L4 larvae both in the absence and presence of the *gly-12* mutation (Table I). Insertion of the *gly-14* mutation into adults also results in 100% protection against killing but only in the absence of both the *gly-12* and *gly-13* mutations (Fig. 3A, B). The protective effect of the *gly-14* mutation in adults is suppressed by either the *gly-12* or *gly-13* mutation (Fig. 3A, B).

When N2 and *gly-14;gly-12 gly-13* L4 larvae were grown on heat-killed *E. coli* OP50 on PGS plates, there was 100% survival over the 72-h length

TABLE I

PERCENT SURVIVAL OF C. ELEGANS L4 LARVAL STRAINS AFTER 5 H ON E. COLI OP50 ON PGS MEDIUM. C. ELEGANS STRAINS A AND B WERE COMPARED BY UNPAIRED t TEST ANALYSIS. MEAN VALUES (N = 3–6), STANDARD DEVIATIONS (SD), AND p VALUES ARE SHOWN

Strain A	Mean % Survival/SD	Strain B	Mean % Survival/SD	p Value (A vs B)	Significant (p < 0.05)
Effect of gly-12 null					
N2	95.1/2.7	**gly-12**	95.0/3.5	0.9711	No
dpy-6 gly-13	85.0/1.4	**gly-12** gly-13	84.4/2.0	0.6728	No
gly-14	99.5/1.2	gly-14;**gly-12**	100		No[a]
gly-14;dpy-6 gly-13	72.2/3.9	gly-14;**gly-12** gly-13	62.9/7.2	0.0786	No
Effect of gly-13 null					
dpy-6	94.9/1.5	dpy-6 **gly-13**	85.0/1.4	0.0003	**Yes**
gly-12	95.0/3.5	gly-12 **gly-13**	84.4/2.0	0.0020	**Yes**
gly-14	99.5/1.2	gly-14;dpy-6 **gly-13**	72.2/3.9	<0.0001	**Yes**
gly-14;gly-12	100	gly-14;gly-12 **gly-13**	62.9/7.2		**Yes**[a]
Effect of gly-14 null					
N2	95.1/2.7	**gly-14**	99.5/1.2	0.0043	**Yes**
gly-12	95.0/3.5	**gly-14**;gly-12	100		**Yes**[a]
dpy-6 gly-13	85.0/1.4	**gly-14**;dpy-6 gly-13	72.2/3.9	0.0016	**Yes**
gly-12 gly-13	84.4/2.0	**gly-14**;gly-12 gly-13	62.9/7.2	0.0017	**Yes**

[a] 100% of gly-14;gly-12 worms survived to 5 h, so that a standard deviation and a p value could not be calculated. The differences between gly-14;gly-12 and gly-14 (99.5 ± 1.2 % survival), gly-14;gly-12 gly-13 (62.9 ± 7.2 % survival), and gly-12 (95 ± 3.5 % survival) are considered, respectively, to be not significant, highly significant, and probably significant.

of the experiment; 100% survival over 72 h was also observed when these worms were placed on PGS plates that had been exposed to *E. coli* OP50 growing on a filter followed by removal of the filter. Both experiments indicate that *E. coli* OP50 on PGS kills the worms by an infectious process, although a heat-sensitive nondiffusible toxin cannot be ruled out.

The *gly-14* promoter is expressed only in gut cells from L1 to adult developmental stages (Chen *et al.*, 1999). Because the mechanism of killing by *E. coli* on PGS medium seems to be an infectious process, the survival data for both larvae and adults (Table I, Fig. 3A, B) suggest that GLY-14-dependent *N*-glycans are required by *E. coli* to adhere to the gut endothelium and/or to cross the endothelial barrier into the body of the worm. Infection is blocked by the protective presence of the GLY-13 enzyme as indicated by the increased killing of larvae with the *gly-13* mutation (Table I) and by suppression of the protective role of the *gly-14* mutation in adults by the *gly-13* mutation (Fig. 3A). The GLY-12 enzyme has no apparent role in larvae but has a minor protective role in adults as can be observed by comparing *gly-14* and *gly-14;gly-12* worms (Fig. 3A, B). The data show that the three GlcNAcTI enzymes in *C. elegans* have different functional roles. GLY-14-dependent *N*-glycans are required for infection by *E. coli*, whereas GLY-12- and GLY-13-dependent *N*-glycans have protective roles. The *N*-glycan structures involved in these effects and their mechanisms of action remain to be determined.

Effects of *Pseudomonas aeruginosa* Strain 14 on PGS Medium on the Survival of *C. elegans*

The molecular mechanisms of bacterial virulence used by the human opportunistic pathogen *P. aeruginosa* strain 14 (PA14) have been studied extensively in *C. elegans* as a model organism (Aballay and Ausubel, 2002; Alegado *et al.*, 2003; Darby *et al.*, 1999; Gravato-Nobre and Hodgkin, 2005; Mahajan-Miklos *et al.*, 2000; Sifri *et al.*, 2005; Mahajan-Miklos *et al.*, 1999; Tan and Ausubel, 2000; Tan *et al.*, 1999a,b). Infection, biofilm formation, and toxins have been identified as mechanisms of worm killing by human pathogens (Mahajan-Miklos *et al.*, 1999; Sifri *et al.*, 2005). When *C. elegans* is fed PA14 on NGM, bacteria accumulate in the lumen of the intestine, and the worms are killed relatively slowly over 2–3 days, a process that has been called "slow killing" (Aballay and Ausubel, 2002; Tan *et al.*, 1999a). However, when L4 larvae are fed PA14 grown on rich media with a high osmolarity such as PGS, the worms are killed quickly (over 80% death within 10 h, Fig. 4A), a process that has been called "fast killing" (Aballay and Ausubel, 2002; Mahajan-Miklos *et al.*, 1999; Tan *et al.*, 1999a). Fast killing has been attributed to secretion of diffusible toxins by the bacteria (Mahajan-Miklos *et al.*, 1999). Fast killing takes

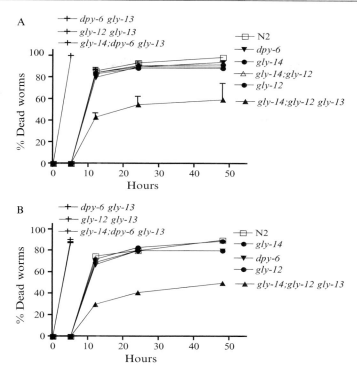

FIG. 4. A. Survival of C. elegans L4 larvae on P. aeruginosa strain 14 on PGS medium. L4 larvae (30–40/plate) were loaded on PA14/PGS plates. Three plates were analyzed for each worm strain. The number of dead worms at various times after loading is plotted as a percent of worms loaded on the plate (error bar = standard error of the mean). The worm strains are identified by the symbols on the figure. (B) Survival of C. elegans L4 larvae on bacteria-free plates previously exposed to PA14. PA14 was spread over a filter placed on a PGS agar plate. The plate was incubated at 37° overnight. The filter was removed, and 40 L4 larvae were transferred to the PGS plate and incubated at 25° in the absence of bacteria. The number of dead worms at various times after loading is plotted as a percent of worms loaded on the plate. Only a single plate was analyzed. The worm strains are identified by the symbols on the figure.

place only when L4 larvae are loaded on the plates; 1-day-old adults are killed much more slowly than larvae (Mahajan-Miklos et al., 1999; Tan et al., 1999a), and, therefore, fast killing studies with PA14 are carried out on L4 larvae (Fig. 4A) (Mahajan-Miklos et al., 1999). One of the most critical factors in fast killing is the osmolarity of the medium; the presence of 0.15 M sorbitol in PGS medium results in a dramatic increase in the killing rate relative to medium free of sorbitol (Mahajan-Miklos et al., 1999).

PA14 on PGS (fast killing conditions) is very toxic for *dpy-6 gly-13, gly-12 gly-13* and *gly-14;dpy-6 gly-13* L4 larvae (100% death at 5 h, Fig. 4A). Adult *dpy-6 gly-13* worms also die rapidly (100% death at 4 h, Fig. 5), whereas adult wild-type worms die much more slowly (50% death between 70 and 90 h) (Mahajan-Miklos *et al.*, 1999; Tan *et al.*, 1999a). The fast killing curves for N2 L4 larvae in our studies consistently show a 5-h delay before killing commences (Fig. 4A), whereas there does not seem to be a delay in previous studies by others (Mahajan-Miklos *et al.*, 1999; Tan *et al.*, 1999a); we have no explanation for this discrepancy other than to suggest that the larvae in this study may have been somewhat more mature than the larvae used in the previous work. *dpy-6, gly-12, gly-14* and *gly-14;gly-12* larvae were also resistant to PA14 for the first 5 h and died rapidly after that time (Fig. 4A). GlcNAcTI-null *gly-14;gly-12 gly-13* larvae showed a significantly reduced killing rate relative to all other worms tested (Fig. 4A).

PA14 bacteria were spread over a filter placed on a PGS agar plate. The plate was incubated at 37° overnight. The filter was removed, and 30–40 L4 larvae were transferred to the bacteria-free PGS agar plate. Remarkably,

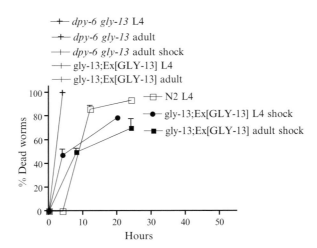

FIG. 5. Rescue of the *gly-13* phenotype by a wild-type *gly-13* extrachromosomal array. Homozygous *dpy-6 gly-13; Ex[rol-6; hs::Myc-gly-13(+)]* L4 and adult worms (abbreviated as gly-13; Ex[GLY-13] in the figure) were subjected to heat shock to activate the wild-type *gly-13* gene in the extrachromosomal array (shocked worms are designated as shock in the figure). Shocked and nonshocked *dpy-6 gly-13; Ex[rol-6; hs::Myc-gly-13(+)]* larvae and adults, as well as N2 and *dpy-6 gly-13* worms, were deposited on PGS agar plates covered with PA14 and analyzed for survival (3 plates/worm strain, ~20 worms per plate). The number of dead worms at various times after loading is plotted as a percent of worms loaded on the plate (error bar = standard error of the mean).

the killing curves in both the presence (Fig. 4A) and absence (Fig. 4B) of PA14 are almost identical, proving that a soluble diffusible bacterial toxin is killing the worms, as previously shown by others for wild-type worms (Mahajan-Miklos *et al.*, 1999). Mutants of PA14 have been isolated that exhibit an attenuated fast killing phenotype (Mahajan-Miklos *et al.*, 1999). Some of these mutant bacteria are deficient in pyocyanin, a tricyclic phenazine pigment, suggesting that pyocyanin exerts a toxic effect on *C. elegans* through the generation of reactive oxygen species (Mahajan-Miklos *et al.*, 1999). However, several PA14 mutants with an attenuated fast killing phenotype have normal levels of pyocyanin, indicating that other toxins are also involved (Mahajan-Miklos *et al.*, 1999).

The susceptibility of *gly-14;gly-12 gly-13* worms to the toxin (Fig. 4A, B) shows that the toxin can work in the complete absence of GlcNAcTI. However, the toxic effect is greatly enhanced if the GLY-13 enzyme is present both in the absence or presence of GLY-12 and/or GLY-14 (Fig. 4A), indicating a role for GLY-13-dependent *N*-glycans in toxin action. Paradoxically, the PA14 toxin is most effective in the absence of GLY-13 and the presence of GLY-12 and/or GLY-14 (Fig. 4A), indicating that GLY-13-dependent *N*-glycans also have a protective role. Although the mechanisms of these effects remain to be determined, it is clear that the action of PA14 toxin on *C. elegans* is controlled by GlcNAcTI-dependent *N*-glycans.

Production of *gly-13* and *gly-14* Phenotypes by RNA-mediated Interference (RNAi)

dpy-6 gly-13 and *dpy-6* L4 larvae that had not been subjected to RNAi showed, respectively, 100% and 0% death at 4 h when placed on PA14/PGS agar plates (Fig. 6A). *dpy-6* larvae exposed to RNAi specific for the wild-type *gly-13* sequence showed an increased killing rate on PA14 (from 0–48% death at 4 h, Fig. 6A). The data confirm that reduction of GLY-13 enzyme activity is responsible for the high toxicity of PA14 toward *dpy-6 gly-13* larvae relative to *dpy-6* larvae (Fig. 4A).

Exposure of *gly-14;gly-12* worms to RNAi specific for the wild-type *gly-13* sequence results in a significant reduction in death rate on PA14 on PGS (from 80-60% death between 12 and 24 h) to a level similar to that of *gly-14; gly-12 gly-13* worms (Fig. 6B). The data confirm that reduction of GLY-13 enzyme activity is responsible for the relatively low toxicity of PA14 toward *gly-14;gly-12 gly-13* worms relative to *gly-14;gly-12* worms (Fig. 4A).

N2 worms are killed by *E. coli* OP50 on PGS (Figs. 3A and 6C). Exposure of N2 worms to RNAi specific for the wild-type *gly-14* sequence resulted in 0% death up to 72 h (Fig. 6C). The data confirm that reduction of GLY-14

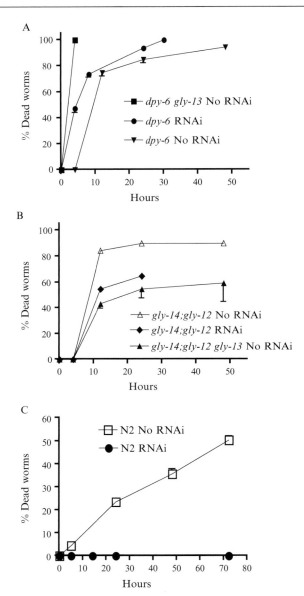

FIG. 6. Production of *gly-13* and *gly-14* phenotypes by RNA-mediated interference (RNAi). (A) *dpy-6* L4 larvae that had been subjected to RNAi specific for the wild-type *gly-13* gene, and untreated *dpy-6* and *dpy-6 gly-13* larvae were placed on PA14/PGS agar plates (about 20 worms/plate) and survival rates were determined. Error bars show standard errors of the mean (3 plates/worm). (B) *gly-14;gly-12* L4 larvae that had been subjected to RNAi

enzyme activity is responsible for the dramatic resistance of *gly-14* worms to infection by *E. coli* OP 50 on PGS (Fig. 3A).

Rescue of the Mutant *gly-13* Phenotype by a Wild-type *gly-13* Extrachromosomal Array

PCR screening was carried out on 13 F3 homozygous dumpy *dpy-6 gly-13* progeny from F2 *dpy-6 gly-13/+ +; Ex[rol-6; hs::Myc-gly-13(+)]* hermaphrodites to detect the presence of the extrachromosomal array. Six homozygous dumpy *dpy-6 gly-13* worms with the array were identified. F4 and F5 homozygous *dpy-6 gly-13; Ex[rol-6; hs::Myc-gly-13(+)]* progeny from four of these six worms were deposited on PGS agar plates covered with PA14 and analyzed for survival (Fig. 5). Approximately 20 L4 or adult worms from each of the heat-shocked and non-heat-shocked groups were placed on each plate. *dpy-6 gly-13* and nonshocked *dpy-6 gly-13; Ex[rol-6; hs::Myc-gly-13(+)]* worms showed 100% death at 4 h when either L4 larvae or adults were loaded on the PA14 plates (Fig. 5). In contrast, homozygous *dpy-6 gly-13; Ex[rol-6; hs::Myc-gly-13(+)]* worms subjected to heat shock showed partial rescue of the *gly-13* larval phenotype to about 50% death between 4 and 8 h when either L4 larvae or adults were loaded on the PA14 plates (Fig. 5). The killing rates of the heat-shocked *dpy-6 gly-13; Ex[rol-6; hs::Myc-gly-13(+)]* worms after 4–8 h was similar to that of the N2 worms. The partial rescue at 4–8 h may be because suppression of the roller phenotype in the F4 and F5 dumpy worms prevents detection of worms that have undergone spontaneous loss of the extrachromosomal array. The data confirm that reduction of GLY-13 enzyme activity is responsible for the high toxicity of PA14 toward *dpy-6 gly-13* worms relative to N2 and *dpy-6* worms (Figs. 4A and 5).

Discussion

Null mutations in the gene encoding GlcNAcTI do not affect the viability of mammalian cells in culture (Chen and Stanley, 2003) but result in embryonic lethality in mice (Ioffe and Stanley, 1994; Metzler *et al.*, 1994)

---

specific for the wild-type *gly-13* gene, and untreated *gly-14;gly-12* and *gly-14;gly-12 gly-13* larvae were placed on PA14/PGS agar plates (about 20 worms/plate) and survival rates were determined. Error bars show standard errors of the mean (3 plates/worm). (C) N2 L4 larvae that had been subjected to RNAi specific for the wild-type *gly-14* gene and untreated N2 larvae were placed on *E. coli* OP50/PGS agar plates (about 20 worms/plate) and survival rates were determined. Error bars show standard errors of the mean (3 plates/worm).

and defective locomotion in *Drosophila melanogaster* (Sarkar *et al.*, 2006). These findings and data from many laboratories (Dennis *et al.*, 1999; Ye and Marth, 2004; Zhou, 2003) indicate that GlcNAcTI-dependent *N*-glycans serve a major function in the interactions of cells with their environment and in the development of multicellular eukaryotes. However, the absence of GlcNAcTI activity seems to have no effect on the viability and development of *C. elegans* (Zhu *et al.*, 2004). We present evidence in this chapter that GlcNAcTI plays a role in the interaction of *C. elegans* with *E. coli* OP50 and *P. aeruginosa* strain 14. *C. elegans* has three GlcNAcTI genes (*gly-12, gly-13, gly-14*) (Chen *et al.*, 1999), each of which plays a different role in the worm's interaction with these bacteria.

   E. coli OP50 on NGM is the standard food for *C. elegans* in the laboratory. When worms are fed *E. coli* OP50 on PGS, an enriched medium with high osmolarity, they are killed at a significant rate by an infectious process. Larvae (Table I) and adults (Fig. 3A, B) differ in their response to infection by OP50. The presence of GLY-13 enzyme in larvae and the absence of GLY-14 enzyme in adults both result in dramatic protective effects on killing of worms by OP50 on PGS. In larvae, GLY-12 enzyme has no effect, and GLY-14 enhances killing moderately. In adults, GLY-12 and GLY-13 must both be present to obtain the protective effect of GLY-14 absence. Because the *gly-14* promoter is expressed only in gut cells (Chen *et al.*, 1999), GLY-14-dependent *N*-glycans may be involved in the ability of *E. coli* to adhere to the gut endothelium and infect the worm. RNA-mediated interference has been used to show that down-regulation of the *gly-14* gene in N2 worms mimics the phenotype of the *gly-14* worms on OP50/PGS, thereby ruling out other mutations in the *gly-14* worm.

   We also studied the response of our mutant worms to fast killing by PA14 on PGS (Aballay and Ausubel, 2002; Mahajan-Miklos *et al.*, 1999; Tan *et al.*, 1999a). Fast killing is caused by a toxin. Only larvae were studied, because fast killing is observed in N2 worms only when larvae are loaded on the plate. The PA14 toxin can work in the absence of GlcNAcTI, but our data show that GlcNAcTI plays important roles in the interaction of toxin with the worm. The effect of GLY-13 enzyme in fast killing is complex. The toxic effect is greatly enhanced relative to the GlcNAcTI-null worm (*gly-14;gly-12 gly-13*) if the GLY-13 enzyme is present. However, the toxic effect is enhanced even further in the absence of GLY-13 and the presence of GLY-12 and/or GLY-14. Therefore, GLY-13-dependent *N*-glycans enhance both protection from the action of the toxin and killing by the toxin, depending on the expression of GLY-12 and GLY-14. RNA-mediated interference has been used to show that down-regulation of the *gly-13* gene in *dpy-6* and *gly-14;gly-12* worms, respectively, mimics the phenotypes of the *dpy-6 gly-13* and *gly-14;gly-12 gly-13* worms on PA14/PGS. We have also been able to rescue the phenotype of the *gly-13* worm on PA14/PGS by an extrachromo-somal array expressing the wild-type *gly-13* gene. These experiments rule out other mutations in the *dpy-6 gly-13* and *gly-14;gly-12 gly-13* worms.

Although the mechanisms responsible for the behavior of mutant worms fed either *E. coli* OP50 or PA14 growing on PGS medium are not under-stood, it is clear that GlcNAcTI-dependent *N*-glycans are involved in these effects. Furthermore, the three GlcNAcTI enzymes in *C. elegans* have different functional roles. The data suggest that GlcNAcTI-dependent *N*-glycans are part of the worm's innate immune system. A great deal of information has been obtained in recent years on bacterial pathogens that infect *C. elegans* or produce lethal toxins, and on *C. elegans* genes that play a role in defense against these pathogens (Alegado *et al.*, 2003; Gravato-Nobre and Hodgkin, 2005; Gravato-Nobre *et al.*, 2005; Kurz and Ewbank, 2003; Millet and Ewbank, 2004; Pujol *et al.*, 2001; Sifri *et al.*, 2005). Worms possess neither a vascular system nor leukocytes and are not able to mount vertebrate-type adaptive humoral or cell-mediated immune responses. However, *C. elegans* dwells in the soil and feeds on bacteria. It is, therefore, not surprising that the worm has developed defenses against pathogenic bacteria. The cuticle is the worm's first line of defense. Access beyond the cuticle can be gained through the mouth, anus, vulva, or sensory openings. Microbes entering through the mouth immediately encounter the grinder that breaks open bacteria as they pass to the intestine (Riddle *et al.*, 1997). Worms also possess a complex immune system involving multiple signaling pathways and antimicrobial proteins and peptides, including responses that resemble those present in vertebrates (e.g., barriers that keep harmful materials from entering the body, and homologs of Toll-like receptors [TLRs] such as TOL-1 and the Toll/Interleukin-1 Receptor [TIR-1]). Worms can even induce the expression of antimicrobial weapons in re-sponse to bacterial pathogens (Aballay and Ausubel, 2002; Alegado *et al.*, 2003; Kurz and Ewbank, 2003; Kurz and Tan, 2004; Millet and Ewbank, 2004).

Glycosylation has been found to play a role in the interaction between *C. elegans* and pathogens or toxins in at least two situations. (1) Recessive *C. elegans* mutants (*bre*) have been isolated that are resistant to the Crystal (Cry) toxins made by *Bacillus thuringiensis* (*Bt*). *Bt* toxin resistance is due to the absence of specific "arthro series" glycosphingolipids presumably needed for toxin binding in the intestine (Griffitts *et al.*, 2003; Schwientek *et al.*, 2002; Wandall *et al.*, 2003). (2) *C. elegans* mutants with altered surface antigenicity (*srf*) (Hodgkin *et al.*, 2000; Hoeflich *et al.*, 2004; Link *et al.*, 1992; Politz *et al.*, 1990) are resistant to the nematode-specific pathogen *Microbacterium nematophilum*. The data strongly suggest that *M. nematophilum* adhesion to wild-type *C. elegans* requires the presence of glycoconjugates on the worm cuticle.

Other worm pathways have been implicated in pathogen resistance: (1) The TGF-β/DBL-1 signaling pathway (Alegado *et al.*, 2003; Kurz and Tan, 2004; Millet and Ewbank, 2004). Two members of this pathway (SMA-6 and DAF-4) are probably *N*-glycosylated. (2) The DAF-2 pathway (Alegado *et al.*, 2003; Kurz and Tan, 2004; Millet and Ewbank, 2004). At least one member of this pathway (DAF-2) is probably *N*-glycosylated. (3) The p38 MAPK pathway (Alegado *et al.*, 2003; Kurz and Tan, 2004; Millet and Ewbank, 2004). This is a cytoplasmic pathway, indicating that probably none of the protein components are *N*-glycosylated. (4) The programmed cell death (PCD) pathway (Aballay and Ausubel, 2001). CED-9 is a transmembrane protein and may be *N*-glycosylated. (5) The Toll signaling pathway (Alegado *et al.*, 2003; Kurz and Tan, 2004; Millet and Ewbank, 2004) is the best characterized conserved signaling pathway implicated in innate immunity. TOL-1 is the only homolog of the vertebrate Toll-like pathogen-associated molecular pattern (PAMP) receptors in the *C. elegans* genome (Pujol *et al.*, 2001). One defining feature of Toll-like receptors is the presence of a conserved intracellular domain, the Toll/Interleukin-1 Receptor (TIR) domain. There are two putative TIR domain-encoding genes in *C. elegans* (Pujol *et al.*, 2001) one of which is a transmembrane protein with at least two proven *N*-glycosylated Asn-X-Ser/Thr sequons (Fan *et al.*, 2004; 2005; Kaji and Isobe, 2003).

This chapter has shown that GlcNAcTI-dependent *N*-glycans are involved in the interaction of *C. elegans* with two bacterial pathogens. The preceding discussion has identified several *C. elegans* genes that encode proven or putative *N*-glycoproteins implicated in innate immunity; other such genes almost certainly remain to be discovered. Future research must be directed toward development of experimental approaches to study the role of *N*-glycans in the functions of these proteins in innate immunity.

Acknowledgments

We dedicate this chapter to the late Dr. Charles Warren, who met an untimely death in mid-career. Charles was an expert in *C. elegans* glycobiology and provided much advice that was of great help to us in the execution of this project. We thank Dr. Andrew Spence, Department of Molecular and Medical Genetics, University of Toronto, for sharing with us his vast expertise in *C. elegans* genetics; Dr. Lori Burrows, Department of Biochemistry and Biomedical Sciences, McMaster University, for providing us with *Pseudomonas aeruginosa* strain 14 and with advice on the use of this pathogen; Dr. Shaoxian Zhu, Department of Paediatric Laboratory Medicine, Hospital for Sick Children, Toronto, for constructing the *C. elegans gly-14(III);dpy-6 gly-13(X)* double mutant; and Drs. Aldis Krizus and James W. Dennis, Samuel Lunenfeld Research Institute, Toronto, for providing the *E. coli* strains used for RNA-mediated interference studies and advice on this procedure. This work was supported by grants to H. S. from the Canadian Institutes of Health Research and the Canadian Protein Engineering Network of Centres of Excellence (PENCE).

References

Aballay, A., and Ausubel, F. M. (2001). Programmed cell death mediated by ced-3 and ced-4 protects Caenorhabditis elegans from Salmonella typhimurium-mediated killing. *Proc. Natl. Acad. Sci. USA* **98,** 2735–2739.

Aballay, A., and Ausubel, F. M. (2002). *Caenorhabditis elegans* as a host for the study of host-pathogen interactions. *Curr. Opin. Microbiol.* **5,** 97–101.

Aebi, M., and Hennet, T. (2001). Congenital disorders of glycosylation: Genetic model systems lead the way. *Trends Cell Biol.* **11,** 136–141.

Alegado, R. A., Campbell, M. C., Chen, W. C., Slutz, S. S., and Tan, M. W. (2003). Characterization of mediators of microbial virulence and innate immunity using the *Caenorhabditis elegans* host-pathogen model. *Cell Microbiol.* **5,** 435–444.

Altmann, F., Fabini, G., Ahorn, H., and Wilson, I. B. (2001). Genetic model organisms in the study of N-glycans. *Biochimie.* **83,** 703–712.

Altmann, F., Schwihla, H., Staudacher, E., Glossl, J., and Marz, L. (1995). Insect cells contain an unusual, membrane-bound beta-N-acetylglucosaminidase probably involved in the processing of protein N-glycans. *J. Biol. Chem.* **270,** 17344–17349.

Atienza-Samols, S. B., Pine, P. R., and Sherman, M. I. (1980). Effects of tunicamycin upon glycoprotein synthesis and development of early mouse embryos. *Dev. Biol.* **79,** 19–32.

Ausubel, F. M., Brent, R., Kingston, R. E., Moore, D. D., Seidman, J. G., Smith, J. A., and Struhl, K. (1993). "Current Protocols in Molecular Biology." John Wiley and Sons, New York, NY.

Brenner, S. (1974). The genetics of *Caenorhabditis elegans. Genetics* **77,** 71–94.

Bulik, D. A., and Robbins, P. W. (2002). The *Caenorhabditis elegans* sqv genes and functions of proteoglycans in development. *Biochim. Biophys. Acta.* **1573,** 247–257.

Burda, P., and Aebi, M. (1999). The dolichol pathway of N-linked glycosylation. *BBA Gen. Subjects* **1426,** 239–257.

Chang, J. Y., and Korolev, V. V. (1996). Specific toxicity of tunicamycin in induction of programmed cell death of sympathetic neurons. *Exp. Neurol.* **137,** 201–211.

Charuk, J. H. M., Tan, J., Bernardini, M., Haddad, S., Reithmeier, R. A. F., Jaeken, J., and Schachter, H. (1995). Carbohydrate-deficient glycoprotein syndrome type II—An autosomal recessive N-acetylglucosaminyltransferase II deficiency different from typical hereditary erythroblastic multinuclearity, with a positive acidified-serum lysis test (HEMPAS). *Eur. J. Biochem.* **230,** 797–805.

Chen, S., Spence, A. M., and Schachter, H. (2003). Isolation of null alleles of the *Caenorhabditis elegans* gly-12, gly-13 and gly-14 genes, all of which encode UDP-GlcNAc: Alpha-3-D-mannoside beta1,2-N-acetylglucosaminyltransferase I activity. *Biochimie.* **85,** 391–401.

Chen, S. H., Zhou, S. H., Sarkar, M., Spence, A. M., and Schachter, H. (1999). Expression of three *Caenorhabditis elegans* N-acetylglucosaminyltransferase I genes during development. *J. Biol. Chem.* **274,** 288–297.

Chen, W., and Stanley, P. (2003). Five Lec1 CHO cell mutants have distinct Mgat1 gene mutations that encode truncated N-acetylglucosaminyltransferase I. *Glycobiology* **13,** 43–50.

Chitwood, D. J., Lusby, W. R., Thompson, M. J., Kochansky, J. P., and Howarth, O. W. (1995). The glycosylceramides of the nematode Caenorhabditis elegans contain an unusual, branched-chain sphingoid base. *Lipids* **30**, 567–573.

Cipollo, J. F., Awad, A. M., Costello, C. E., and Hirschberg, C. B. (2005). N-Glycans of *Caenorhabditis elegans* are specific to developmental stages. *J Biol Chem* **280**, 26063–72.

Cipollo, J. F., Costello, C. E., and Hirschberg, C. B. (2002). The fine structure of *Caenorhabditis elegans* N-glycans. *J. Biol. Chem.* **277**, 49143–49157.

Darby, C., Cosma, C. L., Thomas, J. H., and Manoil, C. (1999). Lethal paralysis of *Caenorhabditis elegans* by Pseudomonas aeruginosa. *Proc. Natl. Acad. Sci. USA* **96**, 15202–15207.

Dennis, J. W., Granovsky, M., and Warren, C. E. (1999). Glycoprotein glycosylation and cancer progression. *Biochim. Biophys. Acta* **1473**, 21–34.

Epstein, H. F., and Shakes, D. C. (1995). *Caenorhabditis elegans*: Modern biological analysis of an organism. *In* "Methods in Cell Biology" (L. Wilson and P. Matsudaira, eds.), Vol. 48. Academic Press, San Diego.

Fan, X., She, Y. M., Bagshaw, R. D., Callahan, J. W., Schachter, H., and Mahuran, D. J. (2004). A method for proteomic identification of membrane-bound proteins containing Asn-linked oligosaccharides. *Anal. Biochem.* **332**, 178–186.

Fan, X., She, Y. M., Bagshaw, R. D., Callahan, J. W., Schachter, H., and Mahuran, D. J. (2005). Identification of the hydrophobic glycoproteins of *Caenorhabditis elegans*. *Glycobiology* **15**, 952–964.

Finnie, J. W. (2001). Effect of tunicamycin on hepatocytes *in vitro*. *J. Comp. Pathol.* **125**, 318–321.

Francis, B. R., Paquin, L., Weinkauf, C., and Jarvis, D. L. (2002). Biosynthesis and processing of *Spodoptera frugiperda* alpha-mannosidase III. *Glycobiology* **12**, 369–377.

Fukuda, M. (2002). Roles of mucin-type O-glycans in cell adhesion. *Biochim. Biophys. Acta* **1573**, 394–405.

Gan, Y. H., Chua, K. L., Chua, H. H., Liu, B., Hii, C. S., Chong, H. L., and Tan, P. (2002). Characterization of Burkholderia pseudomallei infection and identification of novel virulence factors using a *Caenorhabditis elegans* host system. *Mol. Microbiol.* **44**, 1185–1197.

Garsin, D. A., Sifri, C. D., Mylonakis, E., Qin, X., Singh, K. V., Murray, B. E., Calderwood, S. B., and Ausubel, F. M. (2001). A simple model host for identifying Gram-positive virulence factors. *Proc. Natl. Acad. Sci. USA* **98**, 10892–10897.

Gerdt, S., Dennis, R. D., Borgonie, G., Schnabel, R., and Geyer, R. (1999). Isolation, characterization and immunolocalization of phosphorylcholine-substituted glycolipids in developmental stages of *Caenorhabditis elegans*. *Eur. J. Biochem.* **266**, 952–963.

Gerdt, S., Lochnit, G., Dennis, R. D., and Geyer, R. (1997). Isolation and structural analysis of three neutral glycosphingolipids from a mixed population of *Caenorhabditis elegans* (Nematoda:Rhabditida). *Glycobiology* **7**, 265–275.

Granovsky, M., Fata, J., Pawling, J., Muller, W. J., Khokha, R., and Dennis, J. W. (2000). Suppression of tumor growth and metastasis in Mgat5-deficient mice. *Nat. Med.* **6**, 306–312.

Gravato-Nobre, M. J., and Hodgkin, J. (2005). *Caenorhabditis elegans* as a model for innate immunity to pathogens. *Cell Microbiol.* **7**, 741–751.

Gravato-Nobre, M. J., Nicholas, H. R., Nijland, R., O'Rourke, D., Whittington, D. E., Yook, K. J., and Hodgkin, J. (2005). Multiple genes affect sensitivity of Caenorhabditis elegans to the bacterial pathogen *Microbacterium nematophilum*. *Genetics* **171**, 1033–1045.

Griffitts, J. S., Haslam, S. M., Yang, T., Garczynski, S. F., Mulloy, B., Morris, H., Cremer, P. S., Dell, A., Adang, M. J., and Aroian, R. V. (2005). Glycolipids as receptors for Bacillus thuringiensis crystal toxin. *Science* **307**, 922–925.

Griffitts, J. S., Huffman, D. L., Whitacre, J. L., Barrows, B. D., Marroquin, L. D., Muller, R., Brown, J. R., Hennet, T., Esko, J. D., and Aroian, R. V. (2003). Resistance to a bacterial toxin is mediated by removal of a conserved glycosylation pathway required for toxin-host interactions. *J. Biol. Chem.* **278**, 45594–45602.

Grunewald, S., Matthijs, G., and Jaeken, J. (2002). Congenital disorders of glycosylation: A review. *Pediatr. Res.* **52**, 618–624.

Guerardel, Y., Balanzino, L., Maes, E., Leroy, Y., Coddeville, B., Oriol, R., and Strecker, G. (2001). The nematode *Caenorhabditis elegans* synthesizes unusual O-linked glycans: Identification of glucose-substituted mucin-type O-glycans and short chondroitin-like oligosaccharides. *Biochem. J.* **357**, 167–182.

Haltiwanger, R. S., and Lowe, J. B. (2004). Role of glycosylation in development. *Annu. Rev. Biochem.* **73**, 491–537.

Hanneman, A., Lapadula, A., Ashline, D., Zhang, H., Schachter, H., and Reinhold, V. (2005). Aberrant protein *N*-glycosylation in *Caenorhabditis elegans* GnT-I triple knockout worms. *Glycobiology* **15**, 1218.

Hanneman, A., and Reinhold, V. (2003). Abundant and unusual N-linked glycans from the eukaryote, *C. elegans*. *Glycobiology* **13**, 899–900.

Haslam, S. M., and Dell, A. (2003). Hallmarks of *Caenorhabditis elegans* *N*-glycosylation: Complexity and controversy. *Biochimie.* **85**, 25–32.

Haslam, S. M., Gems, D., Morris, H. R., and Dell, A. (2002). The glycomes of *Caenorhabditis elegans* and other model organisms. *Biochem. Soc. Symp.* **69**, 117–134.

Hemming, F. W. (1995). Biosynthesis. The Coenzymic Role of Phosphodolichols.. *In* "Glycoproteins" (J. Montreuil, J. F. G. Vliegenthart, and H. Schachter, eds.) Vol. 29a, pp. 127–143. Elsevier, Amsterdam, The Netherlands.

Herscovics, A. (1999). Importance of glycosidases in mammalian glycoprotein biosynthesis. *Biochim. Biophys. Acta* **1473**, 96–107.

Hirabayashi, J., Hayama, K., Kaji, H., Isobe, T., and Kasai, K. (2002). Affinity capturing and gene assignment of soluble glycoproteins produced by the nematode *Caenorhabditis elegans*. *J. Biochem. (Tokyo).* **132**, 103–114.

Hodgkin, J., Kuwabara, P. E., and Corneliussen, B. (2000). A novel bacterial pathogen, Microbacterium nematophilum, induces morphological change in the nematode *C. elegans*. *Curr. Biol.* **10**, 1615–1618.

Hoeflich, J., Berninsone, P., Goebel, C., Gravato-Nobre, M. J., Libby, B. J., Darby, C., Politz, S. M., Hodgkin, J., Hirschberg, C. B., and Baumeister, R. (2004). Loss of srf-3 encoded nucleotide sugar transporter activity in *Caenorhabditis elegans* alters surface antigenicity and prevents bacterial adherence. *J. Biol. Chem.* **279**, 30440–30448.

Houston, K. M., and Harnett, W. (2004). Structure and synthesis of nematode phosphorylcholine-containing glycoconjugates. *Parasitology* **129**, 655–661.

Hübel, A., and Schwarz, R. T. (2002). Dolichol Phosphate GlcNAc-1-P Transferase. *In* "Handbook of Glycosyltransferases and Related Genes" (N. Taniguchi, K. Honke, and M. Fukuda, eds.) pp. 550–556. Springer-Verlag, Tokyo, Japan.

Ioffe, E., and Stanley, P. (1994). Mice lacking N-acetylglucosaminyltransferase I activity die at mid-gestation, revealing an essential role for complex or hybrid N-linked carbohydrates. *Proc. Natl. Acad. Sci. USA* **91**, 728–732.

Jaeken, J., and Carchon, H. (2004). Congenital disorders of glycosylation: A booming chapter of pediatrics. *Curr. Opin. Pediatr.* **16**, 434–439.

Jaeken, J., Schachter, H., Carchon, H., De Cock, P., Coddeville, B., and Spik, G. (1994). Carbohydrate deficient Glycoprotein syndrome type II: A deficiency in Golgi localised N-acetylglucosaminyltransferase II. *Arch. Dis. Child.* **71**, 123–127.

Kaji, H., and Isobe, T. (2003). Protein database of *Caenorhabditis elegans*. *J. Chromatogr. B Analyt. Technol. Biomed. Life Sci.* **787**, 91–99.

Kaji, H., Saito, H., Yamauchi, Y., Shinkawa, T., Taoka, M., Hirabayashi, J., Kasai, K., Takahashi, N., and Isobe, T. (2003). Lectin affinity capture, isotope-coded tagging and mass spectrometry to identify N-linked glycoproteins. *Nat. Biotechnol.* **21**, 667–672.

Kawar, Z., Karaveg, K., Moremen, K. W., and Jarvis, D. L. (2001). Insect cells encode a class II {alpha}-mannosidase with unique properties. *J. Biol. Chem.* **276**, 16335–16340.

Kurz, C. L., and Ewbank, J. J. (2003). *Caenorhabditis elegans*: An emerging genetic model for the study of innate immunity. *Nat. Rev. Genet.* **4**, 380–390.

Kurz, C. L., and Tan, M. W. (2004). Regulation of aging and innate immunity in *C. elegans*. *Aging Cell* **3**, 185–193.

Laine, R. A. (1994). A calculation of all possible oligosaccharide isomers both branched and linear yields 1.05 x 10(12) structures for a reducing hexasaccharide: The isomer barrier to development of single-method saccharide sequencing or synthesis systems. *Glycobiology* **4**, 759–767.

Lewis, J. A., and Fleming, J. T. (1995). Basic culture methods. *Methods Cell Biol.* **48**, 3–29.

Lin, T. Y., Wang, S. M., Fu, W. M., Chen, Y. H., and Yin, H. S. (1999). Toxicity of tunicamycin to cultured brain neurons: Ultrastructure of the degenerating neurons. *J. Cell Biochem.* **74**, 638–647.

Link, C. D., Silverman, M. A., Breen, M., Watt, K. E., and Dames, S. A. (1992). Characterization of *Caenorhabditis elegans* lectin-binding mutants. *Genetics* **131**, 867–881.

Lochnit, G., Dennis, R. D., and Geyer, R. (2000). Phosphorylcholine substituents in nematodes: Structures, occurrence and biological implications. *Biol. Chem.* **381**, 839–847.

Lowe, J. B. (2003). Glycan-dependent leukocyte adhesion and recruitment in inflammation. *Curr. Opin. Cell Biol.* **15**, 531–538.

Lowe, J. B., and Marth, J. D. (2003). A genetic approach to mammalian glycan function. *Annu. Rev. Biochem.* **72**, 643–691.

Mahajan-Miklos, S., Rahme, L. G., and Ausubel, F. M. (2000). Elucidating the molecular mechanisms of bacterial virulence using non-mammalian hosts. *Mol. Microbiol.* **37**, 981–988.

Mahajan-Miklos, S., Tan, M. W., Rahme, L. G., and Ausubel, F. M. (1999). Molecular mechanisms of bacterial virulence elucidated using a *Pseudomonas aeruginosa–Caenorhabditis elegans* pathogenesis model. *Cell* **96**, 47–56.

Marek, K. W., Vijay, I. K., and Marth, J. D. (1999). A recessive deletion in the GlcNAc-1-phosphotransferase gene results in peri-implantation embryonic lethality. *Glycobiology* **9**, 1263–12671.

Mawuenyega, K. G., Kaji, H., Yamuchi, Y., Shinkawa, T., Saito, H., Taoka, M., Takahashi, N., and Isobe, T. (2003). Large-scale identification of *Caenorhabditis elegans* proteins by multidimensional liquid chromatography-tandem mass spectrometry. *J. Proteome Res.* **2**, 23–35.

Mello, C., and Fire, A. (1995). DNA transformation. *Methods Cell Biol.* **48**, 451–482.

Metzler, M., Gertz, A., Sarkar, M., Schachter, H., Schrader, J. W., and Marth, J. D. (1994). Complex asparagine-linked oligosaccharides are required for morphogenic events during post-implantation development. *EMBO J.* **13**, 2056–2065.

Millet, A. C., and Ewbank, J. J. (2004). Immunity in Caenorhabditis elegans. *Curr. Opin. Immunol.* **16**, 4–9.

Moremen, K. W. (2002). Golgi alpha-mannosidase II deficiency in vertebrate systems: Implications for asparagine-linked oligosaccharide processing in mammals. *Biochim. Biophys. Acta* **1573**, 225–235.

Natsuka, S., Adachi, J., Kawaguchi, M., Nakakita, S.-I., Hase, S., Ichikawa, A., and Ikura, K. (2002). Structure analysis of N-linked glycans in Caenorhabditis elegans. J. Biochem. (Tokyo) 131, 807–813.

Paschinger, K., Rendic, D., Lochnit, G., Jantsch, V., and Wilson, I. B. (2004). Molecular basis of anti-horseradish peroxidase staining in Caenorhabditis elegans. J. Biol. Chem. 279, 49588–49598.

Paschinger, K., Staudacher, E., Stemmer, U., Fabini, G., and Wilson, I. B. (2005). Fucosyltransferase substrate specificity and the order of fucosylation in invertebrates. Glycobiology 15, 463–474.

Pittet, M., Uldry, D., Aebi, M., and Conzelmann, A. (2006). The N-glycosylation defect of cwh8{Delta} yeast cells causes a distinct defect in sphingolipid biosynthesis. Glycobiology 16, 155–164.

Politz, S. M., Philipp, M., Estevez, M., O'Brien, P. J., and Chin, K. J. (1990). Genes that can be mutated to unmask hidden antigenic determinants in the cuticle of the nematode Caenorhabditis elegans. Proc. Natl. Acad. Sci. USA 87, 2901–2905.

Pujol, N., Link, E. M., Liu, L. X., Kurz, C. L., Alloing, G., Tan, M. W., Ray, K. P., Solari, R., Johnson, C. D., and Ewbank, J. J. (2001). A reverse genetic analysis of components of the Toll signaling pathway in Caenorhabditis elegans. Curr. Biol. 11, 809–821.

Riddle, D. L., Blumenthal, T., Meyer, B. J., and Priess, J. R. (1997). C. elegans II. In "Cold Spring Harbor Monograph Series" Vol. 33. Cold Spring Harbor Laboratory Press, Plainview, NY.

Ruiz-Diez, B., Sanchez, P., Baquero, F., Martinez, J. L., and Navas, A. (2003). Differential interactions within the Caenorhabditis elegans–Pseudomonas aeruginosa pathogenesis model. J. Theor. Biol. 225, 469–476.

Sambrook, J., Fritsch, E. F., and Maniatis, T. (1989). "Molecular Cloning: A Laboratory Manual." Cold Spring Harbor Laboratory, Cold Spring Harbor, NY.

Sanger, F., Nicklen, S., and Coulson, A. R. (1977). DNA sequencing with chain-terminating inhibitors. Proc. Natl. Acad. Sci. USA 74, 5463–5467.

Sarkar, M., Leventis, P. A., Silvescu, C. I., Reinhold, V. N., Schachter, H., and Boulianne, G. L. (2006). Null mutations in Drosophila N-acetylglucosaminyltransferase I produce defects in locomotion and a reduced lifespan. J. Biol. Chem. 281, 12776–12785.

Schachter, H. (1986). Biosynthetic controls that determine the branching and microheterogeneity of protein-bound oligosaccharides. Biochem. Cell Biol. 64, 163–181.

Schachter, H. (1991). The "yellow brick road" to branched complex N-glycans. Glycobiology 1, 453–461.

Schachter, H. (1995). Biosynthesis. Glycosyltransferases Involved in the Synthesis of N-glycan Antennae. In "Glycoproteins" (J. Montreuil, J. F. G. Vliegenthart, and H. Schachter, eds.) Vol. 29a, pp. 153–199. Elsevier, Amsterdam, The Netherlands.

Schachter, H. (2000). The joys of HexNAc. The synthesis and function of N- and O-glycan branches. Glycoconj. J. 17, 465–483.

Schachter, H. (2004). Protein glycosylation lessons from Caenorhabditis elegans. Curr. Opin. Struct. Biol. 14, 607–616.

Schachter, H. (2005). Deficient Glycoprotein Glycosylation in Humans and Mice. In "Neuroglycobiology" (M. Fukuda, U. Rutishauser, and R. L. Schnaar, eds.) pp. 157–198. Oxford University Press, Oxford.

Schachter, H., Chen, S., Zhang, W., Spence, A. M., Zhu, S., Callahan, J. W., Mahuran, D. J., Fan, X., Bagshaw, R. D., She, Y. M., Rosa, J. C., and Reinhold, V. N. (2002). Functional post-translational proteomics approach to study the role of N-glycans in the development of Caenorhabditis elegans. Biochem. Soc. Symp. 69, 1–21.

Schwientek, T., Keck, B., Levery, S. B., Jensen, M. A., Pedersen, J. W., Wandall, H. H., Stroud, M., Cohen, S. M., Amado, M., and Clausen, H. (2002). The Drosophila gene brainiac encodes a glycosyltransferase putatively involved in glycosphingolipid synthesis. *J. Biol. Chem.* **277,** 32421–32429.

Sifri, C. D., Begun, J., and Ausubel, F. M. (2005). The worm has turned–microbial virulence modeled in *Caenorhabditis elegans. Trends Microbiol.* **13,** 119–127.

Spiro, R. G. (2002). Protein glycosylation: Nature, distribution, enzymatic formation, and disease implications of glycopeptide bonds. *Glycobiology* **12,** 43R–56R.

Spiro, R. G. (2004). Role of N-linked polymannose oligosaccharides in targeting glycoproteins for endoplasmic reticulum-associated degradation. *Cell Mol. Life Sci.* **61,** 1025–1041.

Stanley, P. (2000). Functions of Carbohydrates Revealed by Transgenic Technology. *In* "Molecular and Cellular Glycobiology" (M. Fukuda and O. Hindsgaul, eds.), Vol. 30, pp. 169–198. Oxford University Press, Oxford.

Surani, M. A. H. (1979). Glycoprotein synthesis and inhibition of glycosylation by tunicamycin in pre-implantation mouse embryos: Compaction and trophoblast adhesion. *Cell* **18,** 217–227.

Tan, J., Dunn, J., Jaeken, J., and Schachter, H. (1996). Mutations in the MGAT2 gene controlling complex N-glycan synthesis cause carbohydrate-deficient glycoprotein syndrome type II, an autosomal recessive disease with defective brain development. *Am. J. Hum. Genet.* **59,** 810–817.

Tan, M. W. (2002). Identification of host and pathogen factors involved in virulence using *Caenorhabditis elegans. Methods Enzymol.* **358,** 13–28.

Tan, M. W., and Ausubel, F. M. (2000). *Caenorhabditis elegans*: A model genetic host to study *Pseudomonas aeruginosa* pathogenesis. *Curr. Opin. Microbiol.* **3,** 29–34.

Tan, M. W., Mahajan-Miklos, S., and Ausubel, F. M. (1999a). Killing of *Caenorhabditis elegans* by *Pseudomonas aeruginosa* used to model mammalian bacterial pathogenesis. *Proc. Natl. Acad. Sci. USA* **96,** 715–720.

Tan, M. W., Rahme, L. G., Sternberg, J. A., Tompkins, R. G., and Ausubel, F. M. (1999b). *Pseudomonas aeruginosa* killing of *Caenorhabditis elegans* used to identify *P. aeruginosa* virulence factors. *Proc. Natl. Acad. Sci. USA* **96,** 2408–2413.

Toyoda, H., Kinoshita-Toyoda, A., and Selleck, S. B. (2000). Structural analysis of glycosaminoglycans in *Drosophila* and *Caenorhabditis elegans* and demonstration that tout-velu, a *Drosophila* gene related to EXT tumor suppressors, affects heparan sulfate *in vivo. J. Biol. Chem.* **275,** 2269–2275.

Varki, A. (1993). Biological roles of oligosaccharides: All of the theories are correct. *Glycobiology* **3,** 97–130.

Verbert, A. (1995). Biosynthesis. From Glc$_3$Man$_9$GlcNAc$_2$-protein to Man$_5$GlcNAc$_2$-protein: Transfer 'En Bloc' and Processing. *In* "Glycoproteins" (J. Montreuil, J. F. G. Vliegenthart, and H. Schachter, eds.), Vol. 29a, pp. 145–152. Elsevier, Amsterdam, The Netherlands.

Veronico, P., Gray, L. J., Jones, J. T., Bazzicalupo, P., Arbucci, S., Cortese, M. R., Di Vito, M., and De Giorgi, C. (2001). Nematode chitin synthases: Gene structure, expression and function in *Caenorhabditis elegans* and the plant parasitic nematode *Meloidogyne artiellia. Mol. Genet. Genomics.* **266,** 28–34.

Vitale, A., and Chrispeels, M. J. (1984). Transient N-acetylglucosamine in the biosynthesis of phytohemagglutinin: Attachment in the Golgi apparatus and removal in protein bodies. *J. Cell Biol.* **99,** 133–140.

Wandall, H. H., Pedersen, J. W., Park, C., Levery, S. B., Pizette, S., Cohen, S. M., Schwientek, T., and Clausen, H. (2003). *Drosophila* egghead encodes a beta 1,4-mannosyltransferase predicted to form the immediate precursor glycosphingolipid substrate for brainiac. *J. Biol. Chem.* **278,** 1411–1414.

Wang, Y., Schachter, H., and Marth, J. D. (2002). Mice with a homozygous deletion of the Mgat2 gene encoding UDP-N-acetylglucosamine:alpha-6-D-mannoside beta1,2-N-acetylglucosaminyltransferase II: A model for congenital disorder of glycosylation type IIa. *Biochim. Biophys. Acta* **1573,** 301–311.

Wang, Y., Tan, J., Sutton-Smith, M., Ditto, D., Panico, M., Campbell, R. M., Varki, N. M., Long, J. M., Jaeken, J., Levinson, S. R., Wynshaw-Boris, A., Morris, H. R., Le, D., Dell, A., Schachter, H., and Marth, J. D. (2001). Modeling human congenital disorder of glycosylation type IIa in the mouse: Conservation of asparagine-linked glycan-dependent functions in mammalian physiology and insights into disease pathogenesis. *Glycobiology* **11,** 1051–1070.

Wood, W. B. (1988). The Nematode. "*Caenorhabditis elegans*". Cold Spring Harbor Laboratory Press, Plainview, NY.

Yamada, S., Van Die, I., Van den Eijnden, D. H., Yokota, A., Kitagawa, H., and Sugahara, K. (1999). Demonstration of glycosaminoglycans in *Caenorhabditis elegans*. *FEBS Lett.* **459,** 327–331.

Ye, Z., and Marth, J. D. (2004). N-glycan branching requirement in neuronal and post-natal viability. *Glycobiology* **14,** 547–558.

Zhang, W., Cao, P., Chen, S., Spence, A. M., Zhu, S., Staudacher, E., and Schachter, H. (2003). Synthesis of paucimannose N-glycans by *Caenorhabditis elegans* requires prior actions of UDP-GlcNAc:alpha-3-D-mannoside beta1,2-N-acetylglucosaminyltransferase I, alpha-3,6-mannosidase II and a specific membrane-bound beta-N-acetylglucosaminidase. *Biochem. J.* **372,** 53–64.

Zhou, D. (2003). Why are glycoproteins modified by poly-N-acetyllactosamine glyco-conjugates? *Curr. Protein Pept. Sci.* **4,** 1–9.

Zhu, S., Hanneman, A., Reinhold, V. N., Spence, A. M., and Schachter, H. (2004). *Caenorhabditis elegans* triple null mutant lacking UDP-N-acetyl-D-glucosamine:alpha-3-D-mannoside beta1,2-N-acetylglucosaminyltransferase I. *Biochem. J.* **382,** 995–1001.

[23]   Detection of Cytoplasmic Glycosylation Associated with Hydroxyproline

By CHRISTOPHER M. WEST, HANKE VAN DER WEL, and IRA J. BLADER

Abstract

A special class of glycosylation occurs on a proline residue of the cyto-plasmic/nuclear protein Skp1 in the social amoeba *Dictyostelium*. For this glycosylation to occur, the proline must first be hydroxylated by the action of a soluble prolyl 4-hydroxylase acting on the protein. Cytoplasmic prolyl 4-hydroxylases are dioxygen-dependent enzymes that have low affinity for their O_2 substrate and, therefore, have been implicated in O_2-sensing in

METHODS IN ENZYMOLOGY, VOL. 417
0076-6879/06 $35.00
DOI: 10.1016/S0076-6879(06)17023-8

Dictyostelium, as well as in vertebrates and invertebrates. The sugar-hydroxyproline linkage has low abundance, is resistant to alkali cleavage and known glycosidases, and does not bind known lectins. However, initial screens for this modification can be made by assessing changes in electrophoretic mobility of candidate proteins after treatment of cells with prolyl hydroxylase inhibitors, and/or by metabolic labeling with [³H]sugar precursors. In addition, cytoplasmic hydroxylation/glycosylation can be assessed by assaying for cytoplasmic glycosyltransferases. Here we describe these methods and examples of their use in analyzing Skp1 glycosylation in *Dictyostelium* and the apicomplexan *Toxoplasma gondii,* the causative agent of toxoplasmosis in humans.

Overview

O-glycosylation of proteins occurs typically at Ser and Thr residues but has also been found on other amino acids such as Tyr, Pro, and Lys (Spiro, 2002). For nonhydroxyl-containing amino acids such as Pro and Lys to be O-glycosylated, they must first be hydroxylated by prolyl- or lysyl-hydroxylases. O-glycosidic linkages are then formed by sugar nucleotide-dependent glycosyltransferases (GTs). In the secretory pathway of plants, 4-HyPro residues formed in the rough endoplasmic reticulum (rER) are subsequently modified by galactose or arabinose (Tan *et al.,* 2003). In animals, as well as the amoebazoan *Dictyostelium* and perhaps other eukaryotes, a distinct class of prolyl 4-hydroxylases (P4Hs), with characteristic low affinities for the substrate O_2, resides in the cytoplasm and nucleus (Kaelin, 2005; van der Wel *et al.,* 2005). The transcription factor subunit hypoxia-inducible factor α (HIFα) is the best-known target of this enzyme in animals. Because of their low affinity, these cytoplasmic P4Hs have been suggested to be important cellular O_2 sensors. In *Dictyostelium,* cytoplasmic HyPro is capped by an α-linked GlcNAc and four additional sugars (Teng-umnuay *et al.,* 1998). Unlike most sugar–amino acid O-linkages, HyPro-sugar linkages are resistant to known methods of chemical and enzymatic cleavage that leave the glycan chain intact. This chapter addresses approaches that can be taken to detect and partially characterize O_2-dependent HyPro glycosylation.

The *Dictyostelium* modification has been described on the small cytoplasmic/nuclear protein *Dictyostelium* Skp1. Skp1 is best known as a subunit of SCF-type E3 ubiquitin ligases (Petroski and Deshaies, 2005). Its glycosylation was first discovered by metabolic labeling of cells with [³H] fucose (Fuc) (Gonzalez-Yanes *et al.,* 1992). The modification was characterized by MS-MS and Edman degradation analysis of peptides from the purified protein, which showed it to be a pentasaccharide attached at

HyPro143 (Teng-umnuay *et al.*, 1998). Further characterization using specific glycosidases, and characterization of the enzymes that catalyze formation of the structure, showed that it consists of Galα1,Galα1,3Fucα1, 2Galβ1,3GlcNAcα1,4HyPro143, with only the linkage of the nonreducing terminal αGal uncertain (West *et al.*, 2004). Approximately 90% of *Dd*-Skp1 is hydroxylated/glycosylated, and most of this is apparently fully glycosylated. Four of the five predicted Skp1 GT activities have been purified to near homogeneity and their functions dissected by analysis of recombinantly produced proteins and in some cases gene disruption (Ercan *et al.*, 2006; West *et al.*, 2004). The validated GT genes have remote similarities to the sequences of the catalytic domains of known GTs but lack the N-terminal signal anchors that direct most eukaryotic GTs to the Golgi. Biochemical complementation of the mutant strain extracts with recombinant GTs suggests that Skp1 is the only soluble *Dictyostelium* protein glycosylated by this pathway.

The evolutionary conservation of HyPro-dependent cytoplasmic glycosylation like that seen on *Dictyostelium* Skp1 remains to be investigated. BLAST searches on the basis of the Skp1 GT sequences yield evidence for homologous GTs in other organisms (West *et al.*, 2004, 2005). Because Skp1s from both unicellular and most multicellular organisms, except for vertebrates, possess the equivalent of Pro143, Skp1 may be modified in other organisms (West *et al.*, 2004). However, the putative GT sequences are sufficiently different that it cannot be predicted whether Skp1 is their target or whether the sugar structure is conserved. In addition, animals and plants have more diverged predicted GT genes that may encode cytoplasmic proteins (West *et al.*, 2002), and considerable circumstantial evidence for other forms of complex (>1 sugar) cytoplasmic glycosylation have been published (Hart *et al.*, 1989; West *et al.*, 2002). Some of these might involve HyPro.

Three methods for screening for cytoplasmic glycosylation that have emerged from our studies on *Dictyostelium* Skp1 are described here. (1) The first tests whether a candidate protein is glycosylated on HyPro by screening for an M_r shift after treatment of cells with inhibitors of P4Hs (Sassi *et al.*, 2001). (2) Metabolic labeling of cells with a tritiated sugar precursor was the method that first detected Skp1 glycosylation (Gonzalez-Yanes *et al.*, 1992), but as a general method presupposes which sugars are present and depends on an effective method for preparing a cytosol free of contamination with vesicle contents and their attendant glycoconjugates. (3) Additional early evidence for Skp1 glycosylation was based on radioactive labeling with GDP-[^{3}H]Fuc in a cytosolic extract of a GDP-[^{3}H]Fuc synthesis mutant, which, therefore, accumulated the Skp1 glycoform precursor (Gonzalez-Yanes *et al.*, 1992). Although this method also presupposes sugar identity, an adaptation of this approach to detect the αGlcNAcT that directly modifies Skp1 HyPro, which is more likely to be conserved, may be more general. Subsequent evidence for the modification was based on purification of the [^{3}H]Fuc-labeled glycopeptide from purified Skp1 and structural analysis by MS-MS methods and Edman degradation (Teng-umnuay et al., 1998). Although necessary for establishing a proposed modification, these latter methods are not practical for an initial screen for HyPro-dependent glycans.

In Vivo Detection of HyPro-dependent Glycosylation by M_r-Shift Analysis

Glycosylation results in reduced mobility of Skp1 during SDS-PAGE that can be detected by Western blot analysis. The dependence of glycosylation of any protein on prolyl hydroxylation can be tested by treatment of cells with P4H inhibitors, which act generally across this class of enzymes including both cytoplasmic and rER types. A side-by-side comparison of extracts from treated and untreated cells on a gradient gel optimized for high-resolution separation in the M_r range of the protein can reveal differences, in optimal separations, as small as one to two sugars. Assignment of an M_r-shift to glycosylation can be confirmed by follow-up studies on the basis of labeling with [^{3}H]sugars, lectin probing, or structural studies on the purified protein. The inhibitor method is general for any kind of glycosyl modification of HyPro of more than one sugar.

Choice of Inhibitors

P4H inhibitors active in cells were originally developed for collagen-type P4Hs that reside in the rER. These chelate the Fe^{+2} cofactor or compete with the α-ketoglutarate (αKG) co-substrate and are effective against Dictyostelium Skp1 P4H1 in cells (Sassi et al., 2001). More recently, additional inhibitors of animal HIFα-type P4Hs have been described (Siddiq et al., 2005) and may also potentially be used.

1. α,α'-dipyridyl functions as an Fe^{+2} chelator. Prepare a 20 mM stock solution of α,α'-dipyridyl (Sigma) in water or an aqueous buffer suitable for the cells to be tested, filter through a 0.2-μm filter for sterility, and store at 4°. Incubate cells at 50–200 μM α,α'-dipyridyl, which is suitable for cultured mammalian cells and Dictyostelium cells.

2. Ethyl 3,4-dihydroxybenzoate functions as a competitive inhibitor of αKG. Prepare a 40 mM stock solution of ethyl 3,4-dihydroxybenzoate (protocatechuic acid ethyl ester, Sigma) in 5% MeOH in water, and filter for sterility. Store at 4° and provide to cells at 200–500 μM.

Treatment of Cells with Inhibitors

Treat cell cultures with the inhibitor for a length of time sufficient for at least two cell doublings, or natural turnover of the protein, to be analyzed. If cells detach, the inhibitor concentration may be too high, because these compounds can affect other αKG-dependent dioxygenases and Fe^{+2}-dependent enzymes. Treatment should not result in >20% inhibition of protein synthesis as measured by uptake of [^{35}S]methionine into total protein (Sassi et al., 2001). It may be helpful to include the P4H cofactor ascorbate at 1 mM in the culture medium. In animal cells, inhibition can be verified by accumulation of HIFα (e.g., Bruick and McKnight, 2001).

Analysis of Target Protein

Any protein for which an antibody is available for Western blotting can be evaluated for HyPro-dependent glycosylation using an SDS-PAGE shift method. Modifications as small as one to two sugars can be detected by use of minimum sample loading consistent with sample detection, and polyacryl-amide gradient gels optimized for the M_r range of interest as shown, for example, in the case of glycosylation of influenza viral hemagglutinin (Hebert et al., 1995); 15–20% linear gradient polyacrylamide gels are suitable for resolving Skp1 ($M_r \sim 20,000$) glycoforms (Sassi et al., 2001), and lower polyacrylamide concentrations or decreased cross-linker percentage are likely to be useful for higher M_r proteins.

Interpretation

A typical result is shown for extracts from Dictyostelium cells treated with 50 μM α,α'-dipyridyl and Western blotted with an anti-Skp1 antibody (Fig. 1). Approximately half of the total Skp1 pool is shifted to a lower M_r

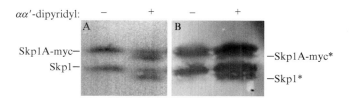

FIG. 1. Effect on Skp1 of treating Dictyostelium cells with a prolyl hydroxylase inhibitor. Strain HW302 cells expressing a myc-tagged isoform in addition to endogenous Skp1 were treated with or without 50 μM α, α'-dipyridyl. Cells were lysed in Laemmli sample buffer, resolved on a 15–20% linear gradient polyacrylamide gel, and Western blotted with mAb 3F9 for Skp1. (A) Growing cells treated for 24 h. (B) Developing cells treated for 18 h. * indicates more rapidly migrating species induced by treatment with the inhibitor.

position consistent with absence of the glycan modification, indicating a dependence on hydroxylation. Longer treatment times or higher concentrations did not result in more of the lower M_r band. This method is applicable for any protein for which an antibody is available and is expected to detect HyPro-dependent glycosylation of secretory proteins, which occurs in algae and plants (Tan *et al.*, 2003) as well. A false-negative result may occur if (1) prolyl 4-hydroxylase is not sufficiently inhibited by the concentration of the inhibitor used, (2) the effect of sugar modification on gel mobility is below the threshold of detection, (3) glycosylation is heterogeneous resulting in dilution of the Western blot signal, or (4) the time of inhibitor treatment does not exceed the half-life of the protein or the doubling time of the cell culture. False-positive results are less likely but may result from indirect effects on other metal- or αKG-dependent cellular processes or modification of HyPro by some other (unknown) substituent. If P4H inhibitor treatment results in reduced apparent M_r, follow-up methods to confirm the suspected modification would include determining whether the immunoprecipitated or purified protein contains sugar by an independent method, such as incorporation of a radioactive sugar precursor, sugar composition analysis (see later), or recognition by a lectin or anticarbohydrate antibody. Ultimately, mass spectrometry (MS) analysis of the purified protein can be done for direct evidence of a covalent linkage between sugars and the protein of interest, as illustrated for *Dictyostelium* Skp1 (Teng-umnuay *et al.*, 1998) and cytoplasmic glycoproteins modified with a single β-linked GlcNAc (Hart, 2006).

Metabolic Labeling of Cells with [³H]fucose

Most cells have uptake mechanisms for sugars that serve as precursors for the synthesis of sugar nucleotides, which are the donor substrates for known cytoplasmic (and other) glycosyltransferases. [³H]Fuc is a relatively specific precursor, because it is minimally interconverted to other compounds *in vivo*. Commercially available [³H]Fuc can be provided in the cell culture medium for a period of hours and cells recovered and fractionated into cytosolic and particulate fractions by differential centrifugation. The cytosolic fraction can be analyzed for radioactive proteins.

Metabolic Labeling and Preparation of the Cytosolic Extract

Methods for metabolic labeling of animal and plant cell lines and microorganisms with radioactive sugars including Fuc are available (Gonzalez-Yanes *et al.*, 1992; Varki, 1994). Tritiated sugars are preferred to ¹⁴C-labeled sugars because of their higher specific activity. After incubation of cells for a period of time, typically for ≥two cell divisions or a time corresponding to twice the turnover time of average cytoplasmic proteins, cells are collected and washed by centrifugation to remove the bulk of unincorporated precursor. Conditions of gentle cell lysis are typically optimized for specific cell lines and cell types. It is best to use a gentle mechanical method of cell lysis to minimize leakage of vesicle contents that are likely to be glycosylated by other mechanisms. If cell walls or other extracellular matrix is present, this material can be predigested enzymatically. Typically, cells are resuspended in ice-cold 0.25 M sucrose, 0.05 M Tris-HCl (pH 7.4), with protease inhibitors appropriate for the cell type (Tris-sucrose buffer) (West *et al.*, 1996). Sucrose provides osmotic balance to protect organelles without contributing ionic strength. Divalent cations and EDTA are avoided to minimize organellar aggregation and nuclear lysis, respectively. Shearing of cells by forced passage through a Nuclepore filter with a 5-μm pore diameter tends to be gentle enough to not disrupt membrane-limited organelles (Das and Henderson, 1983). A cytosolic extract is prepared by centrifugation at 100,000g for 60 min. The purity of the cytosol (supernatant) can be evaluated by monitoring the presence of soluble organelle markers. Normally, it is difficult to exclude contamination of the cytosol by proteins from the nucleoplasm, because some nuclear proteins equilibrate with the cytoplasm during cell lysis by way of nuclear pores (Paine *et al.*, 1983).

Detection of ^{3}H-fucosyl Proteins

A simple and sensitive method to detect radioactive proteins involves submitting the cytosolic sample to SDS-PAGE, which separates proteins according to apparent M_r and removes unincorporated [^{3}H]sugar. If incorporation is low, the gel is fixed in methanol-acetic acid-H$_2$O and stained with Coomassie blue according to standard procedures. Each gel lane is sliced into 30–60 pieces, which are individually equilibrated for 5 days in gel scintillation cocktail and counted in a scintillation counter (Teng-umnuay *et al.*, 1999). The SDS-PAGE method is superior to trichloroacetic acid precipitation, because zero-time controls show that radioactive background attributable to nonspecific adsorption of [^{3}H]Fuc in the gel is very low. If incorporation is sufficient, autoradiography is more convenient.

Interpretation

An example of metabolic labeling of developing *Dictyostelium* cells with [^{3}H]Fuc is shown in Fig. 2. Most of the tritium is incorporated into the P100 fraction. A low level is detected in the S100 (cytosol) fraction. The peak with the highest level of incorporation corresponds to the position of Skp1 and does not represent a region of high incorporation in the P100 fraction, suggesting that it does not result from simple contamination by

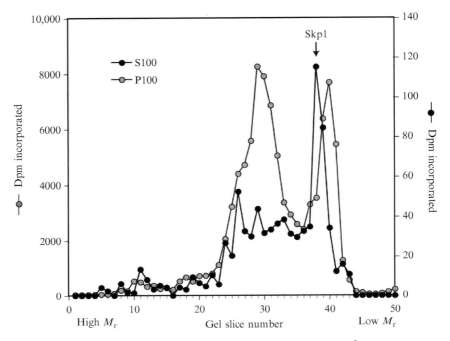

FIG. 2. Detection of Skp1 by metabolic labeling of *Dictyostelium* with [³H]Fuc. Strain Ax3 cells were incubated for 16 h at 2×10^7 cells/ml in the presence of 0.3 mCi/ml [³H]L-Fuc (Dupont) in 10 mM potassium phosphate (pH 6.5). Cells were washed in 10 mM unlabeled Fuc in the same buffer, resuspended in Tris-sucrose buffer, filter lysed, and centrifuged at 100,000g for 60 min. S100 and P100 fractions (150 μg protein each) were separated on the same 7–20% gradient polyacrylamide SDS-PAGE gel. After fixation and staining with Coomassie Blue, the destained gel was sliced, using the bands as a guide to ensure registration between lanes, and individual fragments were counted by liquid scintillation spectrometry. Dpm are shown after correction for background radiation.

vesicle-derived proteins. The lower level of incorporation at higher M_r positions in the S100 fraction (Fig. 2) is variable and may represent P100 contamination (Gonzalez-Yanes *et al.*, 1992). In this example, subsequent purification of the radiolabeled band documented its identity as a cytoplasmic protein, Skp1 (Kozarov *et al.*, 1995). Monosaccharide composition analysis by acid hydrolysis and high pH anion exchange chromatography (HPAEC) (Hardy and Townsend, 1994) confirmed that tritium was incorporated as Fuc and identified the presence of other sugars, and MS-MS studies showed that the sugars were present as a single glycan on HyPro143 (Teng-umnuay *et al.*, 1998). For new investigations, false-negative results can occur in this type of experiment as a result of insufficient labeling time with [^{3}H]Fuc, low abundance or turnover of the protein, or absence of Fuc in the glycan.

Detection of Cytoplasmic Glycosyltransferases

An indirect method of testing for cytoplasmic glycosylation is to assay for the glycosyltransferases that mediate the modification. This approach is sensitive because of the high specific activity of the nucleotide sugar used to assay the enzyme and the catalytic property of enzymes. One approach is to simply supplement an intact or desalted cytosolic extract with tritiated sugar nucleotides and examine for incorporation of the tritiated sugar into protein (Ercan *et al.*, 2006; Gonzalez-Yanes *et al.*, 1992). UDP-[^{3}H]GlcNAc is used in the example shown in the following because this is the first sugar added to *Dd*-Skp1. A variation of the method is to supplement the extract with a known cytoplasmic glycoprotein from another source, such as Skp1 from *Dictyostelium*, and assay for incorporation into *Dd*-Skp1. The addition of recombinant cytoplasmic prolyl 4-hydroxylase may increase the sensitivity for detecting HyPro-dependent glycosylation.

Preparation of Dd Skp1 P4H1

The expression plasmid pET15His$_6$TEV-PH1 is available from *Dictyostelium* Stock Center (http://dictybase.org/StockCenter/StockCenter. html). This plasmid can be used to direct the expression of the protein in *E. coli* (van der Wel *et al.*, 2005), as described in the following for *Dd*-Skp1. P4H1 is purified from *E. coli* extracts on a Ni^{+2}-column, eluted in 0.3 *M* imidazole in 5 μM FeCl$_2$, 0.5 *M* NaCl, 0.05 *M* Tris-HCl (pH 7.5), and stored in aliquots at $-80°$.

Preparation of Dictyostelium His$_{10}$Skp1

The expression plasmid pET19bDdHis$_{10}$Skp1A is available from *Dictyostelium* Stock Center (http://dictybase.org/StockCenter/StockCenter. html). The expression plasmid can be transfected into a strain of *E. coli* suitable for expressing recombinant proteins such as BL21(DE3) (van der Wel *et al.*, 2005; West *et al.*, 1996). Clonal transfectants were grown in 25 ml of LB medium containing 100 μg/ml ampicillin at $22°$ on a shaker until an OD$_{590}$ (1 cm path length) of 0.4–0.6 was attained. Expression was induced by continued shaking in 0.5 m*M* isopropyl-1-thio-β-D-galatopyranoside (Research Products International) for 15 h at room temperature. Cells were cooled on ice for 5 min, centrifuged at 5000*g* for 5 min at $4°$, washed twice by resuspension and recentrifugation in cold 20 m*M* Tris-HCl (pH 8.0), and frozen at $-80°$. Frozen cell pellets were resuspended on ice in 1 ml $E.$ $coli$ lysis buffer (0.1 M Tris-HCl [pH 8.2], 1 mg/ml lysozyme, 5 mM benzamidine, 0.5 μg/ml pepstatin A, 5 μg/ml aprotinin, 5 μg/ml leupeptin, 0.5 mM PMSF) per 25 ml of the original culture volume, lysed in a French Press cell (15,000 psi) or by probe sonication, and centrifuged at 100,000g for 1 h. His$_{10}$Skp1A can be purified directly from the extract on an 1-ml immobilized Ni^{+2}-column and eluted with NaCl and imidazole (van der Wel $et\ al.$, 2005). The eluted protein can be detected using a commercially available anti-His monoclonal antibody from Novagen (70796-3) (van der Wel $et\ al.$, 2005) or antibodies specific for Dd-Skp1 such as mAb 3F9 or 4E1 (Kozarov $et\ al.$, 1995).

Reaction Labeling Conditions

Typically, the reaction consists of 40 μl of cytosol (S100 prepared as in the previous section) and 0.5 μCi of the predicted sugar nucleotide precursor, in this example UDP-[^{3}H]GlcNAc, in a final volume of 50 μl containing 5 μM FeSO$_4$, 1 mM ascorbic acid, 0.5 mM αKG, 5 mM dithiothreitol, 0.02% (v/v) Tween-20, 0.2 mg/ml catalase (Sigma), 10 mM MgCl$_2$ (van der Wel $et\ al.$, 2005). The S100 may be prepared fresh or thawed from aliquots stored at $-80°$. αKG is a co-substrate for P4Hs with O$_2$, FeSO$_4$ is a cofactor, and ascorbic acid and catalase protect the enzyme from oxidation. Assays of crude cell extracts should be supplemented with 1 mM ATP and 3 mM NaF to inhibit sugar nucleotide pyrophosphorylase activity. The reaction proceeds for 0, 1, or 2 h at 29° and is quenched by addition of an equal volume of 2× Laemmli sample buffer. The zero-time trial establishes the blank value, and the time course helps assess the tradeoff between maximal incorporation and potential time-dependent degradation in the complex environment of the cytosol (S100).

If HyPro is the site of addition for the sugar, endogenous prolyl 4-hydroxylase may be rate limiting for the availability. This might occur because, in known examples, prolyl-hydroxylated precursors are unstable (Kaelin, 2005), because it is a signal for polyubiquitylation and, therefore, must presumably be generated in the extract. This is tested by the inclusion of 0.15 μg of Dd-His$_6$P4H1 or a P4H from the organism under evaluation. Nevertheless, target acceptor availability may still be limiting. To test for enzymes that can modify Dd-Skp1, 1 μM unmodified Dd-His$_{10}$Skp1A can be added to the reaction. Another reason for low incorporation might be dilution of UDP-[^{3}H]GlcNAc by endogenous unlabeled UDP-GlcNAc. This can be circumvented by desalting 1 ml of the S100 on a small gel filtration column (e.g., a 9-ml Sephadex G10 column commercially packaged as a PD10 column [Amersham], equilibrated in Tris-sucrose buffer supplemented with 5 mM MgCl$_2$, 5 μM FeSO$_4$, 1 mM ascorbic acid, and

1 mM dithiothreitol to stabilize enzymes). Protein fractions eluting in the void volume are identified using any standard protein assay.

To determine O_2-dependence of the reaction, samples are prepared in 1-ml conical bottom Reactivials (Pierce). The rubber septa are pierced with two syringe needles fitted with plastic tubing. Commercially prepared compressed gasses of different percentages of O_2 in N_2 (Airgas-MidSouth, Tulsa, OK) are each humidified by bubbling through an in-line water reservoir fashioned from an Erlenmeyer flask fitted with a double-hole stopper and introduced through the inlet needle; 21% O_2 should be provided from a compressed air source, because the reaction conducted under ambient atmosphere without flow or agitation yields appreciably lower values (<33%) indicating O_2-starvation (van der Wel *et al.*, 2005). The outlet needle tube is routed into a tube containing H_2O to confirm gas flow.

Interpretation

Incorporation of radioactivity over background at a specific M_r position constitutes preliminary evidence for a soluble cytoplasmic GlcNAcT. Reduced labeling in the absence of added Dd-His$_6$P4H1 would suggest involvement of a HyPro-attachment site, but no difference may be seen if sufficient endogenous P4H activity is present. This can be addressed by the inclusion of a P4H inhibitor (see earlier) and not adding αKG and ascorbate. A dependence on O_2 tension near the ambient range (21%) would also suggest a role for prolyl hydroxylation and would be consistent with a physiological dependence on normal O_2 levels as hypothesized for *Dictyostelium* P4H1 (van der Wel *et al.*, 2005). Evidence for dependence on a P4H would be consistent with a cytoplasmic origin for the αGlcNAcT activity. Unless the identity of the protein is known to confirm its compartmentalization, evidence that the same activity is not found in the particulate (P100) fraction of the cell, as done for the Skp1 modification enzymes (Ketcham *et al.*, 2004; Teng-umnuay *et al.*, 1999; West *et al.*, 1996), will help to rule out an artifactual origin from lysed vesicles of the secretory pathway as discussed previously. Although most GTs are insoluble transmembrane proteins, they are prone to proteolytic release of soluble catalytic domains. The possibility of a nuclear origin for the activity is more difficult to address because of the propensity of some nuclear proteins to dissociate from this organelle during cell lysis (Sassi *et al.*, 2001).

Detection of incorporation near the 20-kD position will suggest a Skp1-like target. This could be further tested by the inclusion of heterologous Dd-His$_{10}$Skp1A, which migrates at a distinct M_r position because of its N-terminal tag. However, this is subject to a false-negative result if species differences affect enzyme recognition. Identification of a radiolabeled band will require follow-up studies to verify that the expected GlcNAc is, indeed, covalently attached to the protein and to HyPro if evidence suggests this to occur. For example, the radioactivity should be characterized after acid hydrolysis as referenced earlier to confirm that it was incorporated as GlcNAc rather than first being converted to another sugar such as GalNAc (as epimerization to UDP-GalNAc). Confirmation of attachment to a HyPro would ultimately require testing dependence of the reaction on a specific P4H gene and on a specific Pro residue by site-directed mutagenesis. Further characterization of the GlcNAcT, including sequence data to confirm cytoplasmic localization, would be required to determine the compartmental mechanism of glycosylation (e.g., van der Wel *et al.*, 2002a). To confirm that the candidate P4H and GlcNAcT directly modify the candidate Pro-residue, biochemical reconstitution of the reaction using purified components (Teng-umnuay *et al.*, 1999; van der Wel *et al.*, 2005) and confirmation of the reaction product (e.g., by MS-MS studies) would be required.

In principle, this approach can be applied to any predicted GT and target glycoprotein. The method has the potential to detect βGlcNAc modification of Ser/Thr residues, a common modification of cytoplasmic and nuclear proteins in animals and plants (Hart, 2006), which is mediated by a βGlc NAcT (distinct from the αGlcNAcT that modifies Skp1). The occurrence of this modification can be tested for by subjecting the proteins to conditions of β-elimination and chromatographic analysis by HPAEC on a Dionex CarboPac MA-1 column and electrometric detection (Weitzhandler *et al.*, 1998) to determine whether the expected [^{3}H]GlcNAcitol product is formed. In contrast, sugar–HyPro linkages are resistant to alkali-catalyzed elimination (Spiro, 1972). Different radioactive sugar nucleotides are commercially available to test for other sugar GTs, and other potential target apoproteins can be added to the reaction. For example, to test for other GTs that modify Skp1, the appropriate glycoform precursor can be prepared and added to the reaction. This method has been used to demonstrate the presence of the Skp1 αGlcNAcT (Teng-umnuay *et al.*, 1999; van der Wel *et al.*, 2002a), the combined Skp1 β3GalT/α2FucT (van der Wel *et al.*, 2002b), and the Skp1 α3GalT (Ercan *et al.*, 2006).

Example in Toxoplasma gondii

These methods were used to test for the occurrence of HyPro-dependent cytoplasmic glycosylation in the obligate intracellular eukaryotic pathogen *Toxoplasma gondii*. Because *Toxoplasma* grows in its host cell within a specialized membrane-bound compartment called the parasitophorous vacuole, metabolic labeling and prolyl hydroxylase inhibitors have less utility. Therefore, the sensitive glycosyltransferase detection method was

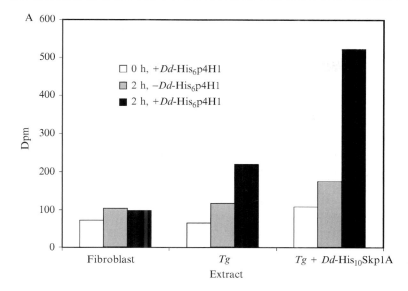

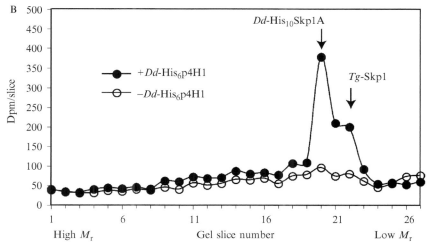

FIG. 3. Detection of Skp1-αGlcNAcT-like activity in S100 extracts of *T. gondii*; 2×10^8 parasites were isolated from human foreskin fibroblasts, and soluble S100 extracts were prepared. As a control, an equal amount of protein prepared from fibroblasts was compared. (A) Fibroblast or parasite S100s (10% of the sample) were supplemented with UDP-[^{3}H] GlcNAc with or without *Dd*-His$_{10}$Skp1 as indicated. Reactions were run for 0 or 2 h at 29° in ambient atmosphere with or without added *Dd*-His$_6$P4H1. Samples were separated on an SDS-PAGE gel, which were then fixed, stained with Coomassie Blue, and sliced into 27 pieces. Incorporation into fibroblast or *Tg* extracts at the expected position of *Tg*-Skp1, or at

used; 2×10^8 parasites of the RH *Toxoplasma* strain were purified from human foreskin fibroblasts (Roos *et al.*, 1994), passed through a 3-μm filter to remove unlysed host cells, and then lysed by probe sonication in 1 ml of Tris-sucrose buffer including protease inhibitors. Sonication was continued in cycles until lysis of >90% of the parasites were confirmed by phase-contrast microscopy. Mechanical disruption is preferred to detergent lysis to reduce the potential for contamination by glycosyltransferases from membranous organelles. The sample was centrifuged at $100,000g$ for 60 min at $4°$.

We first tested whether a P4H/GT pathway existed in *Toxoplasma* by supplementing the *Tg*-S100 fraction with concentrations of substrates and cofactors optimal for *Dd*-P4H1 and *Dd*-αGlcNAcT1 activity. UDP-$[^3H]$ GlcNAc was then added to determine whether this sugar can be conjugated onto parasite proteins in a manner analogous to that of *Dictyostelium* Skp1. After incubation for 2 h at $30°$, the sample, together with no-substrate and zero-time controls, was examined by the SDS-PAGE assay. As shown in Fig. 3A, a low, but significant, level of $[^3H]$GlcNAc incorporation was detected as a single band that interestingly corresponded to the expected M_r of *Tg*-Skp1 (TgTwinScan_0842 at www.ToxoDB.org; predicted apoprotein $M_r = 18780$). Incorporation of the radiolabeled sugar was stimulated by the addition of *Dd*-His$_6$P4H1, suggesting that incorporation was directed to a HyPro residue. Only background incorporation was detected in lysates prepared from uninfected fibroblasts, indicating that it was not due to host cell contamination. These results suggested that *Toxoplasma* expresses a soluble GlcNAcT similar to *Dictyostelium* αGlcNAcT1.

Given that the size of the protein that incorporated $[^3H]$GlcNAc was the same predicted size of *Toxoplasma* Skp1, we tested whether recombinant *Dd*-His$_{10}$Skp1A, which is larger than the *Toxoplasma* protein, could be recognized by the parasite enzyme. As shown in Fig. 3A, *Dd*-His$_{10}$Skp1A was also modified by $[^3H]$GlcNAc in a P4H1-dependent manner. Heterologous Skp1 and the endogenous protein were the major labeled proteins detected (Fig. 3B). These data indicated that similar to *Dictyostelium*, $[^3H]$ GlcNAc modification of Skp1 was HyPro-dependent. These observations are consistent with bioinformatics evidence for P4H1- and αGlcNAcT1-like sequences in the *T. gondii* genome at ToxoDB (not shown). Although these findings are similar to results obtained in *Dictyostelium* extracts, further studies are required to verify that the modification is homologous to that of *Dictyostelium* Skp1.

---

the position of *Dd*-His$_{10}$Skp1A, are shown. (B) Sample from a separate experiment in which the parasite S100 was incubated at 40% O_2 with UDP-$[^3H]$GlcNAc and *Dd*-His$_{10}$Skp1A, in the presence or absence of *Dd*-His$_6$P4H1. All slices from the gel are shown. The position of *Dd*-His$_{10}$Skp1A and the expected position of endogenous *Tg*-Skp1 are indicated.

References

Bruick, R. K., and McKnight, S. L. (2001). A conserved family of prolyl-4-hydroxylases that modify HIF. *Science* **294,** 1337–1340.

Das, O. P., and Henderson, E. J. (1983). A novel technique of gentle lysis of eukaryotic cells isolation of plasma membranes from *dictyostelium discoideum*. *Biochim. Biophys. Acta* **736,** 45–56.

Ercan, A., Panico, M., Sutton-Smith, M., Dell, A., Morris, H. R., Matta, K. L., Gay, D. F., and West, C. M. (2006). Molecular characterization of a novel UDP-Gal:fucoside alpha3-galactosyltransferase that modifies Skp1 in the cytoplasm of *Dictyostelium*. *J. Biol. Chem.* **281,** 12713–12721.

Gonzalez-Yanes, B., Cicero, J. M., Brown, R. D., Jr, and West, C. M. (1992). Characterization of a cytosolic fucosylation pathway in *Dictyostelium*. *J. Biol. Chem.* **267,** 9595–9605.

Hardy, M. R., and Townsend, R. R. (1994). High-pH anion-exchange chromatography of glycoprotein-derived carbohydrates. *Methods Enzymol.* **230,** 208–225.

Hart, G. W., Haltiwanger, R. S., Holt, G. D., and Kelly, W. G. (1989). Glycosylation in the nucleus and cytoplasm. *Annu. Rev. Biochem.* **58,** 841–874.

Hart, G. W. (2006). Identification of O-GlcNAc sites on proteins. *Methods Enzymol.* **417,** 113–133.

Hebert, D. N., Foellmer, B., and Helenius, A. (1995). Glucose trimming and reglucosylation determine glycoprotein association with calnexin in the endoplasmic reticulum. *Cell* **81,** 425–433.

Kaelin, W. G. (2005). Proline hydroxylation and gene expression. *Annu. Rev. Biochem.* **74,** 115–128.

Ketcham, C., Wang, F., Fisher, S. Z., Ercan, A., van der Wel, H., Locke, R. D., Doulah, k.S., Matta, K. L., and West, C. M. (2004). Specificity of a soluble UDP-galactose:fucoside α1,3galactosyltransferase that modifies the cytoplasmic glycoprotein Skp1 in *Dictyostelium*. *J. Biol. Chem.* **279,** 29050–29059.

Kozarov, E., van der Wel, H., Field, M., Gritzali, M., Brown, R. D., and West, C. M. (1995). Characterization of FP21, a cytosolic glycoprotein from *Dictyostelium*. *J. Biol. Chem.* **270,** 3022–3030.

Paine, P. L., Austerberry, C. F., Desjarlais, L. J., and Horowitz, S. B. (1983). Protein loss during nuclear isolation. *J. Cell Biol.* **97,** 1240–1242.

Petroski, M. D., and Deshaies, R. J. (2005). Function and regulation of cullin-RING ubiquitin ligases. *Nat. Rev. Mol. Cell Biol.* **6,** 9–20.

Roos, D. S., Donald, R. G., Morrissette, N. S., and Moulton, A. L.. (1994). Molecular tools for genetic dissection of the protozoan parasite *Toxoplasma gondii*. *Methods Cell. Biol.* **45,** 27–63.

Sassi, S., Sweetinburgh, M., Erogul, J., Zhang, P., Teng-umnuay, P., and West, C. M. (2001). Analysis of Skp1 glycosylation and nuclear enrichment in *Dictyostelium*. *Glycobiology* **11,** 283–295.

Siddiq, A., Ayoub, I. A., Chavez, J. C., Aminova, L., Shah, S., LaManna, J. C., Patton, S. M., Connor, J. R., Cherny, R. A., Volitakis, I., Bush, A. I., Langsetmo, I., Seeley, T., Gunzler, V., and Ratan, R. R. (2005). Hypoxia-inducible factor prolyl 4-hydroxylase inhibition. A target for neuroprotection in the central nervous system. *J. Biol. Chem.* **280,** 41732–41743.

Spiro, R. G. (1972). Study of the carbohydrates of glycoproteins. *Methods Enzymol.* **28,** 3–43.

Spiro, R. G. (2002). Protein glycosylation: Nature, distribution, enzymatic formation, and disease implications of glycopeptide bonds. *Glycobiology* **12,** 43R–56R.

Tan, L., Leykam, J. F., and Kieliszewski, M. J. (2003). Glycosylation motifs that direct arabinogalactan addition to arabinogalactan-proteins. *Plant Physiol.* **132**, 1362–1369.

Teng-umnuay, P., Morris, H. R., Dell, A., Panico, M., Paxton, T., and West, C. M. (1998). The cytoplasmic F-box binding protein Skp1 contains a novel pentasaccharide linked to hydroxyproline in *Dictyostelium. J. Biol. Chem.* **273**, 18242–18249.

Teng-umnuay, P., van der Wel, H., and West, C. M. (1999). Identification of a UDP-GlcNAc: Skp1-hydroxyproline GlcNAc-transferase in the cytoplasm of *Dictyostelium. J. Biol. Chem.* **274**, 36392–36402.

van der Wel, H., Morris, H. R., Panico, M., Paxton, T., Dell, A., Kaplan, L., and West, C. M. (2002a). Molecular cloning and expression of a UDP-GlcNAc:hydroxyproline polypeptide GlcNAc-transferase that modifies Skp1 in the cytoplasm of *Dictyostelium. J. Biol. Chem.* **277**, 46328–46337.

van der Wel, H., Fisher, S. Z., and West, C. M. (2002b). A bifunctional diglycosyltransferase forms the Fucα1,2Galβ,3-disaccharide on Skp1 in the cytoplasm of *Dictyostelium. J. Biol. Chem.* **277**, 46527–46534.

van der Wel, H., Ercan, A., and West, C. M. (2005). The Skp1 prolyl hydroxylase of *Dictyostelium* is related to the HIFα-class of animal prolyl 4-hydroxylases. *J. Biol. Chem.* **280**, 14645–14655.

Varki, A. (1994). Metabolic radiolabeling of glycoconjugates. *Methods Enzymol.* **230**, 16–32.

Weitzhandler, M., Rohrer, J., Thayer, J. R., and Avdalovic, N. (1998). HPAEC-PAD analysis of monosaccharides released by exoglycosidase digestion using the CarboPac MA1 column. *Methods Mol. Biol.* **76**, 71–78.

West, C. M., Scott-Ward, T., Teng-umnuay, P., van der Wel, H., Kozarov, E., and Huynh, A. (1996). Purification and characterization of an α1,2-L-fucosyltransferase, which modifies the cytosolic protein FP21, from the cytosol of *Dictyostelium. J. Biol. Chem.* **271**, 12024–12035.

West, C. M., van der Wel, H., and Gaucher, E. A. (2002). Complex glycosylation of Skp1 in *Dictyostelium*: Implications for the modification of other eukaryotic cytoplasmic and nuclear proteins. *Glycobiology* **12**, 17R–27R.

West, C. M., van der Wel, H., Sassi, S., and Gaucher, E. A. (2004). Cytoplasmic glycosylation of protein-hydroxyproline and its relationship to other glycosylation pathways. *Biochim. Biophys. Acta* **1673**, 29–44.

West, C. M., van der Wel, H., Coutinho, P. M., and Henrissat, B. (2005). Glycosyltransferase Genomics in *Dictyostelium discoideum*. *In* "*Dictyostelium* Genomics" (W. F. Loomis and A. Kuspa, eds.), pp. 235–264. Horizon Scientific Press, Norfolk, UK.

Author Index

A

Aarsman, C. J. M., 221
Aballay, A., 374, 380, 381, 382
Abate-Shen, C., 101
Abdollah, S., 4
Abe, K., 39, 41
Abedin, M. J., 259, 260
Abrous, D. N., 26
Achtman, M., 322
Adachi, J., 361, 362
Adams, J. C., 93, 94
Adang, M. J., 340, 341, 345, 350, 353, 354,
   356, 357, 361
Aebi, M., 25, 365, 366
Ahmad, N., 257, 264, 274
Ahorn, H., 340, 361, 364, 365
Aikawa, M., 93, 94
Aizawa, S., 40, 42, 49
Akahani, S., 261
Akasaka-Manya, K., 146, 148
Akhurst, R. J., 4
Akiyama, S. K., 11
Akiyama, T., 281
Alarcon, T., 294, 316, 322
Albert, M., 54
Albrandt, K., 263
Alegado, R. A., 374, 381, 382
Alessandri, G., 170
Allione, A., 258
Alloing, G., 381, 382
Alon, R., 259, 276
Alonso, C. R., 258
Altman, E., 214, 216, 219
Altmann, F., 340, 361, 364, 365
Altraja, S., 294, 307, 309, 329, 333
Alvarez, M., 258
Alvarez, M. M., 228
Alvarez-Buylla, A. A., 30
Alves, F., 170, 171, 172, 179
Amado, M., 381
Amano, M., 258
Aminova, L., 392

Amselgruber, W., 142
Amsterdam, A., 259, 276
Anazawa, H., 12
Anderson, A. C., 260
Anderson, P., 341
Anderson, R. G., 177
Anderson, W. W., 54, 58
Andersson, M., 328, 329
Andre, S., 16, 257, 259, 264, 274, 275, 277
Angata, K., 25, 26, 27, 31, 32, 33, 34, 35, 54
Ångström, J., 294, 307, 309, 329, 333
Anselmetti, D., 222, 224, 231, 235, 236
Anspach, F. B., 264
Anton, M., 3
Antonenko, S., 195
Aoki, K., 94
Arai, Y., 30
Arakawa, T., 275, 278
Arata, Y., 257, 274
Arbel-Goren, R., 259, 275, 276
Arboleda-Velasquez, J. F., 93, 94, 97, 133
Arbucci, S., 361
Arcaro, A., 286
Arimoto, K., 248
Arnaud, L., 155
Arndt-Jovin, D. J., 17
Arnqvist, A., 293, 294, 307, 309, 316, 322,
   329, 333
Arnsmeier, S. L., 218
Aroian, R. V., 340, 341, 342, 343, 344, 345,
   350, 353, 354, 356, 357, 361, 381
Artavanis-Tsakonas, S., 93, 112, 113,
   127, 128
Artola, A., 54
Arulanandam, A. R., 12
Arvanitis, D. N., 223
Asa, D., 208, 217
Asahi, M., 12, 14, 19
Asano, M., 54, 129
Ascencio, J. A., 228
Ascione, B., 258
Ashkin, A., 329
Ashline, D., 361, 363, 364

Subject Index

A

β1,2-*N*-Acetylglucosaminyltransferase
  activity assays
    brain membrane fraction preparation, 144–145
    glutathione *S*-transferase fusion substrate preparation, 145–146
    human embryonic kidney cell membrane preparation, 142–144
    incubation conditions and high-performance liquid chromatography, 150–151
    materials, 149–150
    principles, 148
  Caenorhabditis elegans and bacterial pathogen defense studies
    double mutant construction, 368
    Escherichia coli OP50 survival studies, 371–374, 380
    function, 365–366
    genes, 366
    Pseudomonas aeruginosa strain 14 survival studies, 374–377, 380
    rescue studies, 369–370, 379, 381
    RNA interference, 368–369, 377, 379
    worm life span in pathogen coculture, 367–368
  dystroglycan as substrate, 138
  mutant production, 151
β1,6 *N*-Acetylglucosaminyltransferase V
  fluorescence imaging studies of growth factor receptor signaling role
    image acquisition, 4–5
    mammary tumor cells from polyomavirus middle T transgenic mice, 5–6
    materials, 4
    nuclear translocation assay, 6–9
    peritoneal macrophages, 6
  growth factor receptors as substrates, 3–4
  knockout mouse phenotype, 3

oncogene activation of expression, 3
Adhesins, *see Helicobacter pylori*
Adipocyte, galectins in signaling, 281
AFM, *see* Atomic force microscopy
Akt, phosphorylation assay of galectin effects, 285
AMPA receptor, *see* Neurotransmitter receptors
Apoptosis
  galectin regulation
    galectin-1, 257–259
    galectin-2, 259
    galectin-3
      expression vectors, 266–267
      overview, 259
      transfection studies, 260–262, 265–268
    galectin-7, 262
    galectin-8, 259
    galectin-9, 259–260
    galectin-12, 262
    T cell assay in Matrigel for galectin-1 effect analysis, 253–254
    transfection studies, 260–263
  pathways, 268
Atomic force microscopy, gold glyconanoparticle analysis of carbohydrate–carbohydrate interactions, 235–239

B

BabA, *see Helicobacter pylori*
Bacillus thuringiensis Cry5B pore-forming toxin, *see Caenorhabditis elegans*
Behavioral testing
  glycosphingolipid synthetic enzyme knockout mice
    foot printing, 47
    formalin test, 47
    hot plate test, 45
    rota-rod test, 46–47
    von Frey test, 45